Nanofibers - Synthesis, Properties and Applications

Edited by Brajesh Kumar

Published in London, United Kingdom

IntechOpen

Supporting open minds since 2005

Nanofibers - Synthesis, Properties and Applications
http://dx.doi.org/10.5772/intechopen.92484
Edited by Brajesh Kumar

Contributors
Lakshmi Swaminathan, Dan Bahadur Pal, Deen Dayal Giri, Harshal Gade, George G. Chase, Sreevalli Bokka, Arshad Khan, Shawkat Ali, Saleem Khan, Moaaz Ahmed, Bo Wang, Amine Bermak, Chandra Sekhar Rout, Sithara Radhakrishnan, Minu Mathew, Anjali Jha, Palak Sondhi, Keith J. Stine, Rudi Dungani, Sri Hartati, Mustika Dewi, Aminudin Sulaeman, Enih Rosamah, Salsabila Mutiara Kanti Muharam, Pingkan Aditiawati, Nand Jee Kanu, Eva Gupta, Venkateshwara Sutar, Gyanendra Kumar Singh, Umesh Kumar Vates, Amos Adeleke Adeleke Akande, Abraham Abdul Adenle, Aderemi Timothy Adeleye, Bonex Wakufwa Wakufwa Mwakikunga, Pablo Rodriguez, Julio Labarga, Daniel Hidalgo, Pedro Navarro Santos, Rafael Herrera-Bucio, Jose Luis Rivera, Judit Aviña-Verduzco, Aparecido Junior de Menezes, Kênia Da Silva Freitas, Thais Eugênio Gallina, Robson Valentim Pereira, Marcelo A. Pereira-da-Silva, Deepthi Sista, Gatut Yudoyono, Diky Anggoro, Lutfi Fitria Ningsih, Rizki Romadoni, Lingayya Hiremath, O. Sruti, B.M. Aishwarya, N.G. Kala, E. Keshamma, Nuray Kizildag

First published in London, United Kingdom, 2021 by IntechOpen
IntechOpen is the global imprint of INTECHOPEN LIMITED, registered in England and Wales, registration number: 11086078, 5 Princes Gate Court, London, SW7 2QJ, United Kingdom
Printed in Croatia

British Library Cataloguing-in-Publication Data
A catalogue record for this book is available from the British Library

Additional hard and PDF copies can be obtained from orders@intechopen.com

Nanofibers - Synthesis, Properties and Applications
Edited by Brajesh Kumar
p. cm.
Print ISBN 978-1-83968-425-8
Online ISBN 978-1-83968-426-5
eBook (PDF) ISBN 978-1-83968-427-2

Meet the editor

Dr. Brajesh Kumar is currently an assistant professor and head in the Post Graduate Department of Chemistry, TATA College, Chaibasa, India. He received a Ph.D. in Chemistry from the University of Delhi, India, in 2009. His research interests include the development of sustainable and eco-friendly techniques for nanoparticle synthesis and their applications for environmental remediation, nanofibers, nanocomposites, nanomedicine, sensors, active films of organic solar cells, natural product extraction, purification and analysis, natural polymers, peptide chemistry, microwave, and ultrasound-assisted organic synthesis, and organic synthesis. Dr. Kumar has several national and international fellowships to his credit and has worked as a faculty member in various universities throughout India, Ecuador, and South Korea. He has also published numerous SCI/SCIE/Scopus research articles and is an active reviewer of more than seventy journals.

Contents

Preface

Nanoscience and nanotechnology offer numerous benefits in various areas, including human health, food processing, environmental safety, and device engineering. Nanofibers are materials in the diameter range of 3–100 nm and length range of 0.1–1000 μm. They are well known for their vast range of applications in sensors, catalysts, conductors, tissue engineering, and so on, owing to their high surface area-to-volume ratio, high porosity, and the ease of tuning their structures, functionalities, and properties. Knowledge of how atoms are united and how nanosized materials are made is essential to understanding nanoscience. In this book, researchers from all over the world highlight the importance of nanofibers in different aspects.

This book includes review and research articles related to the synthesis, characterization, and application of nanofibers. In Chapter 1, Dr. Sista reviews new perspectives of nanofiber synthesis and applications. In Chapter 2, Dr. Pal and Dr. Giri summarize the green synthesis of nanofibers. In Chapter 3, Dr. Yudoyono et al. discuss the electrospinning of carbon nanofibers using PVA polymer. In Chapter 4, Ph.D. Kizildag presents recent advances in applications of ceramic nanofibers. In Chapter 5, Dr. Hiremath et al. introduce antibacterial nanofibers produced by electrospinning. In Chapter 6, Prof. Kanu et al. demonstrate the fabrication of biofunctional curcumin/gelatin nanofibers and their healing process. In Chapter 7, Associate Prof. de Menezes et al. discuss eletrochemical behaviour of cellulose nanofibrils functionalized with dicyanovinyl groups, and their use as electrocatalysts for carbon dioxide reduction. In Chapter 8, Prof. Dungani et al. review the production of nanocellulose using cellulase enzymes through Symbiotic Culture of Bacteria and Yeast (SCOBY). In Chapter 9, Prof. Stine and Dr. Sondhi discuss the progress in synthetic techniques over the last ten years to prepare porous gold films with emphasis on the technique of electrodeposition. In Chapter 10, Dr. Akande et al. report various methods of fabricating and manufacturing micro and nanosensors, membranes, and energy devices.

In Chapter 11, Prof. Jha addresses microwave-assisted synthesis of organic compounds and nanomaterials as a promising area of modern green chemistry. In Chapter 12, Prof. Chase et al. discuss polarization of electrospun polyvinylidene fluoride fiber mats and fiber yarns. In Chapter 13, Prof. Rout et al. summarize recent developments in the field of 2D material-based all-solid-state microsupercapacitors. In Chapter 14, Dr. Rodriguez et al. present research on pulsed electrochemical micromachining of stainless steel. In Chapter 15, Prof. Swaminathan examines the low-cost fabrication processes of radio frequency-micro electro mechanical systems switch and packaging. In Chapter 16, Dr. Khan et. al. presents a comprehensive review of the latest advances of transparent electrodes for flexible and stretchable electronics. Finally, in Chapter 17, Dr. Navarro Santos et al. discuss calculation of the electronic properties and reactivity of nanoribbons and propose a conceptual density functional theory to calculate energetic, electronic, and reactivity of one-dimensional nanomaterials such as carbon nanoribbons. This book is a useful resource for academics, professionals, scientists, and graduate and undergraduate students.

I am indebted to my wife, Mrs. Kumari Smita for her helpful comments on several chapters and excellent support in conceptualizing this book.

Dr. Brajesh Kumar
Department of Chemistry,
TATA College,
Kolhan University,
Chaibasa, Jharkhand, India

Chapter 1

New Perspective of Nano Fibers: Synthesis and Applications

Deepthi Sista

Abstract

Nano fibers are most attractive materials in the scientific world due to their enormous applications in various fields. Their applications start with generation of energy, solution to environmental problems and continues with medical field and many more. Nano materials got much importance from their peculiar electrical, optical, mechanical and thermal properties. Fibrous materials are obtained from several sources and by different mechanisms these materials are converted into nano materials. As of bulk fibers include specific properties compared to other materials, the generation of nano fibers enhance all the properties. The synthesis of nano fibers from natural and synthetic polymers, metals, semiconductors, composite materials, carbon based materials lead to new perspective in science and engineering fields. Most pronouncing techniques that include conventional and modern methods are available to fabricate nano fibers from these materials. Of them some are being used from a long time and some are emerging techniques to generate flexible substrates. Electrospinning, template based synthesis, polymerization, self-assembly, sonochemical synthesis are the conventional methods for the production of nano fibers. New technologies include electro hydrodynamic writing, plasma induced synthesis, centrifugal jet spinning, CO_2 laser supersonic marks a trend in development of nano fibrous materials. This chapter give details about fabrication materials and provides synthetic routes to generate them along with applications. Also this chapter focuses on the challenges in development of nano fiber technology in commercial perspective.

Keywords: Nano fibers, fabrication materials, synthesis techniques-conventional and modern

1. Introduction

Scientific and technological world at present focuses on nano meter range fibrous materials which have excellent physical, chemical, biological and optical properties. Materials made from these fibers have great fundamental importance due to their flexibility and high directional strength. These are light weight with well-regulated pore structures and high surface to volume ratio. Fibrous materials at nano scale have shown excellence in every fundamental property. Nano fibers are well suitable in designing functional materials, used in tissue engineering, filtration, sensors, clothing and can also be used for energy storage. Specific morphological characteristics of nano fibers resembles original cellular matrix impacts living nature.

Unambiguous properties of nano fibers intend to modify or reinforce polymer matrices that have large beneficiations to mankind.

2. Importance of nano fibers to the present day perspective

Today's world is facing many challenges to run things in a smooth manner. Many man made things outburst as a threat to human life. In order to overcome such challenges researchers looked into fabrication of nano fibers that have substantial benefits in various fields. For example, globalization and modernization brought many hazardous things like plastic into day to day life of common man. Plastic is one of major pollutant that cannot decompose easily into Earth. Likewise, many materials are bringing challenges for better livelihood.

Nano fibers are alternative resources for many materials due to their excellent properties. These materials are emerging as substituents for original materials due to their low cost, low density, high porosity, high energy. These unique features enable nano fibers for novel applications.

Nano science and nano technology have been vital applicative for scientific world since antiquity. Impact of nano on the present world empowers and drives one to develop new aspects in many areas. Researchers thrive to dwell the inherence of nano to synthesize from different materials. Fibrous materials have been used as best replacement for many non-renewable sources with less cost. They have been used in day to day applications like mobiles, solar cells, batteries, filtration membranes etc. In many cases these materials are taken from end users or wastes. For example, cellulose nano whiskers can be prepared from coconut fiber [1]. Sea algae is also being used to prepare nano fibers with numerous applications [2]. Different methods like electron spinning, self-assembly, template synthesis, thermal induction and phase separation etc. are used to make nano fibers [3]. These methods include chemical and mechanical techniques. Nano fibers extracted from natural and synthetic polymers are authentic besties to nature.

Flexibility, high tensile strength is the major advantage of these fibers with enhanced properties. Knowing about the various types of nano fibers, their synthesis, properties with basic as well as commercial applications at one glance is the main emphasis of this chapter.

3. Generation of nano fibers

Nano fibers are generated from diverse materials show differences in physical properties as well as application potentials. Natural polymers, synthetic polymers, carbon based materials, metals, ceramics, semiconducting materials and composite materials are used for the preparation of nano fibers [4]. **Figure 1** shows the materials used to synthesize nano fibers. Each of these materials have specific importance and applications.

4. Natural and synthetic polymers

Dominance of renewable sources in the preparation of nano fibrous materials has been increased drastically in this decade due to their environment friendly properties. These renewable sources include polymers were paid much attention from their biocompatible, bio degradable and bio active nature. Starting from

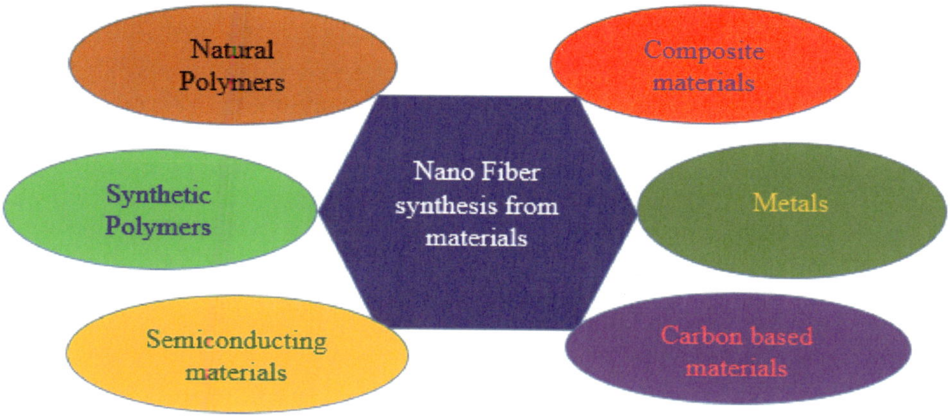

Figure 1.
Synthesis of nano fibers from various materials.

physical properties proliferation, adhesion, migration, cell adhesion are most evitable properties the makes one to rely to produce nano fibers.

Collagen, Cellulose, silk fibroin, keratin, gelatin and polysaccharides (chitosan, alginate) are the natural polymers and can be used to synthesize nano fibers using various techniques [5–7].

4.1 Collagen

Collagen is an excellent protein in the extra cellular matrix found in connective tissues of body. The structure of collagen is helical with amino acids bound together which is elongated fibril also known as collagen helix. Nano fibrous materials generated from this regenerative biopolymer are used in reconstruction of tissues [8].

4.2 Cellulose

Nano structured cellulose fibers are commonly referred as cellulose nano fibers. Cellulose nano crystal, cellulose nano fibers, nano fibrillated cellulose and bacterial nano cellulose are the nano fibered cellulose with different physical, chemical and biological properties [9].

These fibrils are with high aspect ratio that is 5–20 nm width and some micrometers length. Many plant based products, biological products, sea products are the basis of cellulose. As an example the nano fibers from cellulose develops microfiber 3D printing network [10].

4.3 Silk fibroins

Silk fibroin is produced from silkworms and spiders. It has excellent mechanical properties with biological compatibility, morphological flexibility used to produce nano fibers mostly using electrospun technique [11]. The stability of fibers from silk is obtained from chemical treatments such as methanol, ethanol, propanol and water vapor. If viscoelasticity of silk fibroins is increased with blending them with polymers an improvement in mechanical properties were observed while biological properties remain [12].

4.4 Keratin

One of the most abundant non-food protein is Keratin. Components of hair, feathers, nails, horns of mammals and birds are the many sources of keratin [13]. Despite of having so many good characteristics the keratin wastes are being pollutants. If the wastes are brunt they would release toxins due to Sulphur content. Instead of burning them and make as pollutants one can produce nano fibrous out of keratin by electrospinning technique that finds application in tissue engineering and many filtration devices [14].

For example, human hair mixed with polycaprolactone (PCL) in proper proportions to produce nanofibrous membranes that have excellent applications to develop composite materials and also can be used in various biomedical applications [15].

4.5 Gelatin

Gelatin is a natural polymer which is renewable acquired by fractional hydrolysis of collagen. It is most favorable bioengineering material due to its low cost, high biocompatibility and biodegradability [16]. Electrospun gelatin nano fibers face a difficult in water solubility with poor mechanical strength. To overcome this problem crosslinking technique like drying, heating and UV light exposure with some chemical treatments are induced [17].

4.6 Polysaccharides

Chitosan is one of polysaccharide obtained from deacetylation of chitin polymer. It is a natural source found in exoskeleton of insects, crustaceans and fungi. With excellence in properties like biodegradability, biocompatibility, nontoxicity chitosan suits for biomedical applications. Large variety of fungi, yeasts, bacteria can be inhibited by chitosan and through electrospinning technique nano fibers are prepared [18]. These specifications made chitosan ample with opportunities in biomedical and other fields of industry.

5. Synthetic polymers

Polyvinyl alcohol Polycaprolactone (PCL), polyurethane (PU), poly(lactic-co-glycolicacid) (PLGA), poly(3-hydroxybutyrate-co-3-hydroxyvalerate) (PHBV), and poly(ethylene-co-vinyl acetate) (PEVA) are the synthesized polymers that can be used to prepare nano fibers with different techniques [19–21]. Poly Lactic acid (PLA), Poly glycolic acid (PGA) and their copolymer (PLGA) are biodegradable found applications in medical field [22].

These all are water soluble polymers with good mechanical properties. Some of these synthesized polymers are strong enough with good antimicrobial and antifungal activities. Combining with natural polymers these synthesized new scaffolds were prepared for various applications [23].

6. Semiconducting materials

In the present day scenario semiconducting materials looks forward to develop many new technologies that relates to environment and society. Reduction of a material to its nano size exhibit numerous novel properties with a wide range of applications include energy materials, optoelectronic devices, biomedical imaging etc. [23]. The size dependent tuneable band gaps of semiconductors exhibit excellent properties and are used to generate nano fibers. Many fibrous materials that possess semiconducting properties can be prepared from versatile techniques. Three different sources acetylene, ethanol and cotton were used to prepare semiconductor carbon nano fibers. Many electronic functionalities like light emitting diodes, photonic compounds, field effect transistors are used to fabricate nano fibers [24].

6.1 Metals and composite materials

Metals and composite materials have explicit optical, physical and electrical properties. These special features avail the materials amalgamate with nano fibrils to develop new fibrous materials [25]. The newly developed materials exhibit stability, flexibility, bio compatibility, selectivity and improved sensitivity.

6.2 Carbon based materials

Formation of stable organic as well as inorganic molecules is possible from high flexible nature of carbon. Notable mechanical, thermal, electrical properties versatile carbon materials and intrudes the formation of carbon nano fibers [26]. These CNF's also acts as essences for composite materials.

Nano fibers from these materials can be obtained from various physical, chemical or mechanical techniques explained in detail further in this chapter.

7. Synthesis Mechanism

Synthesis of nano fibers include various chemical and mechanical and optical methods. From their early preparation till to date so many techniques keep on coming to generate fibrous materials in nano size. Nano fibers are generated from various technologies electrospinning, self-assembly, template based synthesis, polymerization, sonochemical synthesis [27]. **Figure 2** shows the synthesis mechanism of nano fibers from various techniques. Along with these methods freeze drying or lyophilization is another technique used to produce nano fibers from cellulose materials. There are few ongoing and upcoming new technologies to synthesize nano fibers. Few of them are electro-hydrodynamic writing, plasma induced synthesis, solution blow spinning, centrifugal jet spinning, CO2 laser supersonic drawing.

Of all the availability techniques electrospinning is adaptable mechanism for the production of nano fibers [28–31]. In this technique simple experimental arrangement is used to prepare nano fibers. An electric source, a syringe with nozzle, a counter electrode, target and a pump is the experimental setup to generate fibers at nano scale as shown in **Figure 3**. The principle of this technique is the electrostatic repulsion force produced in a high electrical field. The ejected solution forms into Tylor cone due to the potential difference. The solvent in the solution evaporates that leads to the formation of nano fibers and collected at collector. New improvisation in conventional electrospinning technique is done to generate nano fibers with enhanced properties. There are several types of electrospinning methods that includes multi axial, Co-axial, tri axial electrospinning, bi-component, mutlineedle electrospinning, needle less- bubble, two-layer fluid, splashing electrospinning are the techniques implementing to improve the nano fiber productivity [32].

Majority of fibrous materials from natural and synthetic polymers are generated from electrospinning and its related techniques.

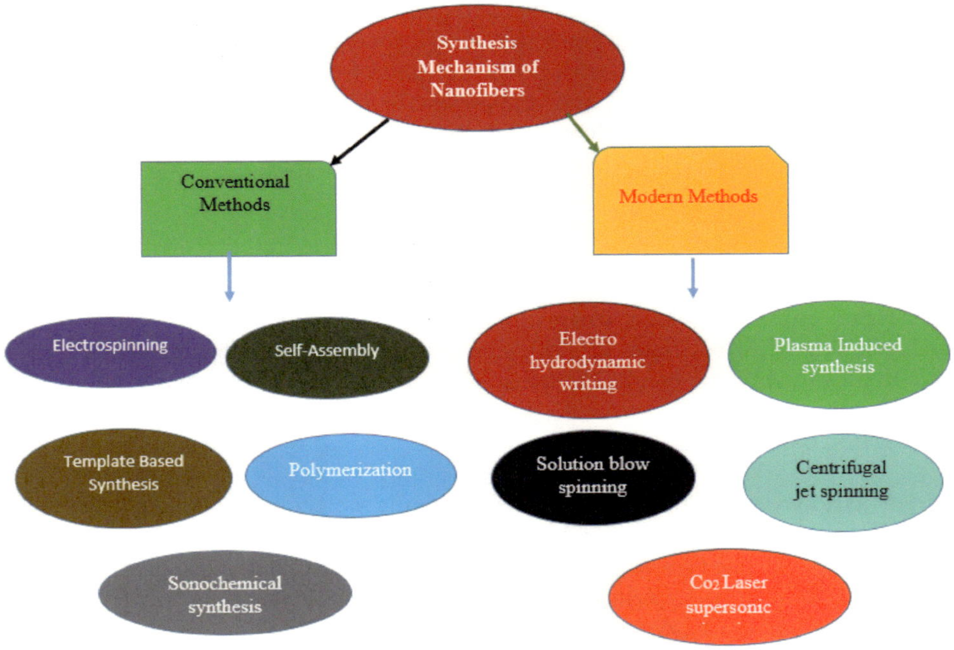

Figure 2.
Representation of Conventional and Modern Methods for synthesis of nano fibers.

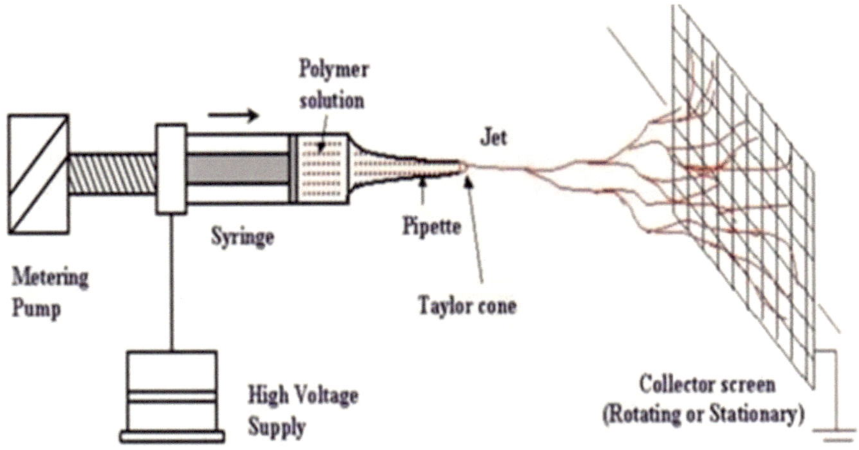

Figure 3.
Representation of nanofiber production with basic electrospinning method [32].

7.1 Self-assembly

Self-assembly is one of the technique used to produce a variety of nano fibers. Self-assembly peptides are capable of producing scaffolds that improvises luminescence efficiency of nano clusters. Natural structural materials are self-assembled by acquiring desirable properties such as mechanical strength, thermal stability and biocompatibility. Peptide and peptide amphiphiles nano fibers are synthesized using self-assembly (**Figure 4**) [33].

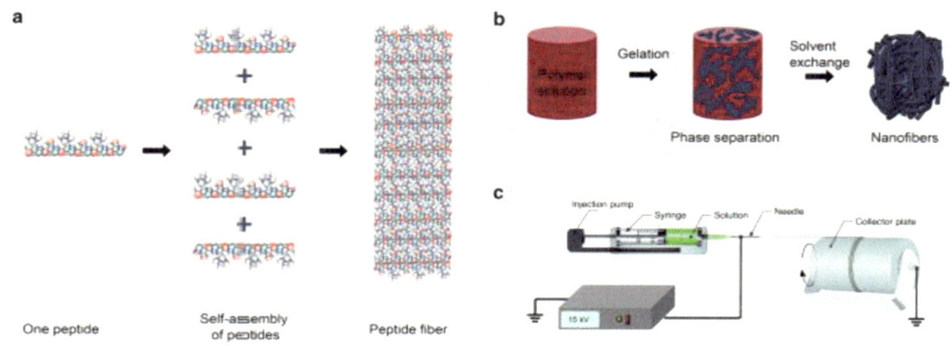

Figure 4.
Generation of nano fibers through self-assembly, phase separation and electron spinning [33].

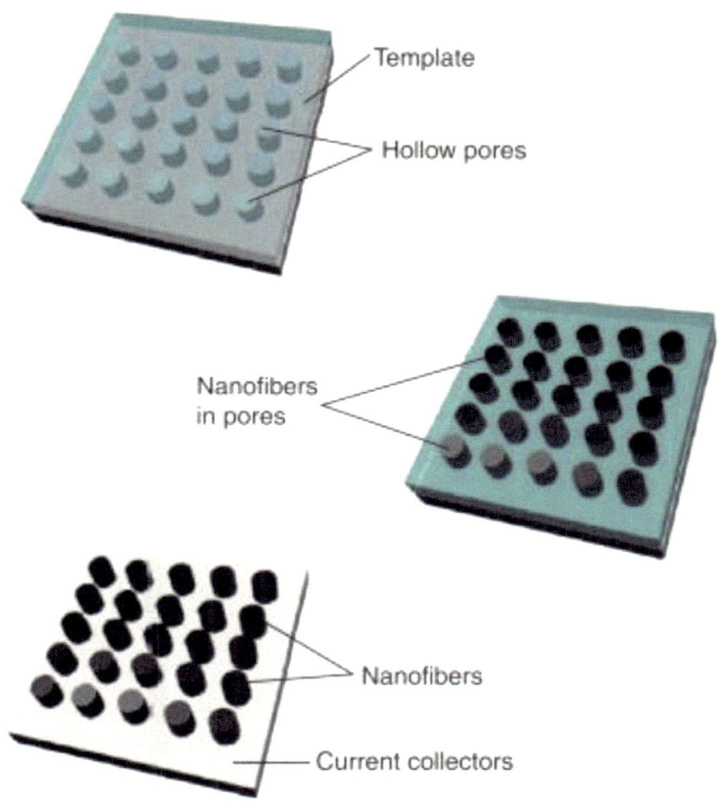

Figure 5.
Nano fibers from template synthesis [34].

7.2 Template synthesis

Nano fibrils and hollow nano fiber are produced by template synthesis mechanism. Nano fibrils are prepared within the microporous membrane or any solid pores. Desired morphology of nano fiber is obtained by pre-configuration. **Figure 5** shows a method for preparation of nanofibers. Sol–gel or electrodeposition are used to fill cylindrical pores at nano scale. Nano fibers are formed below the template [34].

7.3 Polymerization and sonochemical synthesis

Polymerization is a specific method used to synthesize nano fibers from polyaniline [35]. Three major methods are used to produce nano fibers. Chemical oxidative polymerization, interfacial synthesis and rapid mixing reactions [36]. A traditional way of obtaining fibers from polyaniline is the chemical oxidative polymerization. Polymerization takes place with the addition of aniline and an oxidant in acidic solution. By this method nano fibers without fine structure are formed. However, ultrafine nano fibrous material is produced when potassium biiodate is used as oxidant, then crystalline fine structured nano fibers were formed [37]. Polyaniline nano fibers were produced with homogenous nucleation. In this process overgrowth of molecules occurs and is controlled by the formation of nano fibers.

A powerful ultrasound irradiation is utilized for chemical reaction of molecules is sonochemical synthesis. Using this methodology, the molecules undergo high temperature and pressure conditions to produce varied range of nano structured materials. Polyaniline nano fibers can also be obtained from this method [38].

7.4 Electro hydrodynamic writing

Electro hydrodynamic writing is a modern technique that drew attention of many with great potential in rendering nano fibers into highly flexible and controllable substrates. This technology gives highly controllable, flexible aliened micro/nano fibers [39]. Output obtained through electro hydrodynamic writing in developing nano fibers is so popular due to low cost, high flexibility and intense applications in various fields. Direct writing is possible from mechanoelectrospinnig to obtain direct hierarchal nano or microfibers. Unlike the general electrospinning mechanisms this method uses mechanical force that stimulates positioning of fibers with controllable morphology. **Figure 6** shows the mechanism is used to produce nano fibers from assisted air flow with constrained force and additional stress that enhances jet stability. A versatile system is being developed for the production of ultrafine, highly flexible and stretchable electronics using nano fibers and there by applied in the formation of sacrificial structures [40].

7.5 Plasma induced synthesis

Plasma induced synthesis is used to generate nano particles in different shapes and nano fibers in five steps. (i) rapid bombardment of radicals on electrode surface (ii) atomic vapor deposition (iii) plasma expansion (iv) condensation of solution (v) in situ reaction of oxygen and growth of nano fibers [41]. In the production of nano fibers plasma generation is the key role. It is generated from discharge of pulse between electrodes in solution by direct current. Silver nano particles are induced on chitosan nano fibers are to confirm antibacterial activity [42].

7.6 Solution blow spinning

Solution blow spinning is a new emerging modern technique in fabrication of nano fibers. Both electrospinning and melt blowing elements are combined to organize this technique for the production of micro and nano fibers. Using this method production rate increases drastically. **Figure 7** shows the experimental arrangement of solution blow spinning. A syringe pump is used to deliver polymer

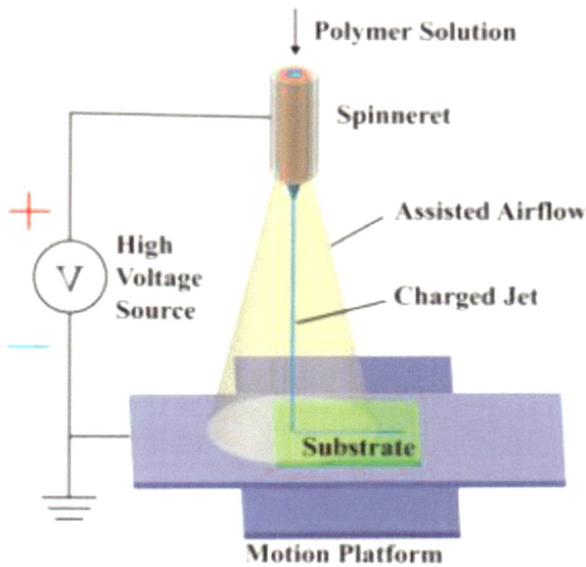

Figure 6.
Electro hydrodynamic writing to produce nano fibers [40].

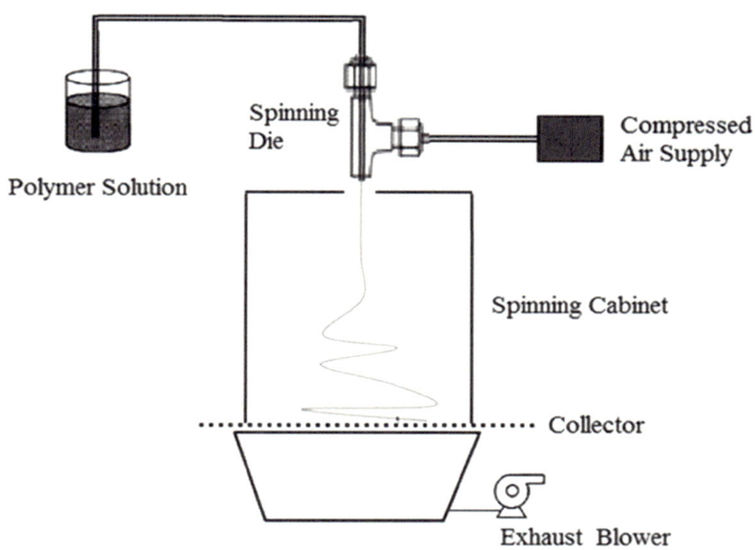

Figure 7.
Representation of solution blowing technique [43].

solution pumped through a nozzle under compressed air supply. This is used in situ deposition of nano fiber mats and mostly used for tissue engineering applications [43]. As an example nano fibers generated from solution blow spinning are used for composite air filter masks [44].

7.7 Centrifugal jet spinning

A new technique with high efficient, low cost and greater throughput to fabricate nano fibers is centrifugal jet spinning. **Figure 8** shows a dc motor with flexible air

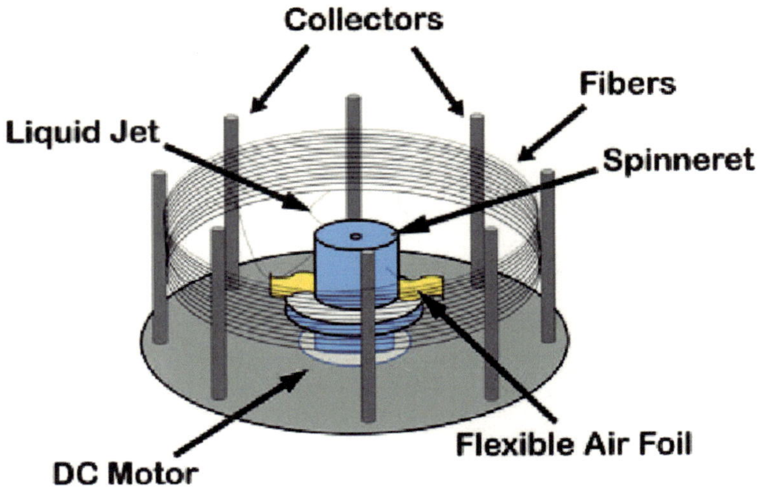

Figure 8.
Schematic diagram shows centrifugal jet spinning [45].

foil with liquid jet and spinneret is used to produce fibers. Two collectors are there to collect the obtained nano fibers [45]. When the centrifugal force overwhelms surface tension of polymer liquid material that stretches out the solution forming nanofibers in solid form. As an example a new spin nano fibers are used to produce nano fibers that is used as a fibrous mat scaffold for bone regeneration [46].

7.8 CO_2 laser supersonic drawing synthesis

This technique is a novel method to develop long nano fibers by irradiation with CO_2 laser at supersonic velocities. Melted fibers are gone through supersonic airflow to draw nano fibers in the range of diameters. Estimation of flow velocity is done by computer program. Through this process several natural and synthetic polymers are being used to yield nano fibers [47].

8. Applications

Numerous applications of nano fibers are fascinating world due to their use in generating energy, many biological, medical field, in defense, food industry, water and environment. Each of these fields have specific developments from nano fibers as shown in **Figure 9**.

8.1 Energy applications

Nano fibers emerged as best replacement for electrodes from anode and cathode materials used in lithium ion batteries. Many factors like electrochemical performance, limited capacity, high cost of materials affect the usage of lithium ion batteries as large scale storage devices [48]. Introduction of nano fibers as electrode materials paved way for generation of energy in batteries and fuel cells. The advantage of nano fibers being used as electrodes came from the properties such as high porosity, large surface area. The capability of nano fibers has extended from lithium-ion batteries to three dimensional interconnected networks. Metal organic frame works derived from metal oxides were intruded

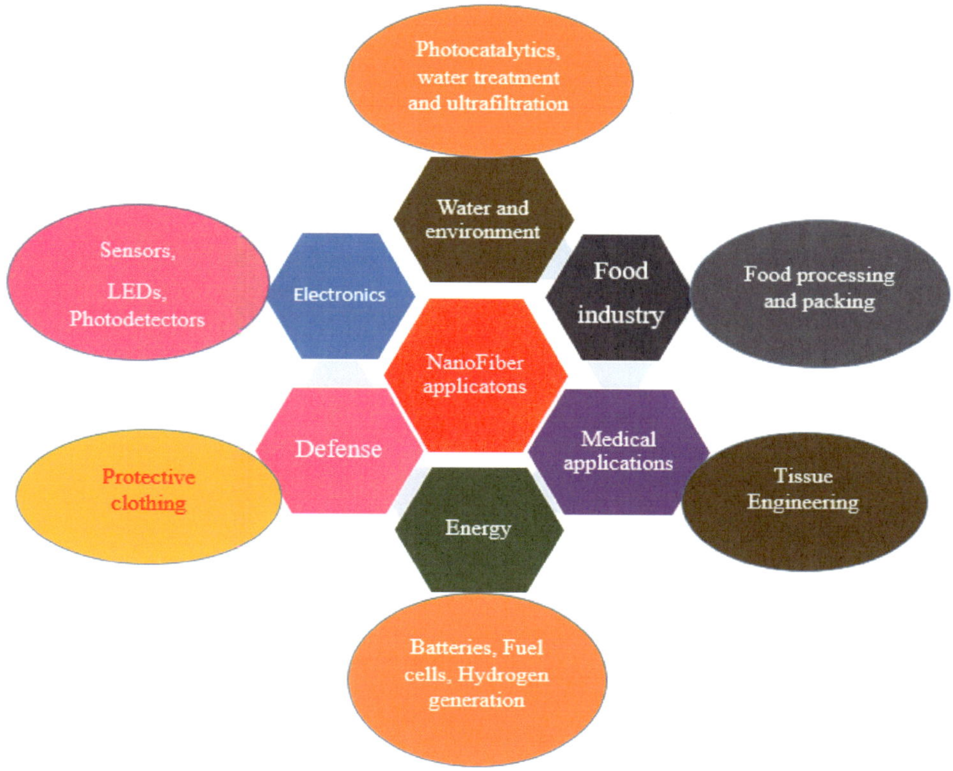

Figure 9.
Applications of nano fibers in various fields.

in carbon nano fibers through electrospinning to improve electrical conductivity with excellent rate capability [49].

Hydrogen is one of the future energy generation sources best carrier for renewable fuels due to its high energy content. Limitation in the availability, economical storage and generation are the challenges in the use of hydrogen as energy carrier. Nano fibered structures are synthesized to improve hydrogen storage.

8.2 Medical applications

Nano fibrous materials acts as best alternative scaffolds for tissue engineering as well as regenerative medicine [50]. Specific properties of nano fibers make them biologically active. Replacement of cells or tissues to exhibit proper body mechanism is the regenerative medicine. Transplantation of cells or tissues is done by various biologically active materials. Of them scaffolds developed from nano fibers are important as these are favorable layout for cellular growth, proliferation [51]. Natural or synthetic fibers are majorly used for tissue engineering as they have biocompatibility, biodegradability. As an example improved angiogenesis (development of new blood vessels) is improved through nano fibers for tissue engineering applications. Hard and soft tissues reconstruction is done by nano fiber scaffolds [52].

8.3 Food industry

The use of nano fibers especially cellulose nano fibers from various materials like fruit peel extracts, sea alga, bacteria, fungi have been utilized in food industry.

The use of nano fibers in food industry starts from packing, protecting aromatic and unstable compounds in beverages, monitors storage and in detection of pesticides [53]. Also nano sensors are used for quality assessment.

8.4 Water and environment

Morphology of nano fiber materials is attracting researchers due to numerous applications in environmental and water related issues in recent developments. Membranes of nano fibers developed from electrospinning technique shows greater prospective in waste water treatment and recycling. Many water treatment functionalities such as separation, adsorption, photo catalysis and antimicrobial activities can be treated effectively by nano fiber membranes Fiber mats prepared from nano fibers act as best filtration membranes that minimizes pressure drop with better efficiency than conventional mats. The contaminants in water and air were absorbed from the membranes of nano fibers with large surface to volume ratio and shows an increment in the life time of these mats [54].

8.5 Electronics and defense

Many micro and nano electrical devices like ultra-light weight space craft materials, electrostatic dissipation, nano solar cells, LCD devices are manufactured with nano fiber materials in electronics [55].

Types of nano fibers	Fabrication techniques	Advantages	Applications
Natural Polymers	Electrospinning, Template based synthesis, Electro hydrodynamic writing, CO_2 laser supersonic, centrifugal jet spinning, plasma induced synthesis	Natural Polymers are obtained from renewable sources, eco-friendly and low cost	Food industry, medical applications, water and environment, protective clothing
Synthetic Polymers	Electrospinning, Electro hydrodynamic writing, self-assembly, polymerization	Synthetic polymers provide optimal support to cell attachment with improved mechanical strength.	Wound dressing, filters, drug delivery, cytotoxicity studies
Semiconducting, metals and composite materials	Electrospinning, centrifugal jet spinning, solution blow spinning	With specific physical, chemical and mechanical properties nanofibers obtained from these materials intrudes into many devices.	Biomedical, optoelectronic, bio imaging and sensors
Carbon based materials	Electrospinning, sonochemical synthesis, template based synthesis	These are the materials which produce three dimensional graphene structures with high surface area, porosity, flexibility.	Super capacitor, air purifier, batteries and sensors

Table 1.
Types of nano fibers advantages and applications.

Nano fibers are the fantastic materials that found applications in defense. Face masks, chemical protective clothing liners, decontamination wipes etc. were prepared from nano fibers to improve their efficiency at low cost. Polyethylene oxide nano fibers serves as detoxifying substances against chemical war fares [56].

Table 1 shows the synthesis of nano fibers from various materials, their fabrication techniques, advantages and applications.

9. Conclusions

Nano fibers are excellent materials designed from various materials like natural, synthetic polymers, metals, semiconductors and many more. These nano fibers have been developing since ages till to date with lot of improvement in their synthesis mechanisms. With a drastic improved properties nano fibers have pronounced applications in different fields. The fabrication methods include conventional methods like electrospun, template based synthesis, polymerization, self-assembly and sonochemical methods and modern methods like electro hydrodynamic writing, plasma induced synthesis, solution blow spinning, centrifugal spinning and CO_2 laser supersonic technologies. New technologies to fabricate nano fibers are being developed day by day to meet the requirement of society. These nano fibers are synthesized from natural, synthetic polymers and from different materials like semiconductors, carbon based materials and many more. Applicative orientation makes these nano fibers a specific alternative to non- renewable sources. Still researchers are keenly looking into the real time applications of these nano fibers at huge.

10. Challenges and future research work

There is a huge demand of nano fibers due to their numerous applications in various fields. These acts as best alternatives for solving many life leading problems that include water and purification, health related issues, electronics, energy derivatives etc. Nano fiber entrustment in all areas with commercial acceptance is the present challenge to the scientific world. Novel development in the fabrication of nano fibrous materials is the ongoing and future research work.

Author details

Deepthi Sista
Department of Science and Humanities, Lendi Institute of Engineering and Technology, Vizianagaram, India

*Address all correspondence to: deepthi.sista@lendi.org

References

[1] M.F. Rosa, E.S. Medeioros, J.A. Malmonge, K.S. Greogo, D.F. Wood, L.H.C Mattoso, Gleen, W.J. Orts, S.H. Imam, Cellulose nanowhiskers: Effect of preparation conditions on their thermal and morphological behavior, Vol.81, no.1, pp.83-92, 2010.

[2] HuiminGao, BoDuan, AngLu, HongbingDeng, YuminDu, XiaowenShi, LinaZhang, Fabrication of cellulose nanofibers from waste brown algae and their potential application as milk thickeners, Vol.79, pp.473-481, 2018

[3] Yaodong Liu and Satish Kumar, Polymer/Carbon Nanotube Nano Composite Fibers–A Review, *ACS Appl. Mater. Interfaces,* No.6, PP.6069–60872014,

[4] Kenry, Chwee Teck Lim, Nanofiber technology: current status and emerging developments, *Progress in Polymer Science*, No. 70, pp. 1-17, 2017

[5] Mina Keshvardoostchokami, Sara Seidelin Majidi, Peipei Huo, Rajan Ramachandran, Menglin Chen and Bo Liu, Electrospun Nanofibers of Natural and Synthetic Polymers as artificial Extracellular Matrix for Tissue Engineering, *Nano materials,* Vol.11, No.1, pp.1-21, 2020

[6] WaruneeTanan, JatePanichpakdee and SayantSaengsuwan, Novel biodegradable hydrogel based on natural polymers: Synthesis, characterization, swelling/reswelling and biodegradability, *European Polymer Journal*, Vol. 112, Pages 678-687, 2019

[7] Sairish Malik, Subramanian Sundararajan, Tanveer Hussain, Ahsan Nazir, Muhammad Ayyoob, Filippo Berto, Seagram Ramakrishna, Sustainable nanofibers in tissue engineering and biomedical applications, *material design and processing communication,* special issue, 2020.

[8] K. S. Silvipriya, K. Krishna Kumar, A. R. Bhat, B. Dinesh Kumar, Anish John, Panayappan lakshmanan, Collagen: Animal Sources and Biomedical Application, *Journal of Applied Pharmaceutical Science* Vol. 5, No.03, pp. 123-127, 2015

[9] Mahuya Das, Rupa Bhattacharyya, Cellulose Nanofibers: Synthesis, Properties and Applications, *Wiley online library*, chapter 1, 2015

[10] Yuanyuan Li, Hongli Zhu, Yibo Wang, Upamanyu Ray, Shuze Zhu, Jiaqi Dai, Chaoji Chen, Kun Fu, Soo-Hwan Jang, Doug Henderson, Teng Li, Liangbing Hu, Cellulose-Nanofiber-Enabled 3D Printing of a Carbon-Nanotube Microfiber Network, *Wiley online library*, 2017

[11] Zhi Liu, Feng Zhang, Jinfa Ming, Shiyu Bie, Jungians Li, Baoqi Zuo, Preparation of Electrospun Silk Fibroin Nanofibers from Solutions Containing Native Silk Fibrils *Journal of applied polymer science,* pp.1-7, 2015.

[12] Weiqin Sheng, Jing Liu, Shanshan Liu, a Qiang Lu, David L. Kaplanac and Hesun Zhud, One-step synthesis of biocompatible magnetite/ silk fibroin core–shell nanoparticles, Journal of materials Chemistry B, Vol. 2, pp. 7394-7402, 2014.

[13] SenaSu, TubaBedir, CevriyeKalkandelen, AhmetOzan Başar, HilalTurkoğlu Şaşmazel CemBulent, Ustundag, MustafaSengor, OguzhanGunduz, Coaxial and emulsion electrospinning of extracted hyaluronic acid and keratin based nanofibers for wound healing applications, *European polymer Journal*, Vol.142, 2021

[14] ChilakamarryChaitanya Reddy, Irshad AhamadKhilji, ArunGupta, PrakashBhuyar SyedMahmood Khater AhmedSaeed AL-Japairai, Gek

KeeChua, Valorization of keratin waste biomass and its potential applications, *Journal of Water Process Engineering,* Vol.40, 2021.

[15] Angela Edwards, David Jarvis, Tracy Hopkins, Sarah Pixley, Narayan Bhattarai, Poly(e-caprolactone)/ keratin-based composite nanofibers for biomedical applications, *Journal of Biomedical Material Research* Part B,2015

[16] Nataliya Babayevska, Łucja Przysiecka, Grzegorz Nowaczyk, Marcin Jarek, Martin Järvekülg, Triin Kangur, Ewa Janiszewska, Stefan Jurga and Igor Iatsunskyi, Fabrication of Gelatin-ZnO Nanofibers for Antibacterial Applications, vol.103, No.14,2021

[17] Y.Z. Zhang, J. Venugopal, Z.-M. Huang C.T. Lim, S. Ramakrishna, Crosslinking of the electrospun gelatin nanofibers, *Polymer,* Vol.47, No.8, pp. 2911-2917, 2006

[18] Rejane C. Goy, Douglas de Britto, Odilio B. G. Assis, A review of the antimicrobial activity of chitosan, *Polímeros,* vol.19 no.3,2009

[19] EmoChiellini, AndreaCorti, SalvatoreD'Antone, RobertoSolaro, Biodegradation of poly (vinyl alcohol) based materials, *progress in polymer science*, Vol.28, No.6, pp.963-1014, 2003.

[20] Yaodong Liu, Satish Kumar, Polymer/carbon nano composite fibers-A review, ACS *Applied Material and Interfaces*, Vol.6, No.9, pp.6069-6087, 2014

[21] Ying Zhao, Yihui Qiu, Huanhuan Wang, Yu Chen, Shaohua Jin, and Shuseng Chen, Preparation of Nanofibers with Renewable Polymers and Their Application in Wound Dressing, *International Journal of Polymer Science*, Vol. 2016, Article ID 4672839, 17 pages, 2016.

[22] Enas M. Elmowafy, Mattia Tiboni, Mahmoud E. Soliman, Biocompatibility, biodegradation and biomedical applications of poly (lactic acid)/poly (lactic-co-glycolic acid) micro and nanoparticles, *Journal of Pharmaceutical Investigation*, Vol. 49, pp.347-380, 2019.

[23] Ibrahim Khan, Khalid Saeed, IdreesKhan, Nanoparticles: Properties, applications and toxicities, *Arabian Journal of Chemistry*, Vol.12, Issue 7, pp.908-931, 2019

[24] Himchan Cho, Sung-Yong Min, Tae-Woo Lee, Electrospun organic nanofiber electronics and photonics, *Macromolecular Materials and Engineering,* Vol.298, No.5, 2013.

[25] Kaushik Mallick, Mike J. Witcomb, Andy Dinsmore, Mike S. Scurrell, Fabrication of a Metal Nanoparticles and Polymer Nanofibers Composite Material by an in Situ Chemical Synthetic Route, Langmuir, Vol.21, No.17, pp.7964-7967, 2005

[26] HyunjinCho, YeonhoKim, Yong JuYun Kyu SeungLee JaehoShim Chil-HyoungLee Jin WonSeo Won G, Hong Hae JinKim Hak YongKim Dong IckSon, Versatile 3D porous recycled carbon garments with fully-loaded active materials in the current collector for advanced lithium-ion batteries, *Composites part B: Engineering,* Vol.179, 2019.

[27] Ibrahim Alghoraibi, Different methods for nanofibers design and fabrication, nano fibers scaffolds for tissue engineering applications, *Springer,* 2018

[28] RuiZhao, XiaofengLu, CeWang, Electrospinning based all-nano composite materials: Recent achievements and perspectives, *Composites Communications*, Vol.10, pp.140-150, 2018

[29] V. Thavasi, G. Singh and S. Ramakrishna, electrospun nano fibers in energy and environment applications, Energy& Environmental Science, No.2, 2008

[30] Jiajia Xue, Jingwei Xie, Wenying Liu, Younan Xia, Electrospun Nanofibers: New Concepts, Materials, and Applications, *Accounts of Chemical Research*, Vol.50, No.8, pp.1976-1987, 2017

[31] Himchan Cho, Sung-Yong Min, Tae Woo Lee, Electrospun Organic Nanofiber Electronics and Photonics, Macromolecular Materials and Engineering, Vol.298, No.5, 201

[32] Hosne Ara Begum, Md. Khalilur Rahman Khan, Study on the Various Types of Needle Based and Needleless Electrospinning System for Nanofiber Production, International Journal of Textile Science, Vol.6, No.4, pp.110-117, 2017

[33] Sorour Nemati, Se-jeong Kim, Young Min Shin and Heungsoo Shin, Current progress in application of polymeric nanofbers to tissue engineering, *Nemati et al. Nano Convergence,* Vol.6, No.36, 2019

[34] B. Meyer, F. Croce, Template Synthesis, Encyclopedia of Electrochemical Power Sources, 2009

[35] Jiaxing Huang, Richard B. Kaner Prof, Nanofiber Formation in the Chemical Polymerization of Aniline: A Mechanistic Study, *Angewandte Chemie,* Vol.116, No.43, 2004

[36] Yibo Zhao, Huige Wei, Moses Arowo, Xingru Yan, Wei Wu, Jianfeng Chen, Yiran Wang and Zhanhu Guo, Electrochemical energy storage by polyaniline nanofibers: high gravity assisted oxidative polymerization vs. rapid mixing chemical oxidative polymerization, Phys. Chem. Chem. Phys., Vol.17, pp.1498—1502, 2015

[37] Abdelaziz Rahy, Duck Yang, Synthesis of highly conductive polyaniline nanofibers, *Material Letters*, Vol.62, No.28, pp.4311-4314, 2008

[38] Xinili Jing, Yangyong Wang, Dan Wu, Jipeng Qiang, Sonochemical synthesis of polyaniline nano fibers, *Ultrasonics Sonochemistry*, Vol.14, No.1, pp.75-80, 2007

[39] Zhenfang Zhanga, Haijun He, Wanlin Fua, Dongxiao Ji, Seeram Ramakrishna, Electro-Hydrodynamic Direct-Writing Technology toward Patterned Ultra-Thin Fibers: Advances, Materials and Applications, *Nano today*, G Model NANTOD-100942; 2020

[40] Jiaxin Jiang, Xiang Wang, Wenwang Li, Juan Liu, Yifang Liu and Gaofeng Zheng, Electrohydrodynamic Direct-Writing Micropatterns with Assisted Airflow, *Micromachines*, Vol.9, 2018

[41] Xiulan Hu, Xin Zahang, Xiaodong, Shen, Hongtao Li, Osamu Takai and Nagahiro Saito, Plasma –Induced Synthesis of CUO Nanofibers and ZNO Nanoflowers in Water, Plasma Chemistry and Plasma Processing, Vol.34, pp.1129-1139, 2014.

[42] Dhyah Annur, Zhi-Kai Wang, Jiunn-Der Liao, Changshu Kuo, Plasma-Synthesized Silver Nanoparticles on Electrospun Chitosan Nanofiber Surfaces for Antibacterial Applications, *Biomacromolecules,* Vol.16, No.10, pp.3248-3255, 2015

[43] Glebert C. Dadol, AliKilic, Leonard D. Tijing, Kramer Joseph A, Lim, Luis K. Cabatingan, Noel Peter B. Tan, ElenaStojanovska, YusufPolat, Solution blow spinning (SBS) and SBS-spun nanofibers: Materials, methods, and applications, *Materialstoday Communications*, Vol.25, 101656, 2020

[44] Noel Peter B. Tan, Shierlyn S. Paclijan, Hanah Nasifa M. Ali,Carl

Michael Jay S. Hallazgo, Chayl Jhuren F. Lopez and Ysabella C. Ebora, Solution Blow Spinning (SBS) Nanofibers for Composite Air Filter Masks, *ACS Applied. Nano Materials,* Vol.2, No.4, pp.2475-2483. 2019

[45] Subhash Singh, Nanofiber electrodes for Biosensors, *Handbook of nano fibers, Springer international publishing,* 2018

[46] AmalorpavaMary, Loordhuswamy, SenthilramThinakaran, Giri DevVenkateshwapuram Rangaswamy, Centrifugal spun osteoconductive ultrafine fibrous mat as a scaffold for bone regeneration, *Journal of Drug Delivery Science and Technology,* Vol.60, 101978,2020.

[47] J. Penidea, F. Quinteroa, J. del Vala, R. Comesañab, F. Lusquiñosa, A. Riveiroc, J. Poua, Laser spinning: a new technique for nanofiber production, *Physics Procedia,* Vol.56, pp.365-370, 2014

[48] ZhengsiHan, FanjunKong, JihuiZheng, JiyunChen, ShiTao, BinQian, MnSe nanoparticles encapsulated into N-doped carbon fibers with a binder-free and free-standing structure for lithium ion batteries, *Ceramics International,* Vol.47, No.1, pp.1429-1438,2021.

[49] Sanjeev Gautam, Harshita Agrawal, Manisha Thakur, Ali Akbari,Hemam Sharda, Rajwant Kaur Mojtaba Amini, Metal oxides and metal organic frameworks for the photocatalytic degradation: A review, *Journal of Environmental Chemical Engineering,* Vol.8, No.3, 103726, 2020

[50] Sakthivel Ngarajan, S.Narayana Kalkura, Sebastien Balme, Celine Pochat Bohatier, Philippe Miele, Mikhael Bechelany, Nano Fibrous Scaffolds for Tissue Engineering Application, *Handbook of Nanofibers*, pp.1-28, 2018

[51] Lisha Zhu, Dan Luo and Yan Liu, Effect of nano/microscale structure of biomedical scaffolds on bone regeneration, *international journal of oral science*, Vol.12, No.6, 2020

[52] Simin Nazarnezhad, Francesco Baino, Hae-Won Kim, Thomas J.Webster and Saeid Kargozar, Electrospun Nanofibers for improved Angiogenesis: Promises for Tissue Engineering Applications, *Nano Materials,* Vol.10, No.8, 2020

[53] JingTian, HongbingDeng, MengtianHuang, RongLiu, YangYi, XiangyangDong, Electrospun Nanofibers for Food and Food Packaging Technology, *Electrospinning: Nanofabrication and Applications, Micro and Nano Technologies*, pp.455-516, 2019

[54] HaishengChen, ManhongHuang, YanbiaoLiu, LijunMeng, MengdieMa, Functionalized electrospun nanofiber membranes for water treatment: A review, *Science of the Total Environment,* Vol.739, 139944, 2020.

[55] Jian Fang, Hao Shao, Haitao Niu, Tong Lin, Applications of Electrospun Nanofibers for Electronic Devices, *Hand book of smart textiles*, pp.617-652, 2015

[56] Seshadri Ramkumar, T.Subbiah, M.M.Hussain, Nano fibers for defence and value added applications, *International Nonwovens Technical Conference*, 2006

Chapter 2

Green Synthesis of Nanofiber and Its Affecting Parameters

Dan Bahadur Pal and Deen Dayal Giri

Abstract

Nanofibers, the widely applied in various field of science research, is one of the important area in nanotechnology research. Nanofibers can be classified into polymeric, ceramic and composite nanofibers depending upon the material used. A variety of nanofibers are applied in field of energy storage, biotechnology and healthcare industry, environmental engineering, as well as security and defense. The wide uses of nanofibers are mainly due to low density, high porosity, tight pore size and large surface area per unit mass. Synthesis of nanofibers depends upon various parameters of solution like molecular weight of polymer, concentration, electrical conductivity, surface tension and viscosity. The process parameters affecting nanofibers synthesis are distance between needle tip and collector, feeding rate of polymer material and electric field.

Keywords: green synthesis, nanofibers, solution parameters, process parameters

1. Introduction

The basic purpose of the green chemistry is to diminish waste generation by complete consumption of material in the processes, or eliminate the generation and avoiding the processes that hazardous waste for human health [1]. Industrial and academic researchers are now more focusing on 'green' polymers from nature, instead of polymers derived from petroleum because of their sustainable environment-friendly nature, easy biodegradability and less energy need for renewing [2]. Fibers having diameters ~100 nm or lower exhibit special features are called nanofibers. These fibers have unique high surface area with respect to the mass compared to conventional fibers. They have high surface area ~ 1000 m^2/g, high porosity, tight pores and low density. These features of materials are desirable for their application in different fields [2]. Some fields in which polymeric, ceramic and composite nanofibers are used include healthcare industry, biotechnological applications, environmental engineering, defense, security and energy storage. Researchers are interest to synthesize nanomaterials with special physical and chemical features for their application in the above mentioned fields. The extensive improvement has been recorded in the area of nanoparticles, nanotubes, nanofibers, nanolayers, nanodevices and nano-structured biological materials.

Providing clean drinking to public and improving their general health status is an important field of research and application of nanomaterials. Nanomaterials adsorb metals from the water on their surface and detoxify it [3]. The nanomaterials with large surface area naturally have high sorption capacities and less disposable waste generation. In nanostructures more unsaturated surface atoms get exposed, proximal functional groups enhance reactivity and show nanoscale effect with decrease in particle size <20 nm diameter. A nanoparticle can have high adsorption for a selected metal [3]. Oxides of aluminum, iron, manganese and titanium have been investigated for their potential heavy metal adsorption from water. Titanium dioxide (TiO2) anatase has been applied in many industries. Bulks as well as nanoparticles of Titanium dioxide (Titanium dioxide (TiO2)) were able to degrade and remove organic compounds in presence of UV light and redox reactions in water [3].

The suitability of the sorbent for treating an inorganic pollutant depends on the cost effectivity and technical applicability. It is always necessary to develop an efficient and cheap adsorbent for the removing heavy metals from the industrial effluents, especially in the economically weak developing countries. Polymeric ceramic nanofibers are good option in this regard. For unit mass ceramic nanofibers harbor extraordinary high surface area and porosity. These low cost fiber show excellent structural and mechanical properties like high axial strength and extreme flexibility for a low basis weight [4].

Iijima (1991) discovered carbon nanotubes (CNTs) having excellent properties and applications. Depending upon layers of carbon atom in the wall, the tubular sheets of graphite are called single-walled CNT or multi-walled CNT [4]. Many heavy metals (Cr, Cu, Cd, Ni and Pb,) have been removed from waste water using CNTs [4]. The actual mechanisms by which metal ions get adsorbed onto CNTs is little bit complicated, however, possible it involves electrostatic attractive forces, sorption-precipitation or other chemical interaction between the metal ions and functional groups on the CNT surface.

The polymers applications are generally limited due to their poor mechanical strength and low antimicrobial resistance. A modification of polymers rectifies the limitations and improves wound healing property. A prominent biopolymer polylactide (PLA) has been exploited immensely in biomedical field due to biodegradability and biocompatibility [5]. Scaffolds of PLA temporary provide structural support to cells and tissue during healing by modulating cellular response that is helpful tissue engineering [6]. Wounds, cuts and damages skin heal rapidly in absence of germ because microbes on such surface trigger the immune response of our body and inflammation damage tissue leading to compromised self-healing. If such polymer scaffolds could contain antimicrobial agents capable to prevent opportunistic microbe on the wound surface, the healing process could be accelerated. When silver nanoparticles (AgNPs) impregnated into the polymer matrix, they impressive impart antimicrobial activity to scaffold in addition to improving mechanical, chemical, catalytic activity. Biomedically Ag-NPs have been applied in wound dressing material and in diagnosis of cancer [7] and sutures [8]. In addition to the antimicrobial activity, non-toxicity AgNPs are utilized in the fabricating non-infectious scaffolds. AgNPs are synthesized by using reducing chemical agents such as sodium borohydride [9], hydrazine etc. However, they have been extensively synthesized by use of medicinally important plant extracts [10]. The colloidal AgNPs has been synthesized using extracts of *Cocciniaindica* [11], *Carica papaya* [12], *Brassica rapa* [13], *Aloe vera* [14], *Melia dubia* [15], *Citrus* sp. [16], *Acalyphaindica* [17], *Prunusamygdalus* [18].

It is the cationic property that gives Chitosan (CHT) the ability to penetrate mucous layers in biomedical applications, perform an antimicrobial function in food preservatives, and trap dyes and metals in waste water [19]. It is also being used in technical applications such as packaging and decontamination, because of its physico-chemical properties, such as hydrophobicity, thermal stability, and mechanical performances [20]. These properties, unique to CHT, combined with high porosity and surface area by its nanosize, render it important for a range of different functional systems.

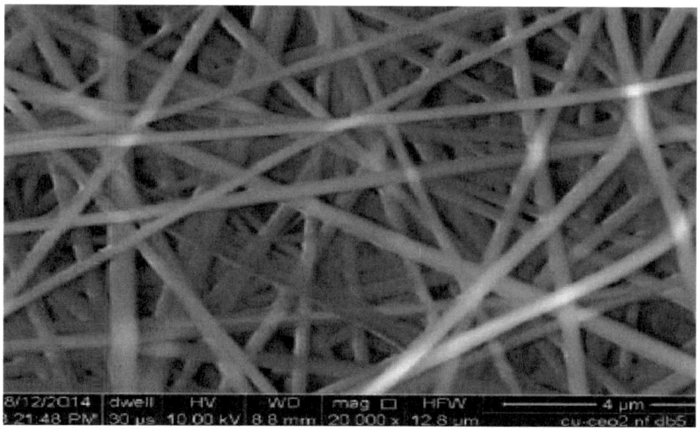

Figure 1.
Scanning electron micrograph of composite nanofibers synthesized from the PVP polymer by electrospinning process.

Hard and inert ceramic materials have excellent mechanical and thermal properties, in addition to superb chemical and corrosion resistance. These characteristics make them suitable for being used in electrodes, photonic devices, electronic and sensors, catalyst supports, drug delivery system and environmental science. The nanoribbons, nanorods, nanotubes, nanowires nanowhiskers and nanofibers are important nanostructures that are predominantly being synthesized in field of nanotechnology. These one-dimensional ceramics are interesting due to their unique optical, thermal, electrical, magnetic, gas sensing and or catalytic property. Such property of nanomaterial develop due to specific surface morphology and very small dimension compared to the same material in bulk. Among these nanostructures, highly porous, low density and high surface area containing nanofibers have been potentially applied in various fields [21]. A scanning electron micrograph of composite copper ceria nanofiber synthesized is given below in **Figure 1**.

2. Preparation methods of nanofibres

A number of methods have been used for fabricating nanofibers. Some of them are template synthesis [22], self-assembly [23, 24], phase separation [23], melt-blown [25] and electrospinning [26, 27]. Each process has its own challenge in preparing nanofibers. Selection of process for producing nanofibers depends on materials, fibers alignment, production rate and most importantly investment cost. Some of them are discussed below and only electrospinning process has been covered in details.

2.1 Template synthesis

The method employs a template or mold of desired material and structure is used to synthesize nanofibers. Generally, a templet of metal oxide membrane of nano-dimension pores is allowed to pass the polymer solution through it by applying water pressure from one side. The nanofibers extrude from the other side of the membrane. The generated fiber comes in contact with solidifying solution. Small sized nanofibers few micrometers length are generated. The fiber diameter is determined by the membrane pore size [28, 29]. The technique is very advantageous in fabricating nanofibers of various diameter by changing templates membrane.

2.2 Phase separation

In this method phases separate because of the physical incompatibility. The phase of the solvent then extracted from the solution while the other phase remain. This method consists mainly of four basic steps:

i. Homogeneous polymer solution is prepared by dissolved the polymer in a suitable solvent.

ii. Gelation of the solution to produce nanofiber matrixes. It is the most difficult step controlling porosity and morphology of the nanofiber. Gelation varies depending upon concentration of polymer and ambient temperature.

iii. Extraction of the solvent from the gel.

iv. Freezing and freeze drying.

The process is not equipment extensive in fabricating nanofiber matrix. The mechanical properties of matrix can be adjusted changing concentration of polymer [30]. By increasing polymer concentration, fiber mechanical properties are improved and porosity gets decreased [29].

2.3 Self-assembly

In self-assembly small building blocks spontaneously organized to build-up the stable nanofibers of very thin diameter. In such nano-material fabrication, building block molecules organize and arrange itself into definite patterns or structures due hydrophobic and electrostatic interactions and hydrogen bonding [31]. The intermolecular forces bring units together to macromolecular nanofiber. The technique produce minute nanofibers of >100 nm to micrometers length. This nanofiber fabrication is complex, long, and extremely elaborate associate with limitation of low yield and uncontrollable fiber dimensions. Further, this method can be used to prepare nanofibers from selected molecules having capability of self-assemble themselves or under an external stimulus [29, 32].

2.4 Freeze drying

The technique is also known as ice segregation-induced self-assembly or solid–liquid phase separation. The technique comprises of following steps-

i. Freezing- polymer solution is frozen at a low temperature

ii. Primary drying- removal of water from frozen material by sublimation in a chamber have reduced pressure of few millibars and

iii. Secondary drying- removal of unfrozen water from the polymer material by desorption.

Important advantages of freeze drying over other techniques of nanofibers fabrication is its ability to produce the porous structures of controlled sizes directly from polymers (e.g. chitin), without addition of structure directing additives and pre-treatment.

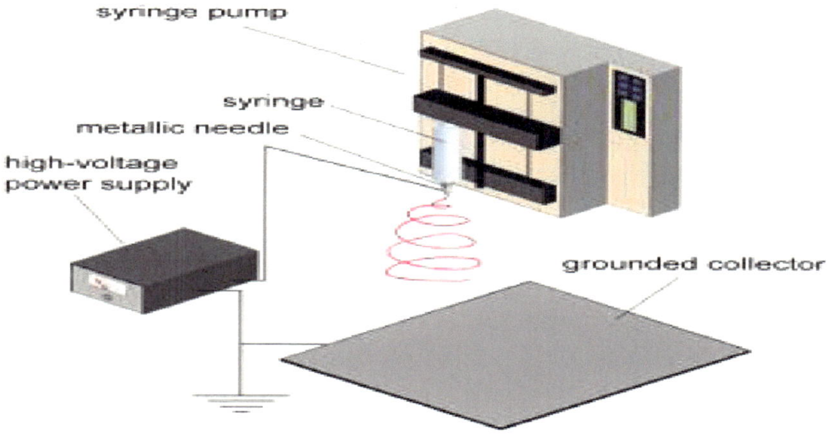

Figure 2.
Schematic representation of the electrospinning.

2.5 Electrospinning (ES)

In the field of nanotechnology electrospinning is the most promising remarkably simple processes for generating nanofibers from polymers solution. In combination with conventional sol–gel process, it offers a versatile technique for producing solid, porous, or hollow structured ceramic nanofibers. The technique has earned enormous attention because of its easy use and flexibility in controlling diameter in range of nanometers to micrometer and alignment of nanofibers as well as continuous nanofibers production capability. Electrospinning is peculiar in the sense that can be useful in producing fine nanofibers from different solution or melt of polymer, ceramic material and composite material for fast synthesis of nanofibers at industrial scale.

Formhals patented a process in 1934 for an experimental setup describing nanofiber production from polymer solution using electrostatic force, termed as electrospinning [33]. Now days, electrospinning is explored as a high efficiency method for the generating ceramic nanofibers [34]. In the generic design, an electrospinning setup consists of a high-voltage power supply, a metallic needle called spinneret, a piece of aluminum foil or silicon wafer acting as an electrically conductive collector and a syringe pump. A plastic syringe loaded with the polymer solution and connected to the metallic needle is often connected to a syringe pump for constant and adjustable feeding rate of the solution. In some cases, especially for electrospinning of ceramic nanofibers, the setup needs to be placed in a closed box for controlled humidity variation.

The collector can be constructed in different configuration from various materials depending on the end use of nanofiber. The electrospinning process is generally performed at room temperature at atmospheric conditions [35]. The ES system continuously need a high voltage (10-40kv) power supply of 40 kV, syringe pump, syringe, metallic needle and an electrically conductive collector plate during nanofiber synthesis. The electrospinning setup is schematically represented in **Figure 2** given below.

3. Parameters affecting the performance of electrospinning

Many parameters affecting ES process can be classified broadly into solution parameters, process parameters, and ambient parameters. Molecular weight, concentration, conductivity, viscosity and surface tension are then solution parameters whereas feeding or flow rate, tip to collector distance and electric field are important process parameters. All these parameters individually and synergistically affect the final fibers morphology. Researcher used to adjust the above mentioned parameters to synthesize nanofibers of desired diameter and morphology and alignment [36]. Further, ambient environmental parameters like the temperature and humidity also affect electrospinning process and nanofiber morphology and diameter [37].

3.1 Solution parameters

3.1.1 Solution concentration

Nanofiber synthesis by ES process needs a minimum concentration of polymer in the solution for continuous nanofiber synthesis, the concentration below it, result in mixture of beads and fibers. The increase in concentration of solution the beads shape changes from spherical to spindle fibers and further increase in concentration synthesis uniform fibers with broad diameters due to the high viscosity resistance [38–41]. The range of concentration synthesizing continuous fibers is determined by surface tension and viscosity of solution. During ES process, solution concentration above optimum prohibits continuous fiber formation because of the inability to maintain the flow of the solution at the tip of the needle leading to the formation of larger fibers.

3.1.2 Molecular weight

Another important parameter is molecular weight of polymer that has strong influence on the morphology of electrospun nanofiber. Molecular weight of polymer affects rheology, electrical properties, conductivity, surface tension, viscosity and dielectric strength [42]. In ES process, generally high molecular weighted polymer offers the desired viscosity for the nanofiber synthesis whereas too low a molecular weighted polymer tends to form beads instead of fibers frequently. But very high molecular weighted polymers synthesize fibers of broad average diameters. Polymer's molecular weight and number polymer chain entanglements affect viscosity of the polymer solution. Polymer chain entanglement significantly determines the fiber synthesis in the electrospinning process.

3.1.3 Solution viscosity

Viscosity of solution is detrimental for the size of fiber and morphology during ES of fiber from polymer solution. The polymer solutions with very low viscosity lack continuous fiber formation whereas polymer solutions having very high viscosity results in difficult ejection of jets of polymer solution from the needle. So, an optimum viscosity is needed for ES process. Viscosity was shown to affect silk nanofibers synthesis by Sukigara and colleagues [43]. In the low viscosity solutions, surface tension becomes dominant factor. The increase in solution viscosity or concentration results in large sized fibers of uniform diameter [38].

3.1.4 Surface tension

Surface tension plays a critical role in the ES process. Surface tension of a solution is more likely a function of solvent compositions. So, the surface tension of polymer solution can be altered by using different solvents. The formation of droplets, bead and fibers depends on the surface tension of polymer solution. If the surface tension of polymer solution is high, jet will be instable and the droplets will be sprayed from the tip of the needle leading to inhibition of ES process [44]. The reduced surface tension of a polymer solution will synthesize nanofibers without beads. The polymer solution with surface tension can of the spinning solution helps in ES process to occur at a relatively low electric field.

3.2 Process parameters

3.2.1 Voltage applied

Applied voltage in the ES process is a crucial process parameter. The applied voltage should attainment of threshold value for starting synthesis of fiber formation. The voltage induces the necessary charge and electric field on the solution to initiates the ES process. In the most cases, a high voltage greatly stretch of the solution because of greater columbic forces in the jet and a strong electric field that result in reduced fiber diameter and rapid evaporation of solvent from the fibers. Thus voltage influences fiber diameter at an extent, but the level of significance varies with the polymer solution concentration and on tip to the collector distance [45].

3.2.2 Feed rate/flow rate

The feed rate of the polymer solution is another important process parameter influencing velocity of jet from needle tip and the material transfer rate. Low feed rate is almost always desirable for getting enough time for evaporation of solvent from the polymer solution during fiber synthesis [46]. The spinning polymer solution should have a minimum flow rate for ES process to occur. Greater polymer solution flow rates forms beaded fibers due to improper smaller drying time period before reaching to the collector.

3.2.3 Types of collectors

A collector in ES process is a conductive substrate on which nanofibers are deposited. Collector is an important process parameter. Aluminum foil is one of the most common collectors in the ES process. To overcome the difficulty to transfer of collected fibers and necessity of aligned fibers for different applications, other collectors could also be conductive paper or cloth, rotating rod or wheel, pin, wire mesh parallel or gridded bar [47].

3.2.4 Distance between tip to collector

Fiber diameters and morphology could be controlled by manipulating the tip to collector distance. However, a minimum distance between collector and needle tip is essential for giving sufficient time for fibers dry before being deposited on the collector. The beads are frequently observed in the cases where tip to collector distance are either too close or too far. The optimum distance between the tip and collector are adjusted depending upon the polymer solution used for proper evaporation of solvent from the nanofibers [37].

3.3 Ambient parameters

In addition to the solution and process parameters, there are some ambient parameters such as humidity, temperature etc. which influence the ES process.

Increase in ambient temperature yield fibers of relatively decreased diameter which can be attributed the decrease in the viscosity of the polymer solutions at high temperatures. In the very low humidity environment, a volatile solvent will rapidly evaporate from synthesized fiber. However, too fast evaporation could be problematic in ES process when solvent get evaporated from the polymer solution just after emission from the tip of the needle. In such condition, ES process in stopped due to clogging of needle tip within few minutes operation of ES process [48]. It has been advised that the high humidity can help in the discharge of the synthesized nanofibers.

4. Preparation of nanofibres using electrospinning method

Well-controlled and high-quality ceramic nanofibers can be generated in ES process by following procedures

1. Preparation of a sol with suitable inorganic precursor and its proper mixing with a polymer solution to get the right rheology for electrospinning

2. Electrospinning of the solution to obtain inorganic/organic composite fibers under appropriate conditions, and

3. Calcination of the as-prepared composite fibers in air to yield pure metal oxide fibers. One of the attractive features associated with this method is that the nanofiber mats thus prepared possess high surface areas and small pore sizes [49].

4.1 The electrospinning solution

Ceramic nanofibers can be obtained by direct electro-spinning of sol of only inorganic precursor metal alkoxides or metal salts dissolved in a solvent. Notable examples are synthesis of nanofibers of CeO_2 [50–52], Titanium dioxide (TiO2) [53] and Al_2O_3/ZnO [54]. However, such synthesis shows inappropriate rheological properties and the rapid hydrolysis rates of metal alkoxides or metal salts, pose difficulties in controlling the ES process. To resolve such problems, one has to introduce a polymer matrix are added into the solution to adjust the rheological properties and catalyst is supplemented to control the hydrolysis rate of the used precursor [54]. Thus a typical spinnable precursor solution composed of metal salt or an alkoxide precursor, a polymer, an additive, and an easily volatile solvent like chloroform, ethanol, iso-propanol, water and dimethyl-formamide. The catalysts added into the solution usually stabilize the precursor and facilitate smooth electrospinning process. The catalyst are required in minute quantity but their addition in spinning solution play an key role in stabilizing the solution and the jet from the needle. The acid catalyst like acetic or hydrochloric or propionic acid is employed for adjusting both the hydrolysis and gelation rates to prevent blocking the needle mouth by solution and ensuring a continuous nanofiber synthesis [49].

At present we are discussing nanofiber synthesis using casting solution comprising precursor cerium nitrate hexahydrate and copper acetate monohydrate, PVP polymer, glacial acetic acid (3-4drops) and a solvent ethanol and water that was used for preparing green nanofibers through ES process. The details of procedure used for preparing the casting solution and green nanofibers are discussed below.

4.2 The precursor and polymer solution

The precursors in the present nanofiber synthesis were cupric acetate monohydrate ($Cu(CH_3COO)_2.H_2O$) and cerium nitrate hexahydrate ($Ce(NO_3)_3.6H_2O$) and polyvinyl pyrrolidone (PVP) was the base polymer. The PVP has a remarkably large molecular weight and high solubility in polar solvents. The solvent used in the process was ethanol and de-ionized water added as co-solvent [55, 56]. The aqueous precursor solutions was prepared in dissolving $Cu(CH_3COO)_2.H_2O$ and cerium nitrate hexahydrate in 4 ml of deionized water and this solution was mixed ethanolic PVP solution(4 ml) for final 10% (w/v) of PVP. The mixed solution stirring continuously till complete dissolution. The 2–3 drops of acetic acid was added during magnetic sterring of solution for about 3 h at room temperature to preparing homogeneous solution for final spinning.

4.3 Electrospinning of composite nanofibers

The prepared polymer spinning solution or sol–gel solution was immediately loaded in plastic syringe of 5 ml capacity and a blunt ended 20-gauge stainless steel needle fitted in it. The nozzle equipped syringe was attached to the syringe pump. The positive electrode of high voltage power supply capable of generating DC voltages up to 40 kV attached with the needle whereas negative electrode was connected to the aluminum foil covered collecting plate. The ground electrode of the power supply was attached to a piece of flat aluminum foil used as the collector plate [57, 58]. The needle tip to collector distance was 10 cm to collect the nanofibers. The ES process was conducted in ambient air at room temperature (25 ± 2°C) having relative humidity of 65 ± 5%. A solution feeding rate was set to 1 ml/h with the help of a syringe pump. The steady deposition of nanofibers on the collector plate was ensured by frequent cleaning of nozzle clogging intermittently. The experiment was conducted by applying positive high voltage of 18 kV across the needle and the collector plate. At such

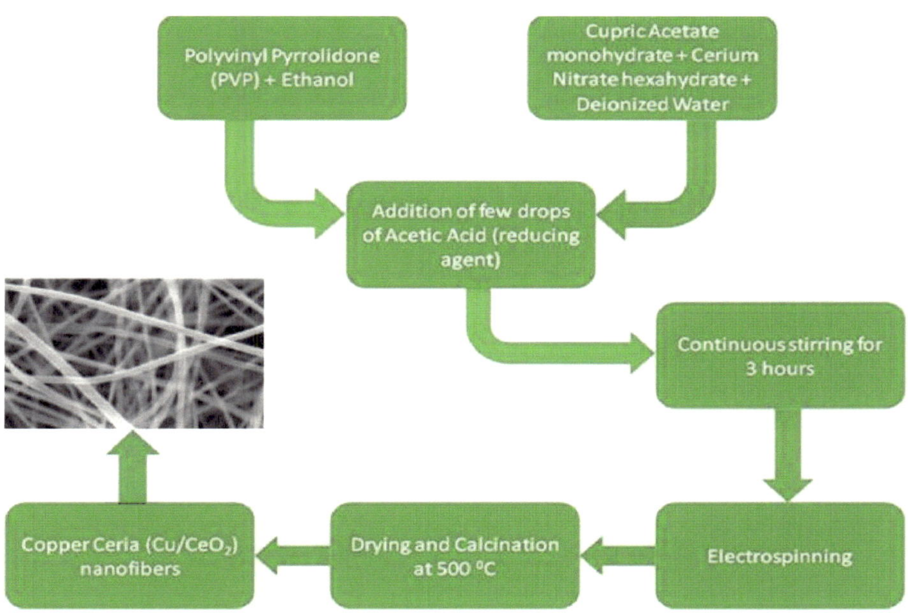

Figure 3.
Process flowchart for the preparation of nanofibers by electrospinning method.

environmental condition fluid jet ejected continuously from the nozzle and accelerated jet moved towards the collector plate. During this movement of jet the solvent get evaporated leaving ultra thin fibers on the collector plate. The ES process continued until all the solution in the syringe was exhausted. The complete sequence of spinning solution preparation and nanofiber synthesis, its calcination and final morphological feature are schematically represented in **Figure 3**.

4.4 Calcinations of green nanofibers

To complete the formation of CuO/CeO_2 nanofibers, the composite fibers prepared as above were left exposed to ambient moisture for 5 hr. to allow complete hydrolysis, then electrospun polyvinyl pyrrolidone (PVP) /cerium nitrate hexahydrate (CN)/copper acetate monohydrate (CA) fiber mats were removed from the aluminum foil. These composite fibers were placed on a ceramic crucible able to withstand high temperature and consequently calcined (500°C, 3 hr) in a muffle furnace in presence of air for eliminating the organic constituents and activating crystallization of Cu-Ce oxide.

5. Potential utilizations of nanofibers

Now a days, scientists and researchers have shown strong interest to develop the electrospun ceramic nanofibers in different field. The most significant characteristic of nanofibers are their long length, high porosity and large surface area. The features are makes them widely applicable in electric and optical devices, optoelectronic components, optical waveguides, gas storage units, fluidic devices, tissue engineering scaffolds and bioreactors. In this century, exciting and important research areas for nanofibers application are in energy technology, catalysis and environmental science [59]. Such applications of nanofibers are briefly discussed below. The basic nanofiber application shown in **Figure 4**.

5.1 Nanofibers application as catalysis

Highly porous and large surface area bearing electrospun nanofibers prominently used as solid supports in catalysts. Nanofibers of some materials having excellent semiconducting property make them suitable in photo-catalysts for removing

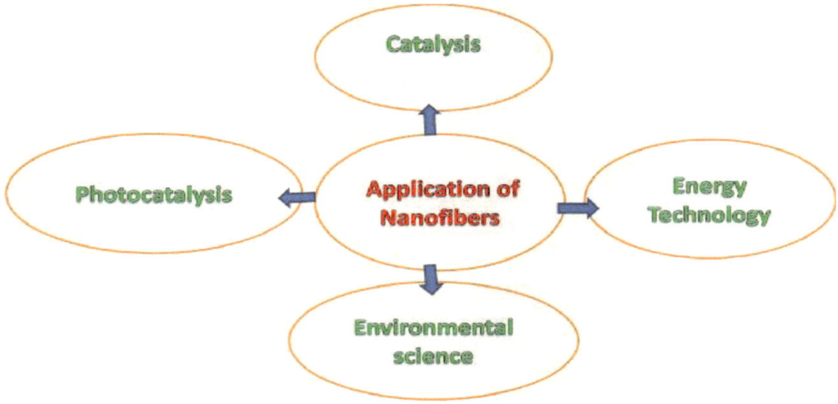

Figure 4.
Basic applications of nanofibers.

organic molecules from air flow or aqueous solution. The extent of catalysis and its efficiency is governed by surface area of the catalyst. Ceramic nanofibers possess extraordinarily high specific surface area are nice materials for their application as catalyst for chemical reaction. The ceramic nanofibers mats of Titanium dioxide (TiO2), zirconium dioxide (ZrO_2) and tin dioxide (SnO_2) have been employed as support material for loading noble metal nanostructures catalysis in different applications [60]. The membranous catalytic system have distinctive benefit over other process in terms of its operation in a continuous flow mode, relatively short reaction time and even more importantly it need not separation of product after completion of reaction. In addition to large surface area, the support material should also be stable and good conductor of electron. The nanoparticles catalyst of noble-metal Pt, Pd, Rh etc. are homogeneous spread on solid support of carbon black for maximizing possible surface site availability [61]. The Pd-coated Titanium dioxide (TiO2) nanofibers performed well in cross-coupling reaction [62]. It is interesting that the ceramic nanofiber substrate can pointedly influence noble-metal catalyst deposition [63].

5.2 Nanofibers application in energy technology

Energy shortage is one of the most serious issues of the 21st century because of limited natural energy resources like crude oil, natural gas, coal and uranium that are fulfilling the energy need for everyday life at presented. For rapid economic growth, will require subsequent increase in energy demand mean but the rate of oil production will no longer be adequate. This is evident from the rising price of crude oil. Presently, large volumes of carbon dioxide emitted by industrial burning of fossil fuels are deteriorating climate of the globe. So, it is necessity of the time to identify alternate environmentally friendly new sources of energy that are able to replace current energy supply. The people are trying to converse energy from renewable sources such as the sun, wind, and tides. The most promising energy conversion/storage devices likely to fulfill the need are based on photovoltaic cells, lithium batteries, and fuel cells. For example, Mai and colleagues demonstrated high performance lithium ion battery electrode based on electrospun vanadium oxide nanofibers [64]. They were able to prepare ultra-long hierarchical vanadium oxide nanowires of fine diameter (100–200 nm) and several millimeters using the low-cost starting materials by ES combined with annealing. The 1D characteristic of electrospun ceramic nanofibers have been widely explored as a new class of promising building blocks for fabricating the devices. Energy storage devices are becoming a very common in everyday life of public due to exponential expansion of digitization. Many devices need stored energy for their functions are electronic mobile, autonomous sensors and various kind of vehicle. So, the reliable energy storage devices with high power density and structural integrity are in demand for their use in various devices [65, 66]. Green flexible network of carbon nanofiber can be deriving from abundantly available but underutilized bioresources lignin. The PVA lignin nanofiber network showed has surface area of $1670m^2$ per gram and excellent specific gravimetric capacitance of ~240 F per gram that was better than many nanostructured carbon or metal oxides [67]. The carbon based nanofibers have high specific electrical double layer capacitance at low cost [68].

5.3 Nanofibers application in environmental science

5.3.1 Nanofiber-based membrane for filtration

Electrospun ceramic nanofibers membranes are able to remove suspended particulate matter from air and other impurities dissolved in water. The highly porous membrane structures, formed by entanglements of nanofibers facilitate material transport without causing resistant to gaseous stream flow or flow of aqueous solution; can be utilized for environmental applications. Dai et al. [50] prepared a hierarchically structured potassium manganese oxide (K_xMnO_2) / Titanium dioxide (TiO2) mats for filtering Congo red dye from waste water. The membranes not only performance well but also showed high filtering efficiency and robustness survival against strong sheer force of solution flowing through it. In another investigation, Song and co-workers reported synthesis of ultrafine porous carbon nanofibers membrane efficiently removing sulfur dioxide (SO_2) from stream of gases. Doping of membranes with minute quantity of nitrogen is likely to improve their capacity, efficiency and durability [69].

5.3.2 Nano-sensors

Detection of threats to the environment is a key feature of environmental management strategy. The contaminants posing environmental threat could be detected with the help of sensors. Some semiconducting materials like Titanium dioxide (TiO2), SnO_2, zinc oxide (ZnO), tungsten oxide (WO_3), molybdenum oxide (MoO_3) have been shown to detect trace level of gaseous species in the parts per million levels [70]. The sensing ability of such materials could be enhanced by increasing specific surface area and porosity. The 1D architectures of nanofiber facilitate fast mass transfer near molecular interaction region and traverse of barriers by charge carriers. Nanofibers of Titanium dioxide (TiO_2), iron oxide ($Fe2O_3$), SnO_2, ZnO/SnO_2, lithium chloride $(LiCl)/TiO_2$, Titanium dioxide $(TiO_2)/ZnO$, potassium chloride (KCl)-doped-ZnO, and Co-doped-ZnO are successfully employed in sensing and enhanced limit of detecting gaseous species. The gases or vapors notably detected are hydrogen (H_2), carbon monoxide (CO), O_2, nitrogen dioxide (NO_2), ammonia (NH_3), H_2O, methanol (CH_3OH), ethanol (C_2H_5OH) and toluene. Wang and colleagues reported very quick response of electrospun orthorhombic-phase WO_3 nanofiber's high sensitivity to different concentrations NH_3 [71]. The superior sensing performance of the material was attributed to high purity, preparation method, high surface area per unit volume and porosity for good accessibility of the gas. The electrospun ceramic nanofibers not only function as sensing materials, but also act as good support material for other sensing bodies derived from noble metal.

5.3.3 As photo-catalysts

Current environmental issue of air and water pollution motivate for sustained fundamental and applied research for remediation of different polluted ecosystem. The steadily growing field of nanoscience research is engaged in synthesizing nanostructured ceramics as photo-catalyst. The low cost and environmental friendliness compounds, ZnO and Titanium dioxide (TiO_2), have showed high catalytic activity are promising photocatalytic material. These small sized and extremely high surface area containing ceramic nanofibers provide channels for quick charge transfer due to its 1D nanostructure. Such ceramics nanomaterial could be potentially used in photo catalysis. Zhang and coworkers synthesized hybrid Fe-Titanium dioxide (TiO_2) /tin dioxide (SnO_2) nanofibers having high photo-catalytic activity and ferromagnetic properties at room temperature. It is supposed that this hybrid nanofiber can act new generation of visible light-excitable photo catalyst with easy recyclable and its potential predictable applications in water purifying and other pollution treatment [72]. Recently, Yoshikawa and colleagues showed Titanium dioxide (TiO_2) nanofibers potentially application as photo-catalysts for H_2 production [73]. The novel structured highly crystalline nanomaterial might be able to reduce lattice defects to facilitate the electron transport for reactions with water adsorbed on their surface. During synthesis and producing of dyes, about 16% of the total global productions are lost with wastewater [74]. A variety of bioremediation techniques are utilize for dye elimination such as Enzymes, Microbiological decolorization. The common pollutants (organic/inorganic) present in aquatic atmosphere are due to the discharge of wastewaters from households as well as from the manufacturing sectors. These contaminants, organic molecules might be found in the land and surface water. The elimination of carcinogenic, non-biodegradable organic dyes and other chemicals from the surroundings is a central environmental problem [75].

5.3.4 Water and wastewater treatment

Pollution of ground and surface water across the world is now critical issue. Heavy metals are the important pollutants that affect physiology of lining entities significantly. The heavy metals frequently reported in the polluted water bodies are As, Cu, Hg, Cd and Pb. The elements are released in industries waste that are discharged and distribution in the environment and finally enter in water. e.g. Smelting of copper release high quantities of Cd in its industrial wastewater. It is nearly impossible to eliminate some metals contaminants from water using conventional water purification procedure. The nanotechnology has greatly advanced water and wastewater treatment potential. They facilitate improved safe use of unconventional water sources. The limited surface area containing conventional adsorbents bear less active sites and also lack of selectivity. In contrast nano sorbents with enhancement specific surface area and large number of available sorption sites, small intra-particle diffusion distance, tunable pore size and surface chemistry facilitate better adsorption. In future, suitable polymer nanofibers functionalized ceramic membranes can be used for fabricating affinity membranes for treating heavy metal containing industrial waste water [76, 77]. The discharge of heavy metals like arsenic, cadmium, chromium, cobalt, nickel, copper, silver, tin, titanium, lead and zinc, into the environment of textile production is an immense concern all over the global because these metals pose unfavorable effects on the human being health, natural atmosphere and aquatic living.

6. Summary

Green synthesis of composite nanofibers can be successfully prepared using sol–gel and ES technique using polymer solutions. Some of the characteristics of the synthesized green composite nanofibers are their controlled average diameter distribution in desired range, remarkably straight fiber over several micrometers of length, uniform and smooth surface. Generally, viscosity of the casting solution plays the most important role in the fiber morphology and diameter. The green syntheses of nanofibers are affected by solution parameters, process parameters and ambient parameters. The nanofibers are at the forefront of nanotechnology due to their merits and suitability to wide range of applications in the field of healthcare, biotechnology, energy storage, environmental engineering, defense and security.

Acknowledgements

The authors acknowledge Birla Institute of technology, Mesra, Ranchi, Jharkhand and Indian Institute of Technology (BHU), Varanasi for characterization of raw materials. The NPIU (TEQIP-III), Govt. of India thankfully acknowledged for the financial support.

Conflict of interest

There are no conflicts of interest between authors. All the co-authors have seen the final manuscript and agreed for submission. No data or figures have been fabricated or manipulated. The authors agree to transfer copy right to book publishers. All the further responsibilities will be undertaken by the corresponding author.

Author details

Dan Bahadur Pal[1*] and Deen Dayal Giri[2]

1 Department of Chemical Engineering, Birla Institute of Technology, Mesra, Ranchi, Jharkhand, India

2 Department of Botany, Maharaj Singh College, Saharanpur, Uttar Pradesh, India

*Address all correspondence to: danbahadur.chem@gmail.com

References

[1] Benelli, G.; Pavela, R.; Maggi, F.; Petrelli, R.; Nicoletti, M. Commentary: Making green pesticides greener? The potential of plant products for nanosynthesis and pest control. J. Clust. Sci. 2017, 28, 3-10.

[2] Fortunati, E.; Peltzer,M.;Armentano, I.; Torre, L.; Jimenez,A.; Kenny, J. Effects ofmodified cellulose nanocrystals on the barrier and migration properties of PLA nano-biocomposites. Carbohydr. Polym. 2012, 90, 948-956.

[3] N. Chitose a, S. Ueta, S. Seino, T. A. Yamamoto. Radiolysis of aqueous phenol solutions with nanoparticles.1. Phenol degradation and TOC removal in solutions containing Titanium dioxide (TiO2) induced by UV, gamma-ray and electron beams. Chemosphere. 50 (2003) 1007-1013.

[4] J. Hu, C.L. Chen, X.X. Zhu, X.K. Wang. Removal of chromium from aqueous solution by using oxidized multiwalled carbon nanotubes. J. Hazard. Mater. 162 (2009) 1542-1550.

[5] R.K. Kulkarni, K.C. Pani, C. Neuman, F. Leonard, Polylactic acid for surgical implants, Arch. Surg. 93 (1966) 839-843.

[6] N. Cai, Q. Dai, Z. Wang, X. Luo, Y. Xue, F. Yu, Preparation and properties of nanodiamond/poly lactic acid composite nanofiber scaffolds, FIBER POLYM. 15(2014) 2544-2552.

[7] S. Vivekanandhan, L. Christensen, M. Misra, A.K. Mohanty, Green process for impregnation of silver nanoparticles into microcrystalline cellulose and their antimicrobial bionanocomposite films, J. Biomater. Nanobiotechnol. 3 (2012) 371-376.

[8] R. Augustine, K. Rajarathinam, Synthesis and characterization of silver nanoparticles and its immobilization on alginate coated sutures for the prevention of surgical wound infections and the in vitro release studies, Int. J. Nano Dimens. 2 (2012) 205-212.

[9] K. Shameli, M.B. Ahmad, W.M.W. Yunus, N.A. Ibrahim, R.A. Rahman, M. Jokar, M. Darroudi, Silver/poly (lactic acid) nanocomposites: preparation, characterization, and antibacterial activity, Int. J. Nanomed. 5 (2010) 573-579.

[10] SK Srikar, DD Giri, DB Pal, PK Mishra, SN UpadhyayGreen synthesis of silver nanoparticles: a review Green and Sustainable Chemistry 6 (1), 34-56

[11] A.S. Kumar, S. Ravi, V. Kathiravan, Green synthesis of silver nanoparticles and their structural and optical properties, Int. J. Curr. Res. 5 (2013) 3238-3240.

[12] D. Jain, H.K. Daima, S. Kachhwaha, S. Kothari, Synthesis of plant-mediated silver nanoparticles using papaya fruit extract and evaluation of their antimicrobial activities, Dig. J. Nanomater. Biostruct. 4 (2009) 557-563.

[13] K.B. Narayanan, H.H. Park, Antifungal activity of silver nanoparticles synthesized using turnip leaf extract (*Brassica rapa* L.) against wood rotting pathogens, Eur. J. Plant Pathol. 140 (2014) 185-192.

[14] S.P. Chandran, M. Chaudhary, R. Pasricha, A. Ahmad, M. Sastry, Synthesis of gold nano triangles and silver nanoparticles using *Aloe vera* plant extract, Biotechnol. Prog. 22 (2006) 577-583.

[15] V. Kathiravan, S. Ravi, S.A. Kumar, Synthesis of silver nanoparticles from Melia dubia leaf extract and their in vitro anticancer activity, Spectrochim.

Acta. Part A: Mol. Biomol. Spectrosc. 130 (2014) 116-121.

[16] P.S. Vankar, D. Shukla, Biosynthesis of silver nanoparticles using lemon leaves extract and its application for antimicrobial finish on fabric, Appl. Nanosci. 2 (2012) 163-168.

[17] D. Kumarasamyraja, N.S. Jeganathan, Green synthesis of silver nanoparticles using aqueous extract of acalyphaindica and its antimicrobial activity, Int. J. Pharm. Biol. Sci. 4 (2013) 469-476.

[18] SK Srikar, DD Giri, DB Pal, PK Mishra, SN UpadhyayLight induced green synthesis of silver nanoparticles using aqueous extract of Prunusamygdalus Green and Sustainable Chemistry 6 (1), 26-33.

[19] Matsuda, A.; Kagata, G.; Kino, R.; Tanaka, J. Preparation of chitosan nanofiber tube by electrospinning. J. Nanosci. Nanotechnol. 2007, 7, 852-855.

[20] Sanmugam, A.; Vikraman, D.; Karuppasamy, K.; Lee, J.Y.; Kim, H.-S. Evaluation of the corrosion resistance properties of electroplated chitosan-Zn1□xCuxO composite thin films. Nanomaterials 2017, 7, 432.

[21] R. Ramaseshan, S. Sundarrajan, R. Jose, S. Ramakrishna. Journal of Applied Physics 102 (2007) 111101.

[22] H. Li, Y. Ke, Y. Hu. Polymer nanofibers prepared by template melt extrusion. J. Appl. Polym. Sci. 99 (2006) 1018-1023.

[23] Z. Yang, B. Xu. Supramolecular hydrogels based on biofunctional nanofibers of self assembled small molecules. J. Mater. Chem. 17 (2007) 2385-2393.

[24] X. Feng, G. Yang, Q. Xu. Self-assembly of polyaniline/Au composites: From nanotubes to nanofibers. Macromol. Rapid Comm. 27 (2006) 31-36.

[25] C.J. Ellison, A. Phatak, D.W. Giles. Melt blown nanofibers: Fiber diameter distributions and onset of fiber breakup. Polymer. 48 (2007) 3306-3316.

[26] B.K. Gu, M.K. Shin, K.W. Sohn. Direct fabrication of twisted nanofibers by electrospinning. Appl. Phys. Lett. 90 (2007) 263902.

[27] P.K. Panda, S. Ramakrishna. Electrospinning of alumina nanofibers using different precursors. J. Mater. Sci. 42 (2007) 2189-2193.

[28] Zhang X, Lua Y (2014) Centrifugal spinning: an alternative approach to fabricate nanofibers at high speed and low cost. Polym Rev 54:677-701.

[29] Kumar P (2012) Effect of collector on electrospinning to fabricate aligned nanofiber. Department of Biotechnology & Medical Engineering National Institute of Technology, Rourkela.

[30] Ramakrishna S, Fujihara K, Teo W-E, Lim T-C, Ma Z (eds) (2005) An introduction to electrospinning and nanofibers. World Scientific Publishing, Singapore.

[31] Beachley V, Wen X (2010) Polymer nanofibrous structures: fabrication, biofunctionalization and cell interactions. ProgPolymSci J 35(7):868-892.

[32] Li W-J, Shanti RM, Tuan RS (2006) Electrospinning technology for nanofibrous scaffolds in tissue engineering. anotechnologies Life Sci 9:135-186

[33] D. L. Jesse, T. McCann, Y. Xia. Electrospinning: A Simple and Versatile Technique for Producing Ceramic

Nanofibers and Nanotubes J. Am. Ceram. Soc. 89 (2006) 1861-1869.

[34] E.J. Chong, T.T. Phan, I.J. Lim, Zhang YZ, Bay BH, Ramakrishna S, et al. Evaluation of electrospun PCL/gelatin nanofibrous scaffold for wound healing and layered dermal reconstitution. Acta Mater. 3 (2007) 321-330.

[35] Li D, Wang Y, Xia Y. Electrospinning nanofibers as uniaxially aligned arrays and layer by layer stacked films. Adv Mater. 16 (2004) 361-366.

[36] Y. Dai, W. Liu, E. Formo, Y. Sun, Y. Xia. Ceramic nanofibers fabricated by electrospinning and their applications in catalysis, environmental science, and energy technology Polym. Adv. Technol. 22 (2011) 326-338.

[37] S. Li, X. Yue, Y. Jing, S. Bai, Z. Dai. Fabrication of zonal thiol-functionalized silica nanofibers for removal of heavy metal ions from wastewater. Colloids and Surfaces A: Physicochem. Eng. Aspects 380 (2011) 229-233.

[38] Y. Gu, F. Jian, X. Wang. Synthesis and characterization of nanostructured Co3O4 fibers used as anode materials for lithium ion batteries. Thin Solid Films 517 (2008) 652.

[39] D. Li, Y. Xia. Fabrication of Titania Nanofibers by Electrospinning. Nano Letters 3 (2003) 555.

[40] Y. Gu, D. hen, X. Jiao. Synthesis and Electrochemical Properties of Nanostructured LiCoO2 Fibers as Cathode Materials for Lithium-Ion Batteries. J. Phys. Chem. B 109 (2005) 17901-17906.

[41] A.F. Lotus, Y.C. Kang, J.I. Walker, R.D. Ramsier, G.G. Chase. Effect of aluminum oxide doping on the structural, electrical, and optical properties of zinc oxide nanofibers synthesized by electrospinning. Mater. Sci.Eng. B 166 (2010) 61.

[42] M. M. V. M. Souza, N. F. P. Ribeiro, M. Schmal. Influence of the support in selective CO oxidation on Pt catalysts for fuel cell applications. International J. Hydrogen Energy 32 (2007) 425.

[43] L. Yang, W. Yang, Q. Cai. Well-Dispersed Pt-Au Nanoparticles Loaded into Anodic Titania Nanotubes. J. Phys. Chem. C 111 (2007) 16613.

[44] E. Formo, P. H. C. Camargo, B. Lim, M. J. Jiang, Y. N. Xia. Functionalization of ZrO2 nanofibers with Pt nanostructures: The effect of surface roughness on nucleation mechanism and morphology control. Chem. Phys. Lett. 476 (2009) 56.

[45] E. Formo, M. S. Yavuz, E. P. Lee, L. Lane, Y. N. Xia. Functionalization of electrospun ceramic nanofibre membranes with noble-metal nanostructures for catalytic applications. J. Mater. Chem. 19 (2009) 3878.

[46] Mai LQ, Xu L, Han CH. Electrospunultralong hierarchical vanadium oxide nanowires with high performance for lithium ion batteries. Nano. Lett. 10 (2010) 4750-4755.

[47] X. Song, Z. Wang, Z. Li, C. Wang. Ultrafine porous carbon fibers for SO2 adsorption via electrospinning of polyacrylonitrile solution. J. Colloid Interface Sci. 327 (2008) 388.

[48] B. Ding, M. Wang, J. Yu, G. Sun. Gas Sensors Based on Electrospun Nanofibers. Sensors 9 (2009) 1609.

[49] G. Wang, Y. Ji, X. Huang, X. Yang, P. I. Gouma, M. Dudley, Fabrication and Characterization of Polycrystalline WO_3 Nanofibers and Their Application for

Ammonia Sensing. J. Phys. Chem. B 110 (2006) 237-77.

[50] X. Yang, C. Shao, Y. Liu, R. Mu, H. Guan. Nanofibers of CeO_2 via an electrospinning technique; Thin Solid Films 478 (2005) 228-231.

[51] C. Qizheng, D. Xiangting, W. Jinxian, Limei. Direct Fabrication of Cerium Oxide Hollow Nanofibers by Electrospinning; J. of Rare Earths, Vol. 26(2008) 664.

[52] S. Xu, D. Sun, H. Liu, X. Wang, X. Yan. Fabrication of Cu-doped cerium oxide nanofibers via electrospinning for preferential CO oxidation; Catalysis Communications 12 (2011) 514-518.

[53] S. Ramakrishna, Wee-EongTeo, R. Ramaseshan. Electrospun nanofibers: solving global issues. Materials today 9 (2006) 3.

[54] R. Baeyens, R. Ebinghous, O. Vasilev. Global and Regional Mercury Cycles: Sources, Fluxes and Mass Balances. Kluwer Academic Publishers 1996.

[55] C. Li, R. Chen, X. Zhang, S. Shu, J. Xiong, Y. Zheng, W. Dong. Electrospinning of CeO_2-ZnO composite nanofibers and their photocatalytic property, Materials Letters, Volume 65, Issue 9, 15 May 2011, Pages 1327-1330.

[56] C. Li, R. Chen, X. Zhang, S. Shu, J. Xiong, Y. Zheng, W. Dong. Mat.Lett. 65 (2011)1327-1330.

[57] D.B. Pal, H Kumar, DD Giri, P Singh, PK Mishra Synthesis and characterization of Cu/CeO2 composite nanofibers by electrospinning method Advanced Science Letters 20 (7-8), 1582-1584.

[58] D. B. Pal, P. Singh, P. K. Mishra. Composite ceria nanofiber with different copper loading using electrospinning method. J. of Alloys and Comp. 694 (2017) 10-16.

[59] Y. Dai, W. Liu, E. Formo, Y. Sun, Y. Xia. Ceramic nanofibers fabricated by electrospinning and their applications in catalysis, environmental science, and energy technology Polym. Adv. Technol. 22 (2011) 326-338.

[60] M. M. V. M. Souza, N. F. P. Ribeiro, M. Schmal. Influence of the support in selective CO oxidation on Pt catalysts for fuel cell applications. International J. Hydrogen Energy 32 (2007) 425.

[61] E. Formo, P. H. C. Camargo, B. Lim, M. J. Jiang, Y. N. Xia. Functionalization of ZrO2 nanofibers with Pt nanostructures: The effect of surface roughness on nucleation mechanism and morphology control. Chem. Phys. Lett. 476 (2009) 56.

[62] E. Formo, M. S. Yavuz, E. P. Lee, L. Lane, Y. N. Xia. Functionalization of electrospun ceramic nanofibre membranes with noble-metal nanostructures for catalytic applications. J. Mater. Chem. 19 (2009) 3878.

[63] DB Pal, H Kumar, DD Giri, P Singh, PK Mishra Enhanced H2 and Reduced CO Level by Use of ElectrospunCuO/CeO2 Nanofibers Catalyst for Water Gas Shift Reaction Advanced Science Letters 22 (4), 967-970.

[64] Mai LQ, Xu L, Han CH. Electrospunultralong hierarchical vanadium oxide nanowires with high performance for lithium ion batteries. Nano. Lett. 10 (2010) 4750-4755.

[65] Wu, Z.; Wang, Y.; Liu, X.; Lv, C.; Li, Y.; Wei, D.; Liu, Z. CarbonNanomaterial-Based Flexible Batteries for Wearable Electronics. Adv. Mater. 2019, 31 (9), 1800716.

[66] (2) Yun, T. G.; Park, M.; Kim, D.-H.; Kim, D.; Cheong, J. Y.; Bae, J. G.;

Han, S. M.; Kim, I.-D. All Transparent-Stretchable Electrochromic-Supercapacitor Wearable Patch Device. ACS Nano 2019, 13 (3), 3141-3150.

[67] Zhang, L. L.; Zhao, X. Carbon-Based Materials as Supercapacitor Electrodes. Chem. Soc. Rev. 2009, 38 (9), 2520-2531.

[68] Jiayuan Wei, ShiyuGeng, Olli Pitkänen, TopiasJärvinen, KrisztianKordas, and KristiinaOksman Green Carbon Nanofiber Networks for Advanced Energy Storage ACS Applied Energy Materials 2020 3 (4), 3530-3540.

[69] X. Song, Z. Wang, Z. Li, C. Wang. Ultrafine porous carbon fibers for SO2 adsorption via electrospinning of polyacrylonitrile solution. J. Colloid Interface Sci. 327 (2008) 388.

[70] B. Ding, M. Wang, J. Yu, G. Sun. Gas Sensors Based on Electrospun Nanofibers. Sensors 9 (2009) 1609.

[71] G. Wang, Y. Ji, X. Huang, X. Yang, P.I. Gouma, M. Dudley, Fabrication and Characterization of Polycrystalline WO3 Nanofibers and Their Application for Ammonia Sensing. J. Phys. Chem. B 110 (2006) 23777.

[72] Zhang R, Wu H, Lin D, et al. Photocatalytic and magnetic properties of the Fe-Titanium dioxide (TiO2) / SnO2 nanofiber via electrospinning. J. Am. Ceram. Soc. 93 (2010) 605-608.

[73] S. Chuangchote, J. Jitputti, T. Sagawa, S. Yoshikawa, Photocatalytic Activity for Hydrogen Evolution of Electrospun Titanium dioxide (TiO2) Nanofibers. ACS Appl. Mater. Interfaces 1 (2009) 1140.

[74] Gupta AK, Pal A, Sahoo C. Photocatalytic degradation of a mixture of Crystal Violet (Basic Violet 3) and Methyl Red dye in aqueous suspensions using Ag+ doped TiO2. Dyes Pigments 2006; 69: 224-32.

[75] Kumar P, Govindaraju M, Senthamilselvi S, Premkumar K. Photo-catalytic degradation of methyl orange dye using silver (Ag) nano- particles synthesized from *Ulva lactuca*. Colloids Surf B Biointer- faces 2013; 103: 658-61.

[76] S. Ramakrishna, Wee-EongTeo, R. Ramaseshan. Electrospun nanofibers: solving global issues. Materials today 9 (2006) 3.

[77] DB Pal, DD Giri, P Singh, S Pal, PK Mishra 2017 Arsenic removal from synthetic waste water by CuOnano-flakes synthesized by aqueous precipitation methodDesal. Wastew 62, 355-359.

Chapter 3

Fabrication of PVA/Carbon-Based Nanofibers Using Electrospinning

Gatut Yudoyono, Diky Anggoro, Lutfi Fitria Ningsih and Rizki Romadoni

Abstract

Nanofibers are widely used in various fields, including water filtration. In the development of nanofibers as water filtration, a mixture of carbon in a polymer solution is often used. Nanofibers can be made by several methods such as multicomponent fiber spinning techniques, melt blowing, electrospinning. Electrospinning is currently a simple development method but can produce nanofibers with a small fiber diameter, it is easy to develop and many parameters can be controlled. Parameters that affect the results of the nanofibers that are formed include flow rate or syringe pump flow rate and high voltage dc high voltage. Various types of nanofibers can be produced from various types of polymers, both natural polymers and synthetic polymers. Generally, because they have properties and characteristics such as high surface area, small pore size, and the possibility to be developed in various applications. Therefore, this chapter discusses the electrospinning of carbon nanofibers using PVA polymer.

Keywords: DC high voltage, electrospinning, nanofibers, PVA/carbon

1. Introduction

Nanofibers are one of the nanotechnology products. Nanofibers are defined as an ultrafine dense fiber having a very small diameter. Nanofibers diameter is tens to hundreds of nanometers, so it is called ultrafine solid fiber (Nanofibers is defined as a fiber with a diameter of 100–500 nm [1]. Research related to nanofibers continues to be carried out for one of the reasons is, has the advantage of a large surface area per unit mass and small pore size and has superior mechanical properties [2], but the quality of nanofibers does not only depend on the chemical properties of the solution but rather the size and mechanical properties also very important [1, 3, 4]. Nanofibers size, density, mechanical properties, and orientation are essential for a reliable product application. The properties and characteristics of the fibers will change drastically when their size subside from micrometers to nanometers, one of which is the increase in the surface area of the fiber to its volume ratio, examples of these mechanical properties are stiffness and strength. With these best characteristics, nanofibers have a very extensive application [5], and nanofibers applications are widely used in various industries. Making nanofibers can be done by several methods, is a multicomponent spinning technique, melt blowing, and electrospinning [6, 7].

The type of nanofibers that is currently developing very rapidly in the field of research, material synthesis systems and product applications is carbon nanofibers (CNF). Carbon nanofibers (CNF) has applications as a promising material and has great potential in various fields, in the chemical field, carbon nanofibers (CNF) have been widely applied to gas and water membranes, by utilizing the advantages of CNF in porosity, surface area high as well as good higher chemical resistance [8]. In the field of physics, the good thermal and electrical conductivity properties make CNF very potential to be applied to electrical devices, batteries in the electrode material, energy storage and as a sensor [9]. While in the field of materials science, CNF has been applied to the strengthening of composites and supercapacitor materials [8, 10].

The precursor for forming carbon nanofibers consists of Polyvinyl Alcohol (PVA) and Carbon, PVA is a polymer that has flexible properties, can form hydrogel bonds, is easily broken down naturally and is often used in the formation of nanofibers [11]. The chemical structure of PVA is shown in **Figure 1**, the degree of hydrolysis of PVA is around 98.5% so that it can dissolve in water with a temperature of 70 ° C [12]. In addition, PVA has optical properties, a quite good load storage capacity but poor conductivity values. Therefore, to overcome the bad conductivity properties can be done by means of doping. The nature of PVA is colorless, odourless, tasteless, and soluble in water [13].

Because PVA has biodegradable properties, this is what makes this polymer widely used for its applications in the medical, food industry and electronics. The physical properties of PVA are presented in **Table 1** below [6]. Anita and Harsojo, in their research, explained that the morphological results of PVA fabrication using electrospinning owned by nanofibers which were formed at a concentration of 10% were continuous [15, 16].

Carbon is a material that has various advantages in terms of physical and chemical properties, so many researchers have developed it today. This advantage of carbon makes it a material with the extensive application. The performance of this carbon is influenced by morphology. This morphological difference will result in the wide application of the carbon, such as catalyst supports, adsorbents, gas storage, separation technology, battery electrodes, porous template materials, fuel cells, and biological cells. In addition, several carbon particles with certain morphologies will have different applications [17], besides carbon material is also an amorphous

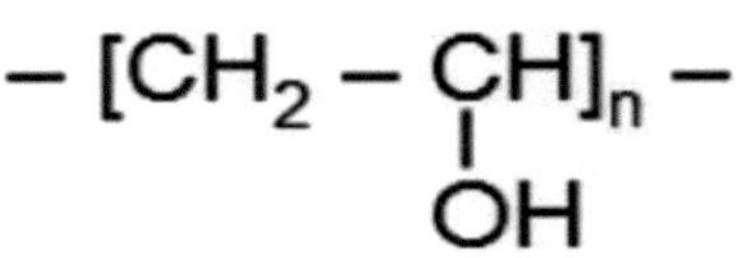

Figure 1.
PVA Chemical Structure.

Character	Value
Density	(1.19–1.31) gr/cm^3
Melting point	180–240 °C
Boiling point	228 °C
Decomposition temp	180 °C

Table 1.
Physical Properties of PVA [14].

compound which is produced from materials containing carbon or charcoal which are specially treated to obtain high adsorption power. Carbon can adsorb certain gases and chemical compounds, or its adsorption properties are selective, depending on the size or volume of pores and surface area. The absorption capacity of activated carbon is very large, namely 25–100% by weight of activated carbon [18].

The electrospinning method is a method that provides many advantages among the existing methods. This advantage is that the electrospinning technique can produce nano-sized fibers. The formation of jet polymer in the electrospinning method affects the morphological shape of the nanofibers, the polymer jet is influenced by environmental conditions, one of which is humidity [19], but most of the existing studies do not consider the relative humidity of the electrospinning spinning environment. Apart from relative humidity parameters, collector rotational speed affects fiber morphology. Collector rotating speed will affect fiber continuity.

2. Synthesis of carbon nanofibers

This section discusses the process of obtaining carbon nanofibers or the synthesis process to obtain carbon nanofibers which can be done, such as electrospinning (plate and drum collector), drawing methods and template methods. Each of these carbon nanofibers synthesis methods has its own advantages and disadvantages of the resulting material.

2.1 Preparation of carbon nanofibers (CNF)

The polymer solution in this study was made from polyvinyl alcohol (PVA), distilled water and carbon powder precursors with a size of 500 mesh. In the process of forming a polymer synthesis preparation material, PVA (molecular weight 60000, Merck Co) and carbon as a solute and distilled water as a solvent. The process scheme in making polymer solutions is by determining the concentration of the solution. The concentration of PVA solution that can be used in this study is 13 wt% with the solvent, and 2% wt carbon with distilled water as a solvent. After being measured, PVA and distilled water were mixed in one beaker. Then the magnetic stirrer is inserted into the reaction glass, then it is placed on the magnetic stirrer hotplate which has been activated. The temperature is set to reach 90 °C. After the temperature is right, this stirring process is carried out for one hour. Then the carbon and distilled water are mixed in one beaker glass, with the same steps as the PVA solution, the carbon and the solvent are placed on a hotplate magnetic stirrer which has been activated and the temperature is set to 30 After the temperature has been adjusted, the stirrer is turned on and the stirring process is carried out for one hour after the two solutions have dissolved well then, the PVA and carbon solutions are mixed with a volume ratio of PVA and carbon 2:1, then sonication is carried out by ultrasonic bath for 5 hours later. Stirred back at 30 °C for one hour.

2.2 Electrospinning

The electrospinning technique is a technology for making nano-sized fiber materials derived from materials in the form of solutions or liquids, as well as an efficient nanofibers manufacturing system by utilizing the influence of electrostatics in producing a solution (jet) of electrically charged polymer solutions or melts [20]. The electrostatic effect is generated by using a high voltage source. The voltage source that can be done in the use of electrospinning is between 7 kV to 32 kV [21]. Apart from the high voltage, the other important parts controlling the process are a

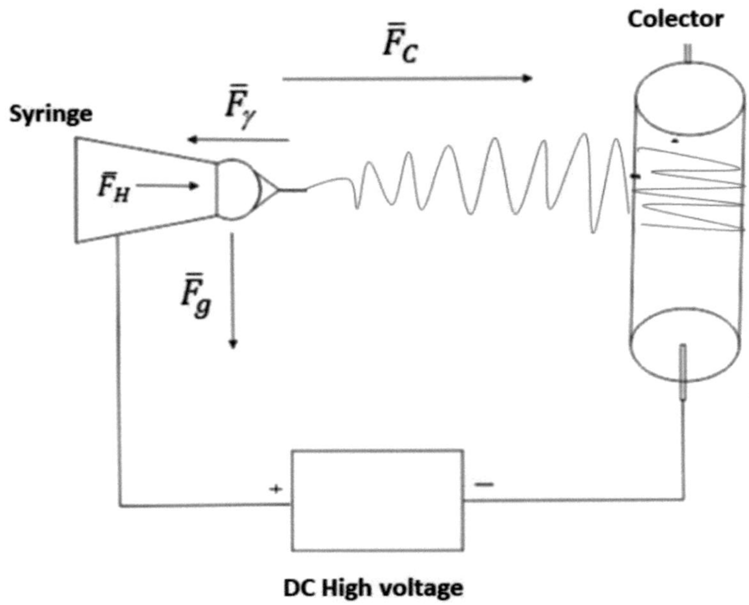

Figure 2.
The forces that appear in the electrospinning process.

syringe pump as a solution sprayer with a precise flow rate, and a collector as a place to collect the nanofibers that are formed [22]. When a high voltage is applied to the needle tip and collector, an electric field is formed around it. The positive pole is connected to the needle, the negative pole is connected to the collector. As the electric field around the needle increases, the hemisphere of the solution droplets at the tip of the needle will expand further and form a cone (also known as Taylor cone). When a high voltage is applied to the needle tip, the electric field will affect the surface tension of the solution droplet. Due to the influence of the electric charge on the needle tip on the solution, the solution is polarized and attracted towards the collector [23]. On the way from the tip of the needle to the collector, the solution undergoes thinning and evaporation of fibers or fibers that form and collect on the collector surface [21].

The forces acting on the electrospinning process can be described as in **Figure 2** which shows the modelling of the forces acting on the electrospinning process. From this figure, it can be written the equation of the forces acting on the electrospinning process as follows,

$$\overline{F}\gamma - \overline{F}_C + \overline{F}_H = 0 \tag{1}$$

when $\overline{F}\gamma$ is the surface tension force of the solution, $\overline{F}_C$ is the Coulomb force that arises because of the electric field, and $\overline{F}_H$ is the hydrodynamic force that occurs when the solution is pushed/pressed by the syringe pump [24].

2.3 Electrospinning with a plate collector

PVA and Carbon precursor solutions that have gone through the preparation stage will then be fabricated using an electrospinning system to produce carbon nanofibers, the electrospinning system used can be done with various schemes, but the schemes used here use electrospinning with a stationary plate collector and a rolling drum collector.

As shown in **Figure 3**, is an electrospinning scheme with a syringe containing a polymer solution that includes a spinneret (needle), a direct current (DC) high-voltage power generator and a stationary collector plate. In the electrospinning method, a high voltage over a certain range is applied between two electrodes to obtain the desired type and quality of carbon nanofibers. The positive electrode is made in contact with the PVA + carbon fluid via a spinneret to produce a charged liquid when subjected to an external electric field, and the negative electrode is attached to a collector which acts as a fiber collector. Due to the electrostatic force, the solution will be attracted towards the collector.

Although such a simple plate-collector electrospinning scheme is sufficient to obtain fibers, it does not produce a homogeneous and evenly distributed CNF layer, as more CNF collects in the center of the collector, resulting in variations in thickness through the layers, which can also affect fiber morphology, as well as carbon nanofibers the resulting system tends to easily form beads as shown in

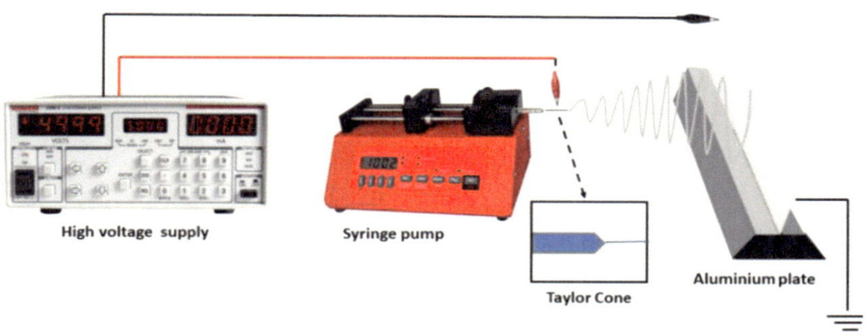

Figure 3.
Schematic illustration of an electrospinning system using a plate collector type.

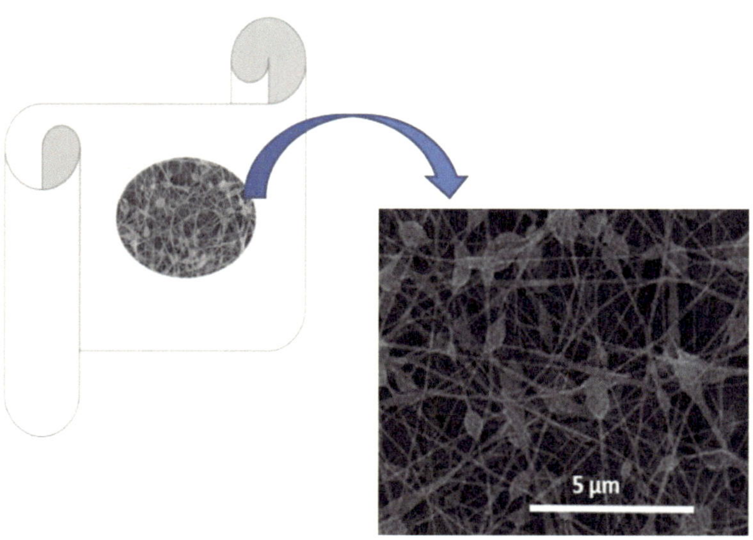

Figure 4.
SEM images of fabricated CNF using an electrospinning system with a stationary plate collector and most of the fibers are formed only in the center of the collector.

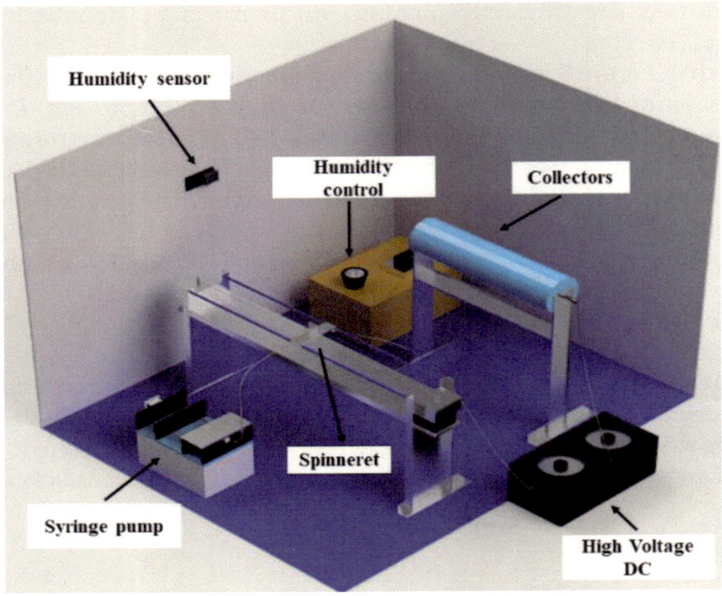

Figure 5.
Schematic illustration of an electrospinning system that uses a rotating drum collector type, with more complex controls and parameters.

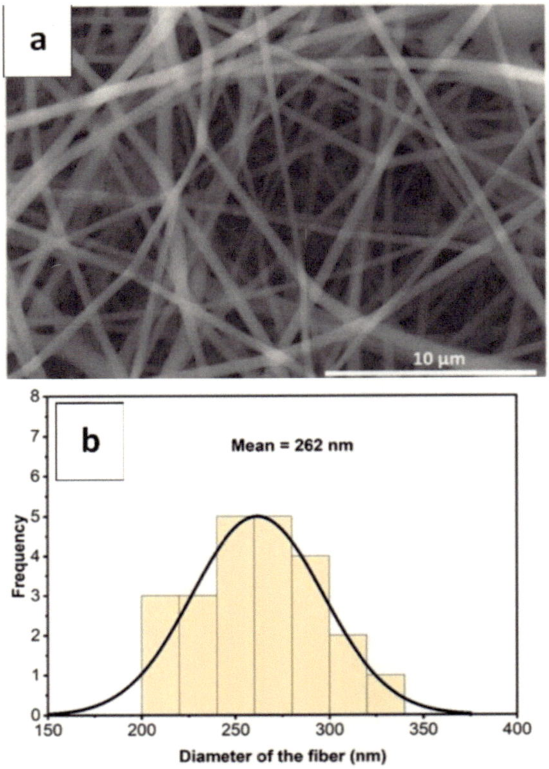

Figure 6.
(a) SEM image results, and (b) size distribution of CNFs from electrospinning fabrication results with rotating drum collectors, with a rotation speed of 130 rpm, a given high DC voltage of 10 kV and a relative humidity of about 30%.

Figure 4. The results were not homogeneous, and beads appeared on the CNF because the plate collector was used in a stationary or stationary position so that the effect of spinning fibers after the Taylor cone and jet polymer processes was very weak.

2.4 Electrospinning with a rolling drum collector

In this regard, electrospinning with a rotating drum collector has been developed to allow the formation of carbon nanofibers homogeneously and thoroughly to all areas on the drum surface, resulting in a CNF of uniform thickness as shown in **Figure 5.** When the drum collector rotates, the fibers will be attracted towards the collector and subjected to a spinning effect on the rotating drum, then the spinneret moves right and left, as well as the influence of the electric field between the needle and the collector which causes the CNF to form evenly throughout and formless beads in the layer collector.

Carbon nanofibers (CNF) formed with this system has an interesting material morphology as shown in **Figure 6** which is attractive in the sense that no beads are formed on the CNF, and the fibers are evenly distributed with a homogeneous thickness. The fiber that is formed enters the nanometer scale area, from direct measurements the diameter of the CNF formed is at 262 nm, as we all know that the limitations of fiber or composite materials are said to be in the nanoscale if they have a size of 50 nm to 500 nm.

3. Electrospinning control parameters

The electrospinning method has many parameters that must be controlled to produce nanofibers. The parameters that influence are high voltage, field, electricity, nozzle to collector distance, solution concentration, and humidity. The formation of jet polymer in the electrospinning method results in the morphological shape of the nanofibers. The polymer jet itself is influenced by environmental conditions, one of which is humidity. Humidity parameters greatly affect the diameter of the nanofibers, at high humidity, the fiber diameter will increase (Medeiros et al., 2018). The application of high voltage to electrospinning is very important in influencing the diameter and morphology of the nanofibers. The increase in high voltage causes an increase in the electric field as well as this affects the decrease in the diameter of the nanofibers and shortens the time of the solution from the tip of the needle to the collector. The flow rate in electrospinning is the flow of fluid from the syringe pump to the collector. The rate of solution (flowrate) affects the formation of fiber diameter and morphology. This process affects the material transfer rate and jet speed. The diameter of the fiber will increase as the rate of solution used increases [25].

Viscosity is the thickness of a solution; this viscosity is influenced by the concentration of the solution. High viscosity is difficult to force the solution out of the syringe so that the control on the needle is unstable, the higher the viscosity, the higher the fiber diameter. The diameter and morphology of the nanofibers are basically influenced by the distance between the needle tip and the collector. Distance affects fiber diameter and morphology because distance can determine the deposition time, evaporation rate, and polymer jet instability [26]. Therefore, an optimum nozzle to collector distance is needed to form carbon nanofibers with the desired diameter and fiber morphology. Several studies have studied the effect of the nozzle to collector distance and concluded that increasing the distance makes the fiber diameter decrease but the polymer jet instability increases [25].

The effect of relative humidity on the morphology of carbon nanofibers polymers is about the size of the diameter of the nanofibers which is strongly influenced by humidity during the spinning process seen from **Figure 7**. An environment with high humidity helps skin form rapidly with clear boundaries, whereas if there is less moisture, solvents will easily evaporate. When the polymer is hydrophobic, water acts as a non-solvent so that the fiber shell is more easily formed. This makes PMMA, PVC,

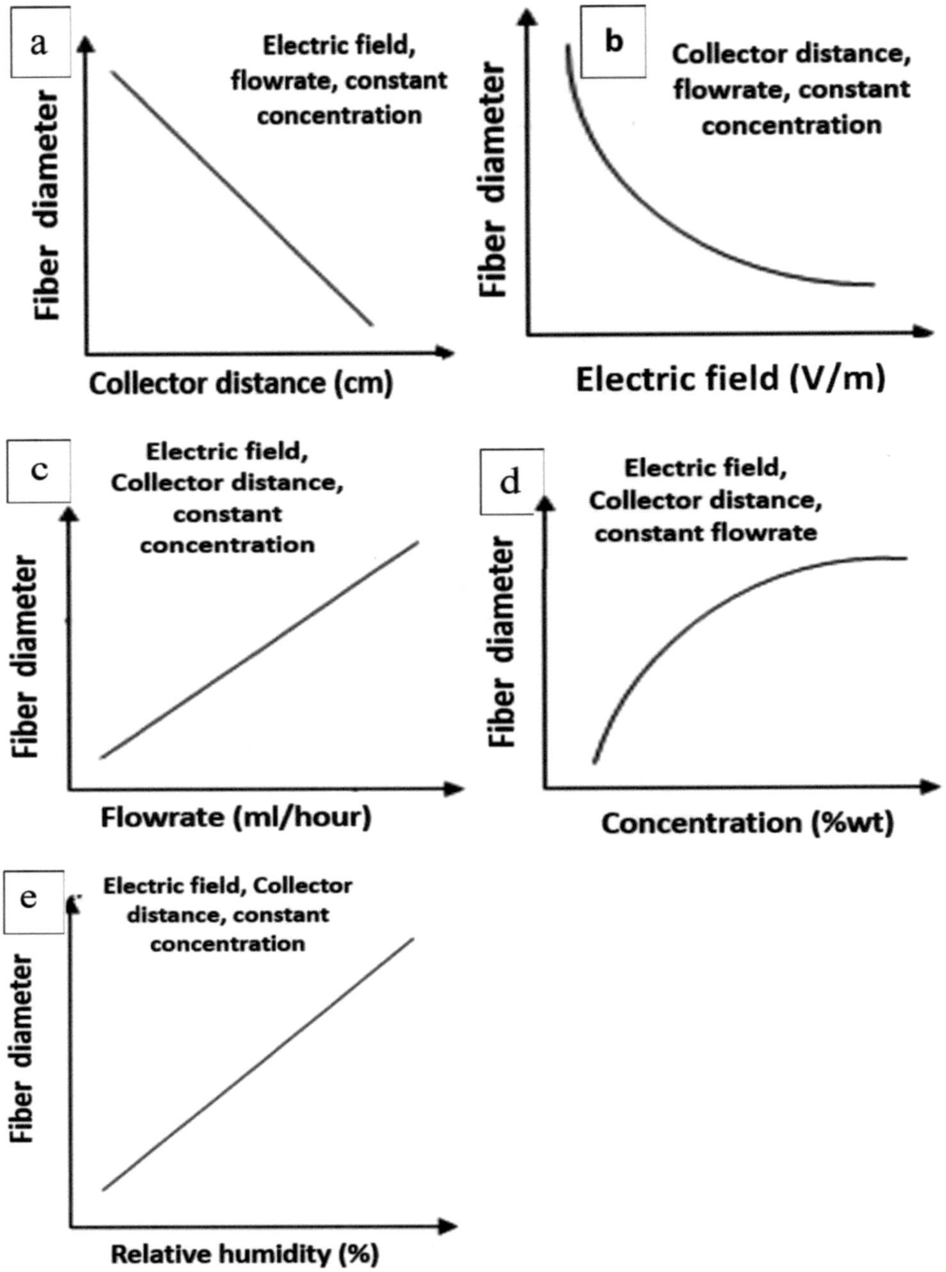

Figure 7.
Parameters affecting the size of the diameter of the nanofibers. (a) collector Distance (cm), (b) electric field (V/m), (c) flowrate (ml/hour), (d) concentration (%wt), (e) Relative humidity (%).

or PS in DMF, PMMA or PS in toluene have porous fibers when electros are in an environment with a relative humidity of more than 30%, while PVA is a hydrophilic polymer solution, water acts as a solvent so that the formation skin is easy to form humidity 20%. and there was no pore formation at all at a relative humidity value of 20% -80%. PVA nanofibers were successfully made in the relative humidity range of 20% -80%, but at high relative humidity, the morphology of the fibers contained more beads [27].

4. Characterization of carbon nanofibers (CNF)

The characterization carried out on the CNF depends on the application to be performed on it, the CNF which has been fabricated is researched as a sensor and as a Capacitive Deionization (CDI) electrode. To find out and measure the morphology and diameter of the nanofibers, a Scanning Electron Microscope or SEM is used for short. Also from SEM we get EDX data which shows the number of elements present in the CNF sample. To determine the electrical properties of CNF, a four-point probe characterization method was used, namely by using the I-V meter four probes. The measurement results using the I-V meter show the

Figure 8.
Carbon nanofibers (CNF) electrospinning results with a flow rate of 0.1 ml/hour and relative humidity (a) 30% d and (b) 40% and an average fiber diameter of 262 nm (a) and 309 nm (b), (c) distribution of Carbon nanofibers elements at a flow rate of 0.5 ml/hour and relative humidity of 40% and (d) CNF sheets that are ready to be applied.

Ohmic curve on the I-V graph according to Ohm's law. The resulting Voltage and Current data are then processed according to Ohm's law to determine the conductivity value of the material.

Figure 8 shows the electrospinning results with a flow rate of 0.1 ml/hour and relative humidity of 30% and the parameters still produce fibers with an average diameter of 262 nm, the fibers in this parameter have a morphological shape with minimum and almost no beads. Whereas the electrospinning carbon nanofibers with a flow rate of 0.1 ml/hour and relative humidity of 40% and the parameters still produce a fiber with the smallest size of 200 nm and the largest diameter of 400 nm and an average diameter of 309 nm, the fiber in this parameter has a morphological shape with minimum and almost non-existent beads and more uniform diameter sizes and a narrower size range when compared to 30% humidity parameters. The CNF fabrication results were carried out by SEM–EDX to determine the

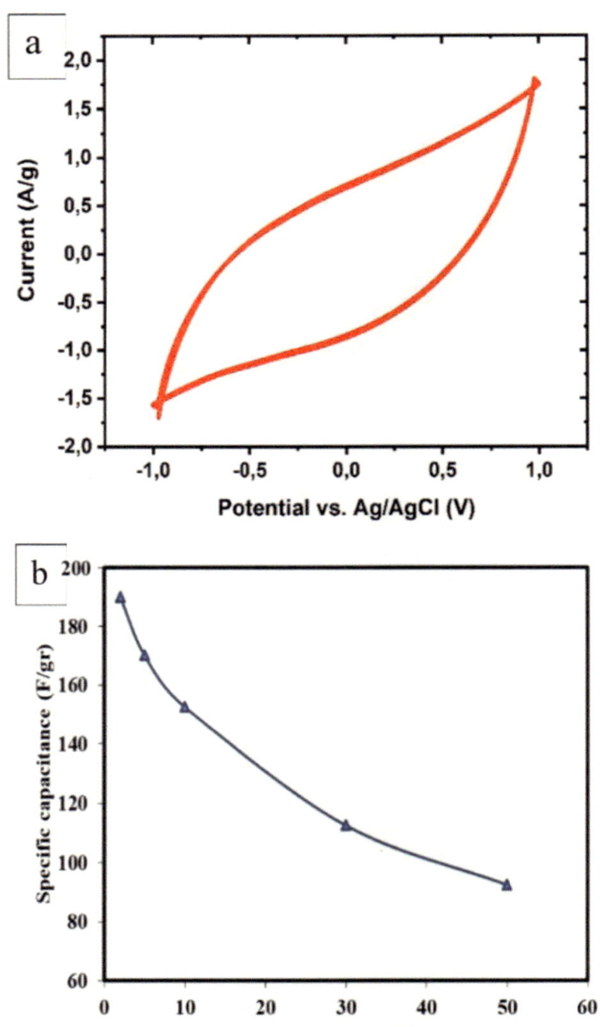

Figure 9.
Electrochemical characterization results. (a) diagram of the cyclic voltammogram (CV) data on the CDI electrode with a sweep rate of 5 mV/s, and (b) the specific capacitance of the carbon electrode from the cyclic voltammogram data.

Sample	CNF diameter (nm)	Voltage (mV)	Sheet resistance (Ω/sq)
A	417	28.7510	19.9822
B	358	26.8532	18.6633
C	309	26.5402	18.4457
D	262	25.6280	17.8117

Table 2.
The Average Value of Voltage and Resistance Sheet of Nanofibers from measurements using the I-V meter four probes.

elements of the nanofibers fabrication as in **Figure 8c**, the distribution of carbon elements has been evenly distributed, as well as the results of nanofibers fabrication with flow rate parameters of 0.1 ml/hour and 40% humidity and fixed parameters. Based on this characterization, carbon nanofibers consists of several chemical elements. The chemical elements, namely, O (Oxygen) by 24.57%, and C (Carbon) by 75.43%.

The cyclic voltammogram (CV) curve was obtained by scanning at a potential sweep rate of 5 mV/s on CNFS which had undergone temperature treatment and became a CDI electrode as shown in **Figure 9**. The CV curve shows the ideal behavior of the capacitor, the ideal CV curve with an almost square shape at different scanning sweep rates. Cyclic voltammetry testing on the three types of electrodes used in this study was carried out in a potential range of −0.5 V to 0.5 V and a potential sweep speed of 5 mV/s. The electrodes were immersed in an electrolyte solution of 0.5 M KCl with a submerged surface area of 1 cm2. Cyclic voltammetry measurement experiments were carried out at room temperature of 25 ° C. The results of the cyclic voltammetry test can be seen in **Figure 9**. On the voltammogram graph, a redox reaction is formed in an up-current pattern that shows the transfer of electrons from the electrode to the electrolyte solution, which involves the transfer of electrolyte ions based on the change in potential applied to the electrochemical cell, so this pattern increases in current indicating an increase in ion absorption capacity and ion absorption rate at potential given.

Based on **Table 2**, there is the largest sheet resistance value owned by CNF with the largest fiber diameter, but the difference is not too significant, the smallest sheet resistance value is owned by CNF with the smallest fiber diameter. The resistance value is directly proportional to the resistivity, but the resistivity is inversely proportional to the conductivity. Thus, the sample with the highest sheet resistance value has the greatest resistivity but the lowest conductivity.

5. Applications of carbon nanofibers

The application of CNF continues to develop today, based on the characteristics of CNF in the form of morphology and electrical properties, CNF has the potential to be applied to various fields, as shown in **Figure 10**, applications of CNF can be developed in the field of sensors, environmental applications and fields of electronics and electronics. Optical devices including energy storage fields.

Based on the characterization on the morphology and electrical properties of CNF, it is very possible to apply to sensor devices, the principle of chemiresistor sensors is more widely used for gas sensors because these sensors can be made easily and at relatively low cost. The chemiresistor mechanism is the reaction that occurs between the layers on the electrode and the gas which will result in a change in the value of resistance or conductivity.

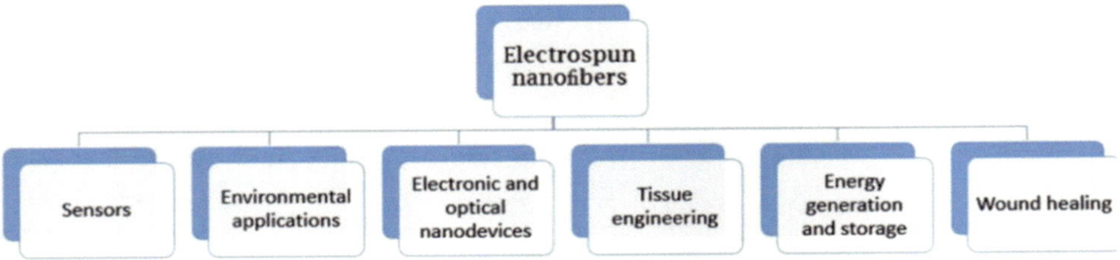

Figure 10.
General application areas of Carbon nanofibers [1].

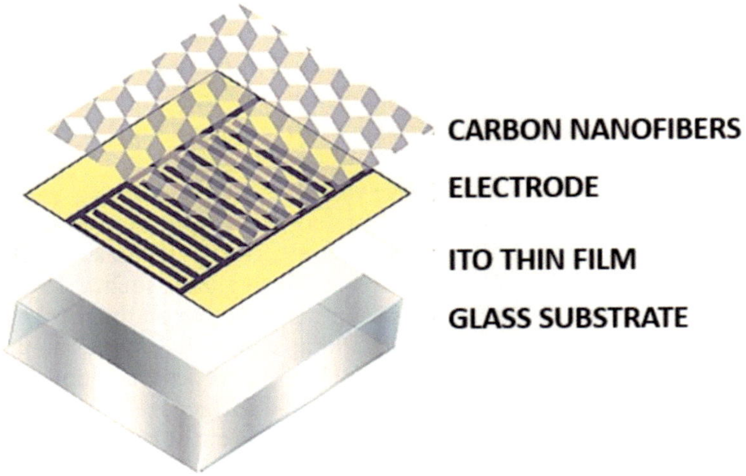

Figure 11.
Schematic of CNF application in sensor device fabrication.

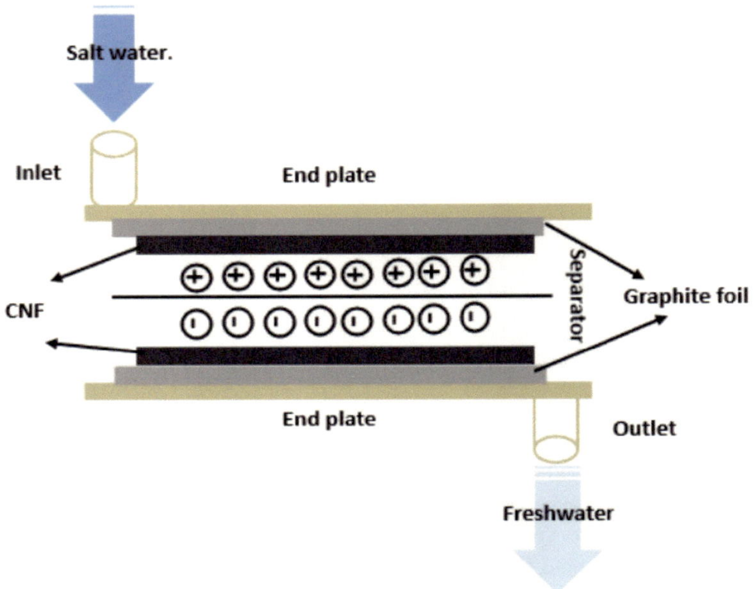

Figure 12.
Schematic of CNF application in electrode capacitive deionization CDI) [20].

This change occurs due to the transfer of valence electrons to the atoms of the sensor material due to the reaction with the reactant gas. The changes that occur are decreased resistance or increased conductance. Conductivity indicates the ability of a material to conduct electric current. The conductivity value can be determined by the equation below.

$$\sigma = \frac{1}{\rho} \tag{2}$$

Where σ is electrical conductivity (1/Ω.m) and ρ is electrical resistivity (Ω.m), the CNF application scheme for sensors can be seen in **Figure 11**. Whereas CNF which has been fabricated has high porosity and very high capacitance values. Good for capturing salt ions, it is very potential to be applied to Electrodes For Capacitive Deionization (CDI) with the scheme that can be seen in **Figure 12**.

6. Conclusions

Synthesis of CNF using electrospinning is an important part of this research, the synthesis process plays an important role in the quality of the CNF produced. In addition, the results of characterization of the different morphological structures of the CNFs formed show different properties, so that surface modification can help make CNF compatible for various applications. The characterization results show that the electrospinning result of the CNF composite diameter is in the range 200–450 nm, this shows that the electrospinning process with a rotating drum collector has advantages compared to a stationary plate type collector. Based on the electrical properties of the CNF I-V meter measurement results, a good conductivity value is obtained, which is due to the larger surface area of the CNF making it easier for electrons to move freely, so that CNF has the potential to be a good and modern sensor material. Whereas from the capacitive properties measured by the cyclic voltammogram (CV) curve with 5 mV/s and a measuring temperature of 25 °C shows the ideal behavior of the capacitor, the ideal CV curve with an almost square shape at different scanning sweep rates, so the use of CNF as a capacitive electrode is proper application.

Acknowledgements

This work was supported by Department of Physics and Lembaga Penelitian dan Pengabdian Kepada Masyarakat (LPPM), Institut Teknologi Sepuluh Nopember.

Conflict of interest

The authors declare no conflict of interest.

Author details

Gatut Yudoyono, Diky Anggoro*, Lutfi Fitria Ningsih and Rizki Romadoni
Department of Physics, Institut Teknologi Sepuluh Nopember, Surabaya, Indonesia

*Address all correspondence to: anggoro@physics.its.ac.id

References

[1] Mohamed A. Synthesis, Characterization, and Applications Carbon Nanofibers. Carbon-Based Nanofillers and Their Rubber Nanocomposites, Elsevier; 2019, p. 243-57. https://doi.org/10.1016/B978-0-12-813248-7.00008-0.

[2] Anggoro D, Saefuddin M, Fatimah I, Indrawati S. Optimization of high temperature furnace system as one of the spray pyrolysis subsystems based on R type thermocouples and PID control. vol. 1153, IOP Publishing; 2019, p. 012037.

[3] Fatimah I, Sari TI, Anggoro D. Effect of Concentration and Nozzle-Collector Distance on the Morphology of Nanofibers. vol. 860, Trans Tech Publ; 2020, p. 315-9.

[4] Akhtar K, Khan SA, Khan SB, Asiri AM. Scanning Electron Microscopy: Principle and Applications in Nanomaterials Characterization. In: Sharma SK, editor. Handbook of Materials Characterization, Cham: Springer International Publishing; 2018, p. 113-145. https://doi.org/10.1007/978-3-319-92955-2_4.

[5] han2014.pdf n.d.

[6] Goodship V, Jacobs D. Polyvinyl alcohol: materials, processing and applications. vol. 16. Smithers Rapra Technology; 2009.

[7] Feng L, Xie N, Zhong J. Carbon Nanofibers and Their Composites: A Review of Synthesizing, Properties and Applications. Materials 2014;7:3919-3945. https://doi.org/10.3390/ma7053919.

[8] Song H, Shen W. Carbon nanofibers: synthesis and applications. Journal of Nanoscience and Nanotechnology 2014;14:1799-1810.

[9] Vilaplana JL, Baeza FJ, Galao O, Zornoza E, Garcés P. Self-sensing properties of alkali activated blast furnace slag (BFS) composites reinforced with carbon fibers. Materials 2013;6:4776-4786.

[10] Acharya S, Adam J, Adamová D, Adolfsson J, Aggarwal MM, Aglieri Rinella G, et al. Production of deuterons, tritons, He 3 nuclei, and their antinuclei in p p collisions at s = 0.9, 2.76, and 7 TeV. Phys Rev C 2018;97:024615. https://doi.org/10.1103/PhysRevC.97.024615.

[11] Kumar M, Hietala M, Oksman K. Lignin-Based Electrospun Carbon Nanofibers. Front Mater 2019;6:62. https://doi.org/10.3389/fmats.2019.00062.

[12] Wang G, Pan C, Wang L, Dong Q, Yu C, Zhao Z, et al. Activated carbon nanofiber webs made by electrospinning for capacitive deionization. Electrochimica Acta 2012;69:65-70. https://doi.org/10.1016/j.electacta.2012.02.066.

[13] Zargham S, Bazgir S, Tavakoli A, Rashidi AS, Damerchely R. The Effect of Flow Rate on Morphology and Deposition Area of Electrospun Nylon 6 Nanofiber. Journal of Engineered Fibers and Fabrics 2012;7:155892501200700. https://doi.org/10.1177/155892501200700414.

[14] Goodship V, Jacobs DK. Polyvinyl alcohol: materials, processing and applications. vol. Vol.16. Shrewsbury, Shropshire: Smithers Rapra Technology; 2009.

[15] Mohamed A, Osman TA, Toprak MS, Muhammed M, Uheida A. Surface functionalized composite nanofibers for efficient removal of arsenic from aqueous solutions. Chemosphere

2017;180:108-116. https://doi.org/10.1016/j.chemosphere.2017.04.011.

[16] Inagaki M, Yang Y, Kang F. Carbon Nanofibers Prepared via Electrospinning. Advanced Materials 2012;24:2547-2566. https://doi.org/10.1002/adma.201104940.

[17] Rahman IA, Padavettan V. Synthesis of Silica Nanoparticles by Sol-Gel: Size-Dependent Properties, Surface Modification, and Applications in Silica-Polymer Nanocomposites—A Review. Journal of Nanomaterials 2012;2012:1-15. https://doi.org/10.1155/2012/132424.

[18] Nayak R. Nano fibres by electro spinning, properties and applications n.d.:13.

[19] Pelipenko J, Kristl J, Janković B, Baumgartner S, Kocbek P. The impact of relative humidity during electrospinning on the morphology and mechanical properties of nanofibers. International Journal of Pharmaceutics 2013;456:125-134. https://doi.org/10.1016/j.ijpharm.2013.07.078.

[20] Anggoro D, Muhlas I, Fatimah I. Carbon Nanofibers Form by Electrospinning with Flowrate Variations as Electrodes for Capacitive Deionization. vol. 860, Trans Tech Publ; 2020, p. 351-6.

[21] Patil JV, Mali SS, Kamble AS, Hong CK, Kim JH, Patil PS. Electrospinning: A versatile technique for making of 1D growth of nanostructured nanofibers and its applications: An experimental approach. Applied Surface Science 2017;423:641-674. https://doi.org/10.1016/j.apsusc.2017.06.116.

[22] Li L. Electrospun hollow nanofibers for advanced secondary batteries. Nano Energy 2017:29.

[23] Nguyen - Vu VL. TAYLOR CONE–JET MODE IN THE FABRICATION OF ELECTROSPRAYED MICROSPHERES. JST 2018;55:216. https://doi.org/10.15625/2525-2518/55/1B/12112.

[24] Zhu J, Ge Y, Jasper S, Zhang X. Physical characterization of electrospun nanofibers. Electrospun Nanofibers, Elsevier; 2017, p. 207-38. https://doi.org/10.1016/B978-0-08-100907-9.00009-X.

[25] Doustgani A, Vasheghani-Farahani E, Soleimani M, Hashemi-Najafabadi S. Optimizing the mechanical properties of electrospun polycaprolactone and nanohydroxyapatite composite nanofibers. Composites Part B: Engineering 2012;43:1830-1836.

[26] Yuniar RA, Widiyastuti W, SetyawanH,PurwaningsihH,MachmudahS, Anggoro D. Formation of Carbon Fibres From Polymer Poly(vinyl alcohol)/Acetylene Black using Electrospinning Method. IOP Conf Ser: Mater Sci Eng 2019;543:012030. https://doi.org/10.1088/1757-899X/543/1/012030.

[27] Huang L, Bui N-N, Manickam SS, McCutcheon JR. Controlling electrospun nanofiber morphology and mechanical properties using humidity. J Polym Sci B Polym Phys 2011;49:1734-1744. https://doi.org/10.1002/polb.22371.

Chapter 4

Recent Advances in Applications of Ceramic Nanofibers

Nuray Kizildag

Abstract

Ceramic materials are well known for their hardness, inertness, superior mechanical and thermal properties, resistance against chemical erosion and corrosion. Ceramic nanofibers were first manufactured through a combination of electrospinning with sol–gel method in 2002. The electrospun ceramic nanofibers display unprecedented properties such as high surface area, length, thermo-mechanical properties, and hierarchically porous structure which make them candidates for a wide range of applications such as tissue engineering, sensors, water remediation, energy storage, electromagnetic shielding, thermal insulation materials, etc. This chapter focuses on the most recent advances in the applications of ceramic nanofibers.

Keywords: applications, Li-ion battery, catalysts, ceramic nanofibers, electromagnetic shielding, electrospinning, nanofiber, sensors, thermal insulation, tissue engineering, water remediation

1. Introduction

Nanofibers are 1D nanostructured materials with some distinctive characteristics, such as large surface areas, well-controlled composition, flexibility, tunable porosity, ease of surface functionalization, and high mechanical/thermal properties which make them suitable for a number of different applications such as energy harvesting and storage [1, 2], filtration [3, 4], sensors [5, 6], tissue engineering [7–9], wound healing [10, 11], drug delivery [12, 13], polymer reinforcement [14, 15], and so on.

The possibility of producing nanofibers from a wide range of materials, including polymers, metals and metal oxides, carbon-based, and composite nanomaterials forms the significant impact of nanofiber technology [16]. The development of ceramic nanofibers has attracted a significant interest over the recent years. Ceramic nanofibers (CNFs) have been developed by the application of different methods such as magnetron sputtering [17], solution blowing [18], laser spinning [19], chemical vapor deposition [20], template synthesis [21], phase separation [22], hydrothermal treatment [23], and electrospinning [24–26]. Various kinds of ceramic nanofibers classified roughly into two groups such as oxide nanofibers (SiO_2, SnO_2, TiO_2, α-Al_2O_3, WO_3, $BaTiO_3$) and non-oxides (carbides (B_4C, SiC), borides, nitrides, silicides, sulfides) have been fabricated and are proven to be used in many different applications with high performance and efficiency.

2. Electrospinning of ceramic nanofibers

Electrospinning is the most used nanofiber production technique. It uses electrostatic forces to produce nanofibers. The electrospinning setup consists of a pump, high-voltage power supply, and a collector. The pump feeds the solution to the tip of the needle, the high-voltage power supply charges the solution, and the grounded collector collects the nanofibers. A jet occurs as the electrostatic force overcomes the surface tension of the solution droplet, undergoes bending and whipping instability during its flight in the electrical field between the tip of the needle and the collector. As the solvent evaporates, nanofibers accumulate on the collector in the nanoweb form. Electrospinning is a very simple and relatively inexpensive method used to fabricate ceramic nanofibers and it allows to control the morphology, average diameters, and compositions of the nanofibers (CNFs) [27–30].

Electrospinning enables the production of CNFs with very small diameters, high surface areas, extremely long length, and small pore size. The fabrication of CNFs by electrospinning is usually achieved in three main steps (**Figure 1**): (i) preparation of an electrospinning solution containing a polymer and sol–gel precursor; (ii) electrospinning of the solution to generate precursor NFs; and (iii) conversion of precursor NFs into the final CNFs by calcination, sintering, or chemical processes [31–36]. CNFs requires the presence of a polymer phase in the electrospinning solution as the ceramic phase on its own is not suitable for electrospinning. For successful electrospinning of CNFs, a suitable combination of ceramic precursor, polymer and solvent that can form a viscous homogeneous solution should be selected. Polyvinyl alcohol (PVA) [37], polyethylene oxide (PEO) [38], and polyvinyl pyrrolidone (PVP) [39] are widely used as the polymer phase in the electrospinning of CNFs.

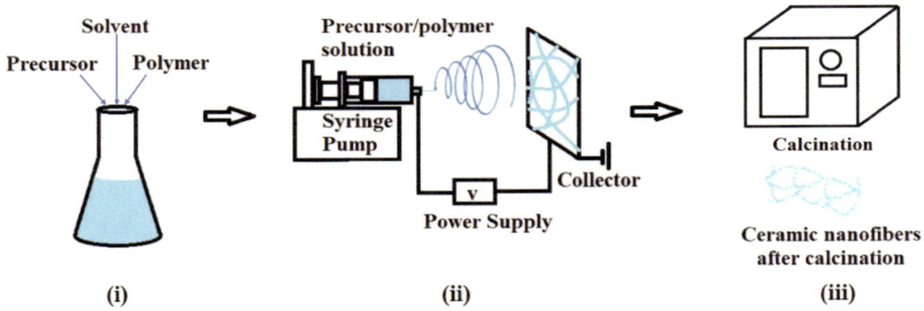

Figure 1.
Schematic showing the preparation of ceramic nanofiber membranes.

3. Applications of ceramic nanofibers

The unique properties of CNFs such as high surface area, extraordinary length, low density, high porosity, and thermo-mechanical properties [40, 41] qualify them for many different applications. This chapter covers a review of their applications related to tissue engineering [35, 36, 42–47], sensors [32–34, 48, 49], water remediation [50–56], batteries [57–63], catalyst supports/catalysts [64–71], electromagnetic interference (EMI) shielding [72–79], and thermal insulation materials [26, 79–82], etc. A list of the recent studies about ceramic nanofibers by electrospinning method is presented in **Table 1**.

Ceramic nanofiber composition	Processes applied	Potential applications	Important results	Ref.
$Ca_{10}(PO_4)_6(OH)_2$	electrospinning, calcination at 600°C	tissue engineering	uniform nanofibers with rough surface	[36]
$Ca_3(PO_4)_2$	electrospinning, calcination at 600°C	tissue engineering drug delivery	solid and micro-porous nanofibers	[35]
$Ca_{10}(PO_4)_6(OH)_2$	electrospinning, heat treatment at 350°C	tissue engineering	structural analysis	[43]
TiO_2	electrospinning, calcination at 700°C	tissue engineering	supported osteoblast viability	[45]
TiO_2	electrospinning, calcination at 500°C bioceramic preparation using TiO2 nanofibers as reinforcement	hard tissue engineering applications	bulk density, compressive strength and microhardness increased, porosity and water adsorption capacity decreased, bone-like apatite formation in simulated body fluid	[46]
BN reinforced gelatin	electrospinning, glutaraldehyde cross-linking	bone tissue engineering applications	nontoxic, biodegradable bioactive materials with enhanced mechanical properties	[47]
$8CeO_2 - 0.5Y_2O_3 - ZrO_2$	electrospinning, calcination at 1300°C	artificial muscle applications	Aligned nanofibers with shape-memory strain and repeatable shape memory actuation behavior	[42]
CeO_2	electrospinning, calcination at 1000°C	O_2 and CO monitoring	good morphological and structural stability in high temperature environment, sensitive, reversible, and reproducible response in real-time O_2 and CO monitoring	[32]
Ln FeO_3 (Ln = La, Nd, Sm)	electrospinning, calcination at 700°C	acetone sensor	lower optimum operating temperature (140°C) and good response	[49]
$PrFeO_3$	electrospinning, calcination at 600°C	acetone sensors	high response value, good selectivity, and good long-time stability at a low operating temperature of 180°C	[34]
silica, silica/PVP, silica/PMMA	electrospinning, drying at 50°C	optical gas sensor	flexible and porous nature, reliable and reproducible visual sensing performance	[48]
$SmFeO_3$	electrospinning, calcination at 700°C	ethylene glycol sensor	good response and distinct selectivity to ethylene glycol	[33]

Ceramic nanofiber composition	Processes applied	Potential applications	Important results	Ref.
α-Fe_2O_3	electrospinning, calcination at 500°C	chromate removal in water treatment	adsorption capacity increased with decreasing diameter, outperformed the commercially available Fe_2O_3 nanoparticles	[50]
CuO-ZnO composite	electrospinning, calcination at 500°C	water treatment and purification	high adsorption capacity, antibacterial activity	[53]
TiO_2 Au coated TiO_2	electrospinning, calcination at 800°C, Au coating	water treatment and purification	enhanced photoactivity	[52]
NiO nanofibers	electrospinning, calcination at 500°C	photocatalyst in water treatment	highly stable and ultrafine nanofibers with photocatalytic activity against congo red model dye	[83]
ZnO ZnO-SnO_2 composite	electrospinning, calcination at 600°C	photocatalyst in water treatment	ZnO–SnO_2 nanofibers showed higher photocatalytic activity than ZnO nanofibers, which was dependent on Sn/Zn molar ratio	[54]
ZnO/fly ash	electrospinning, calcination at 500°C	photocatalyst and adsorbent in water treatment	enhanced adsorption capacity and photocatalytic efficiency	[55]
TiO_2-coated YSZ/silica NF	electrospinning, calcination at 500°C	photocatalyst in water treatment	stable and reusable membranes with improved photocatalytic degradation efficiency and self-cleaning functionality	[51]
SiO_2 nanofibers/Al_2O_3 nanoparticles	electrospinning, calcination at 500, 600, 700, and 800°C	separator for LIBs	Nanofiber membranes with smooth surface, polymer-free composition, high porosity (79%), high electrolyte uptake (876%), and excellent thermal stability Full cell with reversible discharge capacity of 165 mAh g – 1 after 100 cycles at a current density of 50 mA g – 1	[57]
ZrO_2	electrospinning, calcination at 800°C	separator for LIBs and NaIBs	remarkable mechanical flexibility, ample porosity, excellent electrolyte wettability and infiltration, outstanding heat and flame-resistance, and high electrochemical inertness, in lithium or sodium batteries, higher current densities, longer cycling lives	[59]

Ceramic nanofiber composition	Processes applied	Potential applications	Important results	Ref.
CuO nanofibers	electrospinning, calcination at 600°C	anode for LIBs	specific capacity of 310 mAh g − 1 at 1C rate for 100 cycles and stabilized capacity of about 120 mAh g − 1 at 5C rate for 1000cycles	[60]
Co_3O_4@rGO nanofiber	electrospinning, calcination at 600°C, rGO coating	anode for LIBs	efficient stress relaxation and fast Li + ions and electron transport during discharge/charge cycling, in a half cell, displayed high Coulombic efficiency, enhanced cyclic stability, and high-rate capability (~900mAh/g at 1A/g, and ~ 600mAh/g at 5A/g) in half cell	[61]
3D SiO_2 nanofibers in polyethylene oxide–lithium (bis trifluoromethyl) sulfate–succinonitrile (PLS)	impregnating nanofiber membranes by PLS	solid polymer electrolyte for batteries	specific capacity of 167.9 mAh g − 1, suffered only 3.28% capacity degradation after 100 cycles, high safety at elevated temperatures by automatic shut down	[58]
aluminum-doped $Li_{0.33}La_{0.557}TiO_3$ (LLTO) nanofiber network in a polyvinylidene fluoride-hexafluoropropylene (PVDF-HFP)	electrospinning, calcination at 900°C, impregnation	solid polymer electrolyte for batteries	improved ionic conductivity and electrochemical cyclic stability, a full cell with excellent cycling performance and rate capability	[62]
γ-Al_2O_3	electrospinning, calcination	catalyst supports	sinter-resistant performance up to 500°C, higher reaction rates in the catalytic reduction of p-nitrophenol, compared to free nanocrystals	[68]
$MgTiO_3$	electrospinning, calcination at 600°C	photocatalyst for H_2 generation	enhanced efficiency and stability in photocatalytic H_2 generation under ultraviolet light	[71]
CuO-NiO composite	electrospinning, calcination at 600°C	catalyst for hydrazine oxidation	Ni content dependent electro-catalytic activity when used for hydrazine oxidation in alkaline media	[69]
CeO_2/Al_2O_3	electrospinning, calcination at 500°C	catalyst supports	homogenous elemental distribution, displayed good sinter-resistant performance and exhibited 13-times great catalytic activity than that of Pt@Al_2O_3 catalyst towards the hydrogenation of p-nitrophenol	[70]

Ceramic nanofiber composition	Processes applied	Potential applications	Important results	Ref.
TiO_2 ZrO_2	electrospinning, calcination at 510, 550, and 800°C	catalytic system for Suzuki cross-coupling reactions	calcination temperature dependent phase, Pd on anatase TiO2 could be used to catalyze the Suzuki coupling reactions and operated in a continuous flow fashion, nitric acid treatment provided reactivation	[67]
SiC	electrospinning, calcination at 1350°C	EMI shielding	flexible, hydrophobic, corrosion resistant, and thermally stable nanofibers with excellent EMI shielding properties with an effective absorption bandwidth of 4–18 GHz	[75]
SiC-Si_3N_4/graphite	electrospinning, calcination at 800, 1300, 1400, and 1500°C	EMI shielding	nanofibers annealed at 1300°C in Ar showed a RL_{min} of −57.8 dB at 14.6 GHz with EAB of 5.5 GHz, nanofibers annealed at 1300°C in N_2 exhibited a RL_{min} of −32.3 dB with EAB of 6.4 GHz over the range of 11.3–17.7 GHz	[76]
SiC/ZrC/SiZrOC nanofibers	electrospinning, calcination at 1400 and 1500°C	EMI shielding	reasonable electrical conductivity microwave-absorbing capability RL_{min} of about −40.38 dB at 14.1 GHz, antioxidant properties at 600°C	[72]
SiC/C	electrospinning, calcination at 1400°C	EMI shielding	lightweight and high EMI shielding efficiency, an ultrathin paraffin matrix with 30 wt.% SiC/C nanofibers exhibited superior EM wave absorption performance ($RL < -10$ dB) over the range of 12.6 and 16.7 GHz	[73]
ZrC/SiC	electrospinning, calcination at 1300-1500°C	EMI shielding	Flexible, lightweight nanofiber mats, 3-layer ZrC/SiC nanofiber mats with a thickness of 1.8 mm showed EMI shielding effectiveness (SET) of 18.9 dB, further improved to 20.1 dB at 600°C	[74]
SiO_2	electrospinning, calcination at 600, 800, 1000, and 1200°C	thermal insulation materials	ultra-softness, enhanced tensile strength, ultra-low thermal conductivity of 0.0058 W m − 1 K − 1	[77]
mullite	electrospinning, calcination between 900 and 1500°C	thermal insulation materials	diphasic nanofibers with average diameter of 216 ± 40 nm	[79]
SiO_2/aluminoborosilicate nanofibrous aerogels	electrospinning, calcination at 900°C	thermal insulation materials	flyweight densities of >0.15 mg cm − 3, zero Poisson's ratio, rapid recovery from 80% strain, and temperature-invariant superelasticity up to 1100°C, robust fire resistance and thermal insulation performance	[78]

Ceramic nanofiber composition	Processes applied	Potential applications	Important results	Ref.
SiO_2/SiO_2 nanofibrous aerogels	electrospinning, calcination at 900°C	thermal insulation materials	ultralow density of ~0.2 mg cm-3, ultralow thermal conductivity (23.27 mW m-1 K-1), negative Poisson's ratio, temperature-invariant superelasticity from −196 to 1100°C, and editable shapes on a large scale	[81]
SiZrOC	electrospinning, calcination at 800°C	thermal insulation materials	high-temperature stability (~1200°C in Ar) and low thermal conductivity (~0.1392 W m − 1 K − 1 at 1000°C in N_2)	[26]
ZrO_2-Al_2O_3/$Al(H_2PO_4)_3$ nanofibrous aerogels	electrospinning, calcination at 800°C	thermal insulation materials	ultra-strong, superelastic, and high temperature resistant, high fatigue resistance, thermal insulation performance with low thermal conductivity (0.0322 W m − 1 K − 1)	[82]
yttria-stabilized zirconia mixed silica (YSZ/SiO_2)	electrospinning, calcination at 800-1300°C	thermal insulation materials	enhanced use temperature and mechanical properties by stabilization, good thermal insulation performance and hydrophobic property	[80]
TiO_2 nanofibers AgNP coating	electrospinning, calcination at 500°C, coating by impregnation	separation of nuclear waste and recycling of nuclear fuels	high porosity, permeability, loading capacity, stability in extreme conditions	[84]
$ZnTiO_3$	electrospinning, calcination at 300-700°C	detoxification of chemical warfare agents	satisfactory reactivity against chemical warfare simulants	[85]
TiO_2	electrospinning, calcination at 800°C	reinforcement in Mg composite	increased compressive strength	[86]
TiO_2	electrospinning, calcination at 450°C	capacitive pressure sensor	high sensitivity (≈4.4 k Pa − 1), fast response speed (<16 ms), ultralow limit of detection (<0.8 Pa), low fatigue over 50000 loading/unloading cycles, high temperature stability	[87]

Table 1.
Ceramic nanofibers by electrospinning method and their applications.

3.1 Tissue engineering applications

Tissue engineering scaffolds are often made of biodegradable polymeric materials. The biodegradable polymeric scaffolds are not widely used in regeneration of load-bearing bones due to their limited mechanical strength. There have been many efforts invested to enhance the mechanical properties of the scaffolds, i.e., CNFs, reinforced by hydroxyapatite (HA), have provided mechanical support and osteoconductivity to the growing cells in bone regeneration.

Wu et al. [36] electrospun a precursor mixture of $Ca(NO_3)_2{\cdot}4H_2O$ and $(C_2H_5O)_3PO$ with a polymer additive, then applied thermal treatment at 600°C for 1 h and prepared HA ($Ca_{10}(PO_4)_6(OH)_2$) fibers. The pure fibers obtained were 10–30 μm in diameter and up to 10 mm in length. The HA grain size was ~1 μm [36]. Kim and Kim produced HA nanofibers and also their fluoridated forms for dental restoration applications to stimulate bone cell responses and provide protection against the formation of dental caries [44]. Xiaoshu and Shivkumar [35] electrospun nanofibers from polyvinyl alcohol (PVA) solution containing calcium phosphate-based sol and then calcined them at 600°C for 6 h to obtain an inorganic, fibrous network, which was suggested for use in the tissue engineering and drug delivery [35]. Franco et al. used phosphorus pentoxide (P_2O_5) and calcium nitrate tetrahydrate ($Ca\ (NO_3)_2.4H_2O$) as precursors of phosphorus and calcium and polyvinylpyrrolidone (PVP) as the polymer to electrospin CNFs [43]. Wang et al. [45] electrospun pure titanium dioxide (TiO_2) nanofiber meshes with different surface microroughness and nanofiber diameters and investigated the osteoblast differentiation on these meshes by analyzing the cell number, differentiation markers and local factor production for MG63 cells seeded on TiO_2 meshes. Cells with similar morphology were observed to grow throughout the entire surfaces. While the cell number was found to be sensitive to surface microroughness, the cell differentiation and local factor production were observed to be regulated by both surface roughness and nanofiber diameter. These results indicated that the TiO_2 nanofiber meshes could be used to create an osteogenic environment without using exogenous factors [45]. Aly et al. [46] developed wollastonite glass ceramic composites reinforced by electrospun TiO_2 nanofibers for use in hard tissue engineering applications. The composite material exhibited greater densification and better mechanical characteristics in comparison to pure wollastonite. The composites having 0, 10, 20 and 30 wt.% metal oxide nanofibers were sintered at 900, 1100 and 1250°C. While the compressive strength, bulk density, and microhardness increased, the water adsorption capacity and porosity decreased with the increase in the TiO_2 nanofiber content. When the wollastonite and wollastonite/TiO_2 nanofibers were soaked in simulated body fluid, bone-like apatite was formed on their surface. The characteristics of wollastonite were improved with incorporation of TiO_2 nanofibers while its in-vitro bioactivity was preserved. The developed composite was suggested for use as a bone substitute in high load bearing sites [46]. Nagarajan et al. [47] produced boron nitride-reinforced gelatin nanofibers as a new class of two-dimensional biocompatible nanomaterials, showing enhanced mechanical properties, stability to the glutaraldehyde cross-linking, high bioactivity in forming bone-like HA, and biodegradability. Depending on the analysis of osteoblast gene expression and the measurement of alkaline phosphatase activity, they were proven to be suitable for bone tissue engineering applications [47].

Apart from the use of CNFs in bone tissue engineering applications, Du et al. [42] recently fabricated highly aligned, zirconia-based, shape memory nanofiber yarns and springs by electrospinning for artificial muscle applications at elevated temperatures. The nanofiber yarns displayed a recoverable strain of up to ~5% and short recovery time (0.16 s) at actuation temperatures of 328–388°C. When heated by a Bunsen burner, the shape memory ceramic springs could lift up to 87 times their own weight with a stroke of ~3.9 mm. The ceramic yarns/springs exhibited an output stress of 14.5–22.6 MPa, a work density of ~15–20 kJ m^{-3}, and a tensile strength of ~100–200 MPa, which were much higher than those of human muscles and some other polymer-based artificial muscles [42].

3.2 Gas sensors

Sensors, which are used to monitor and quantify volatiles related to environmental monitoring, analyze food quality, and diagnose illnesses, have attracted great interest in the recent years. Sensors designed are required to display high selectivity, low power consumption, fast response/recovery rate, low detection limit and a low humidity dependence [31].

The ceramic nanofibers have been extensively studied for gas sensing applications due to their advantages such as good directional carrier transport, high surface energy, large surface-to-volume ratio, high chemical stability, great sensing performance. Ceramics are inherently resistant to aggressive physical and corrosive chemical circumstances and they offer significantly minimized hysteresis with increased relaxation time, which improves the stability, performance, and response time of pressure sensors [87]. Many researchers showed the applicability of electrospun ceramic nanofibers in gas sensing applications for detection of many different gases such as acetone [34, 49, 88–97], ethanol [98–101], formaldehyde [102], ammonia [103–107], hydrogen sulfide [108], nitrogen dioxide [109–111], acetic acid [112], carbon monoxide [113, 114], hydrogen [115], and toluene [116]. The sensing properties of metal oxides depend on their shape, size, size distribution, surface area, structure, phase, the grain size, crystallinity, the presence of crystal lattice defects, the type of the charge carriers, and the oxidizing or reducing nature of the target gas [31, 117–120]. Besides, the composition of the sensor is another important factor. The most effective methods for improving response time and sensitivity are (i) doping with different metals such as Ce, Cu, Pr, La, Pd, Mn [91, 94, 106, 112, 114, 121], (ii) formation of composites by coupling of two or more oxide metals [118, 119, 122–124], and (iii) addition of graphene [125–127]. For gas sensors made up of ceramic nanofibers, the composition and the nanofiber configuration are other important characteristics that can be controlled to improve the sensor performance.

Liu et al. [32] prepared polycrystalline CeO_2 nanofibers by combination of electrospinning and calcination. The average diameter of the nanofibers was measured as 376 ± 55 nm. They displayed good morphological and structural stability at high temperatures (800–1000°C) and showed reversible, sensitive, and reproducible response when used for real-time oxygen (O_2) and carbon monoxide (CO) monitoring at 800°C and 1000°C, respectively [32]. Tong et al. [49] prepared acetone sensors based on $LnFeO_3$ (Ln = La, Nd, and Sm) nanofibers produced by electrospinning and investigated the effect of lanthanide on acetone sensing properties of the nanofibers at different temperatures and acetone concentrations. The results indicated that the lanthanides significantly affected the sensing properties of $LnFeO_3$. When exposed to 100 ppm acetone at 140°C, the $SmFeO_3$ sensor exhibited the largest sensing response (Response = 9.98). The response and recovery times for the $SmFeO_3$ sensor were about 17 and 15 s, respectively [49]. Ma et al. [34]prepared hollow perovskite praseodymium ferrite ($PrFeO_3$) nanofibers via electrospinning followed by calcination. The samples had a large specific surface area (33.74 m^2 g^{-1}) with mesoporous characteristics. They showed good selectivity, long-time stability at 180°C, and high response value. While the response time of the sensor to 10 ppm acetone was about 4 s, the recovery time was measured as 4 s. [34]. Teli and Nadathur [48] prepared reusable, reversible dye doped nanofibrous silica and silica/PVP, and silica/PMMA membranes by combination of sol–gel and electrospinning and demonstrated their use as an effective optical gas sensor. A durable, sensitive, reversible, and visually detectable response to HCl and NH_3 was observed when the silica composite NF membranes was doped with a pH sensitive dye Bromothymol blue. Eliminating the need for electronic instrumentation, the regeneration of the doped NF membranes' color under mild thermal treatment and their thermal stability, permitted their repeated use for naked-eye sensing. The magnitude of color change was affected by the presence of copolymer in the NF membrane structure due to the copolymer's effects on the fiber diameter, surface area, porosity, and polarity of nanofiber membranes. The reliable and reproducible visual sensor performance along with its flexible and porous nature offered many advantages for different applications such as detecting volatiles relevant to environmental safety, tracing stability to spoilage in fresh food storage, air quality monitoring. Besides, they could be used to detect biogenic amines, or volatiles released by biological materials, to monitor plant life or human health, and to confirm the safety of work environments in chemical and nuclear plants [48]. Han et al. [33] fabricated perovskite samarium ferrite ($SmFeO_3$) nanofibers by electrospinning route and calcination process for use as a sensor for ethylene glycol, which is on one hand one of the most significant raw materials, and on the other hand one of toxic pollutants for animals, humans, and environments. The roughness of $SmFeO_3$ nanofibers contributed to gas sensing properties by increasing the contact area between gases and material surface. They displayed high response value (18.19) to 100 ppm EG at 240°C, fast response/recovery times, good stability and selectivity [33].

3.3 Water remediation applications

Water resources being contaminated by many different types of pollutants such as heavy metal ions, organic dyes, pesticides, bacterial pathogens, etc. generates a major challenge worldwide. Many of these contaminants are skin sensitizers and mutagens responsible for different types of cancers in human beings. On the other hand, bacterial pathogens are responsible for several severe health problems. It has become very critical to develop efficient water purification technologies. Among many different methods such as electrochemical treatment, ozonation, membrane filtration, flocculation, ceramic nanofibers have gained significant interest because of their potential in water remediation due to their adsorption properties and photocatalytic activity. The major advantages of nanofibers in water remediation applications are their high aspect ratio, enhanced surface area, higher surface activity, higher porosity, continuous structure, easy handling and retrieving compared to nanoparticles [53].

Many attempts have been devoted to the use of ceramic nanofibers in the removal of contaminants from water sources. Fe_2O_3 [50], ZnO/TiO_2 [51], TiO_2 [52], CuO/ZnO [53], ZnO/SnO_2 [54], ZnO [55], NiO [83], and many other nanofibers have been developed and proved to be successful in water remediation applications.

Nalbandian et al. [50] produced α-Fe_2O_3 nanofibers via electrospinning and investigated their applicability in chromate removal from water. Adsorption capacity of the nanofibers increased as the average nanofiber diameter decreased. Based on ($CrO_4^{\ 2-}$) adsorption isotherms at pH 6, the nanofibers with 23 nm average diameter exhibited an adsorption capacity of 90.9 mg g^{-1} [50]. Malval et al. [53] fabricated CuO-ZnO composite nanofibers and explored their adsorption capacity and antibacterial properties. The composite ceramic nanofibers displayed excellent adsorption capacity for congo red dye. Depending on the adsorption isotherms and kinetic studies composite nanofibers performed better than their single counterparts. The ceramic nanofibers were also very active inhibitors against the growth of *S. aureus* and *GFP-E. Coli*. With their excellent adsorption capacity and adequate antibacterial properties, the composite ceramic nanofibers were suggested for use in water treatment and purification processes [53].

Nalbandian et al. [52] produced titanium dioxide (TiO_2) nanofibers and tailored their structure and composition to optimize their photocatalytic treatment efficiency. Nanofibers with controlled diameter (30–210 nm), crystal structure (anatase, rutile, mixed phases), and grain size (20–50 nm) were manufactured. Besides, composite nanofibers with either surface-deposited or bulk-integrated Au nanoparticle cocatalysts were developed. When the nanofibers' reactivity against some model pollutants such as phenol and emerging pollutants such as pharmaceuticals, personal care products were analyzed, optimized TiO_2 nanofibers displayed superior performance than the traditional nanoparticulate photocatalysts. The photoactivity increased by 5 to 10 fold after Au deposition onto the surface of the nanofibers independent of the solution concentration which was attributed to the improved charge separation [52]. Malval et al. [83] produced nickel oxide (NiO) nanofibers for use as photocatalysts. The NiO nanofibers were tested for their photoactivity against model dye congo red (CR). The concentration of the catalyst was observed to be a significant factor. Due to their non-aggregating nature in aqueous solution, NiO nanofibers performed better than NiO nanoparticles. Additionally, reusability and stability were other advantages NiO nanofibers provided [83]. Pascariu et al. [54] fabricated $ZnO–SnO_2$ nanofibers by electrospinning technique combined with calcination at 600°C. The composite ceramic nanofibers showed photocatalytic activity against Rhodamine B (RhB) dye and the highest efficiency was obtained for nanofibers with Sn/Zn molar ratio of 0.030 [54]. Pant et al. [55] produced fly ash incorporated zinc oxide nanofibers and investigated their ability to remove methylene blue (MB) from the water. Adding fly ash, which is a waste material from thermal power plants to ZnO nanofibers resulted in increased adsorption and photocatalytic removal of MB from water. Huh et al. [51] prepared TiO_2-coated yttria-stabilized zirconia/silica nanofiber (an) by coating TiO_2 on the surface of YSZ/silica NF using a sol–gel process. The coating layer improved the separation ability of the membrane as well as the photocatalytic degradation ability. With its smaller pore size, TiO_2-coated YSZ/silica NF membrane rejected over 99.6% of the 0.5 μm polymeric particles. Furthermore, the TiO_2-coated YSZ/silica NF membrane showed excellent adsorption/degradation of humic acid (HA, 88.2%), methylene blue (MB, 92.4%), and tetracycline (TC, 99.5%) [51].

Some polymer-ceramic composites have been suggested for the removal of organic pollutants from water. Allabashi et al. [56] impregnated Al_2O_3, SiC and TiO_2 ceramic nanofibers by triethoxysilylated derivatives of poly (propylene imine) dendrimer, polyethylene imine and polyglycerol hyperbranched polymers and β-cyclodextrin and subsequently sol–gel reaction led to their polymerization and chemical bond formation with the ceramic substrates. The resulting organic–inorganic filters were found to remove the polycyclic aromatic hydrocarbons (up to 99%), of monocyclic aromatic hydrocarbons (up to 93%), trihalogen methanes (up to 81%), pesticides (up to 43%) and methyl-tert-butyl ether (up to 46%) [56].

3.4 Batteries

Electrospun ceramic nanofibers, with their extremely high surface area and fast charge-transfer channels along its 1D nanostructure, have been considered as ideal materials for batteries. The use of lithium-ion batteries (LIBs) has attracted great interest for use in the smart micro-electronic devices and electric vehicles because of their potential for high energy density and power. However, their energy density is usually restricted by the working potentials and specific capacity of the electrodes. There is great effort to develop alternative electrodes with higher specific capacity that will be able to replace commercially available graphite anode. Due to their high theoretical specific capacity, transition metal oxides have attracted considerable attention.

Recently, Zaidi et al. [57] prepared a separator film by electrospinning to develop high-performance LIBs. The film had a hybrid morphology of SiO_2 nanofibers (SNFs) and Al_2O_3 nanoparticles (ANPs). It had a polymer-free composition, smooth surface, good thermal stability with high electrolyte uptake (876%) and high porosity (79%). Compared to some commercial products, higher ionic conductivity, lower bulk resistance at elevated temperature (120°C), lower interfacial resistance with lithium metal, and a wider electrochemical window was obtained. When full cells were fabricated, the specific capacity of the full cell with the SNF-ANP separator film was measured as 165 mAh g^{-1}; the cell was stable for 100 charge/discharge cycles and exhibited a capacity retention of 99.9% at room temperature [57]. Jing et al. [59] used electrospun ZrO_2 membrane as separator in lithium or sodium batteries. The membranes displayed remarkable mechanical flexibility, excellent electrolyte wettability, and infiltration, ample porosity, high electrochemical inertness, and outstanding heat and flame-resistance. The separator could withstand high current densities and showed longer cycling life than the state-of-the-art separators [59].

Different types of CNFs were also developed as electrodes for batteries. Gangaja et al. [60] electrospun CuO nanofibers for use as LIB anode. The nanofibers were made up of CuO nanoparticles, which formed a good inter-connected network. Fabricated half cells maintained specific capacity of 310 mAh g^{-1} at 1°C rate for 100 cycles and stabilized capacity of about 120 mAh g^{-1} at 5°C rate for 1000 cycles. While SEI layer content remained the same, its thickness increased at the end of 10th charge according to the ex situ x-ray photoelectron spectroscopy [60].

Incorporation of carbon materials into ceramic nanofibers has been used as an effective strategy to improve electrical conductivity and act as a buffer to suppress the volume variation of the anodes. For example, addition of graphene into Co_3O_4 anodes has improved the electrochemical performance because of the high capacity of Co_3O_4 together with the high surface area, excellent flexibility of graphene, and good electrical conductivity [128, 129]. Hu et al. [61] synthesized porous Co_3O_4@rGO nanofiber anode materials via electrospinning, post thermal and hydrothermal treatments. The interconnected structure of Co_3O_4 nanocrystals and presence of reduced graphene oxide networks were effective in accommodating volume change, deterring aggregation of Co_3O_4 fibers and facilitating rapid Li^+ ion transport during charge/discharge cycling. The developed anode material exhibited high reversible capacity, good rate capability, and excellent cycling stability when it was tested in half and full cells. The full cell constructed from a $LiMn_2O_4$ cathode and a Co_3O_4@rGO anode displayed a stable capacity with operation voltage of ~2.0 V, which was promising for the electronic devices working at low voltages [61].

CNFs with their high ionic conductivity are promising candidates for use as electrolytes in all-solid-state batteries, which are among the most promising technologies to replace conventional lithium-ion batteries [130]. Cui et al. [58] prepared an electrolyte using polyethylene oxide–lithium (bis trifluoromethyl) sulfate–succinonitrile (PLS) and frameworks of three-dimensional SiO_2 nanofibers (3D SiO_2 NFs), which had a conductivity of 9.32×10^{-5} S/cm at 30°C. While the Li/$LiFePO_4$ cells assembled with PLS and 3D SiO_2 NFs (PLS/3D SiO_2 NFs) delivered an original specific capacity of 167.9 mAh g^{-1}, they only suffered 3.28% capacity degradation after 100 cycles. The solid lithium batteries based on composite electrolytes offered high safety at elevated temperatures since the cells automatically shut down with the decomposition of PLS above 400°C [58]. Yang et al. [62] developed a solid-state ceramic/polymer composite electrolyte by embedding a three-dimensional (3D) electrospun aluminum-doped $Li_{0.33}La_{0.557}TiO_3$ (LLTO) nanofiber network in a polyvinylidene fluoride-hexafluoropropylene (PVDF-HFP) matrix and reported an enhanced interfacial Li-ion transport along the nanofiber/polymer interface. The chemical interaction between the nanofibers and the polymer was further enhanced by coating lithium phosphate onto the LLTO nanofiber surface, which also promoted the lithium-ion transport along the polymer/nanofiber interface, improved the ionic conductivity and electrochemical cycling stability of the nanofiber/polymer composite. The full cell consisted of a lithium metal anode, a $LiFePO_4$-based cathode and the composite electrolyte in between exhibited excellent cycling performance and rate capability [62]. Zhang et al. [63] embedded $Li_7La_3Zr_2O_{12}$ (LLZO) nanofibers into a poly (vinylidene fluoride-co-hexafluoropropylene) (PVDF-HFP)/ionic liquid (IL) matrix to construct a solid polymer electrolyte (SPE). The developed SPE, containing 15 wt. % LLZO nanofibers, exhibited an improved ionic conductivity (6.5×10^{-3} S cm^{-1}) at room temperature, a wide electrochemical window (5.3 V vs. Li/Li^+), excellent flexibility and mechanical strength. SPE-based half and full cells delivered impressive electrochemical properties [63].

3.5 Catalyst supports and catalysts

Ceramic nanofibers with their unique properties have been extensively explored as either catalyst supports or catalysts for various types of heterogeneous reactions, such as sinter-resistant catalysts [68], photocatalytic reaction [71], electrocatalytic reaction [69], hydrogenation reaction [70], oxidation reaction [66], and Suzuki coupling reactions [67].

Fu et al. [68] prepared γ-Al_2O_3 nanofibers with a loofah-like surface using single-needle electrospinning for use as Pt supports. After sintering at 500°C, the Al_2O_3 nanofiber supported Pt catalysts were employed in catalytic reduction of p-nitrophenol and 4-times higher reaction rate constant (6.8 s^{-1} mg^{-1}) was observed compared to Pt nanocrystals. The high performance of the Al_2O_3 nanofiber supported Pt catalysts was attributed to the special surface structure and the strong metal–support interactions between Pt and γ-Al_2O_3 [68]. There are many studies on the use of ceramic nanofibers as photocatalysts in water remediation studies, as already explained previously. Other than water remediation studies, ceramic nanofibers are also employed in the photocatalytic H_2 evolution from water splitting, which is a promising renewable energy generation process. Using electrospinning method, Wang et al. [71] fabricated $MgTiO_3$ nanofibers and compared their photocatalytic H_2 generation ability with the $MgTiO_3$ nanoparticles and P25. The $MgTiO_3$ nanofibers showed high efficiency and stability in photocatalytic H_2 generation under ultraviolet light. Attributed mainly to their large specific surface area, special 1D structure, unique mesh morphology, and pure phase, photoelectrochemical measurements showed that the $MgTiO_3$ nanofibers facilitated the transport and separation of the photoinduced charge carriers [71]. Ceramic nanofibers are also utilized in electrocatalysis to speed up the charge transfer reaction between electrodes and electrolytes. Hosseini et al. [69] fabricated CuO/NiO composite nanofibers and investigated their photocatalytic performance as anode catalyst for hydrazine oxidation in alkaline media. The best catalytic performance was observed when the proportion of $Cu(OAc)_2$:$Ni(OAc)_2$ was 25:75 in polymeric solution [69]. Liu et al. [70] electrospun mesoporous CeO_2-based ultrathin nanofibers in fibril-in-tube configuration. The fibril-in-tube configuration was achieved by choosing two metal precursors with different decomposition rates. $Al(acac)_3$, which rapidly led to the growth kinetics varied along the radial direction of nanofibers by releasing gaseous pieces, was selected as Al_2O_3 precursor, and made critical contribution to the formation of fibril-in-tube structure. The novel fibril-in-tube CeO_2 nanofibers with different amount of homogenous Al_2O_3 elemental distribution were investigated as Pt supports. The developed catalytic system exhibited sinter-resistant catalytic activity in the hydrogenation of p-nitrophenol, which was 13-times higher than that of Pt@Al_2O_3 catalyst [70]. Electrospun ceramic nanofibers are also employed as heterogeneous catalysts for oxidation reactions. Formo et al. [67] reported the use of TiO_2 and ZrO_2 nanofibers decorated with Pt, Pd, and Rh nanoparticles as a catalytic system for Suzuki cross-coupling reactions. The new catalytic system offered an efficient process, which was cost-effective as it could be fully regenerated and repeatedly used [67]. Formo et al. deposited Pt nanoparticles on the TiO_2 porous nonwoven mats, which acted as superior catalyst toward methanol oxidation reaction [66].

3.6 EMI shielding applications

Intensive effort is being spent to develop high-performance EMI shielding materials to protect people from the potential damages of high frequency electromagnetic waves generated by the electronic devices. Among many different electromagnetic wave absorbers such as carbon nanotubes, hollow carbon microspheres, graphene, Fe_3O_4 microspheres, composite spheres, needle-like metal oxides, and carbon nanofibers, specifically SiC nanofibers are considered as an important candidate because of their high temperature stability, mechanical strength, and resistance to oxidation and corrosion [75].

Wang et al. [75] fabricated flexible, hydrophobic, corrosion resistant, and thermally stable SiC ceramic nanofibers, which showed excellent EMI shielding properties with an effective absorption bandwidth of 4–18 GHz. 90% of electromagnetic waves below −10 dB were absorbed. The maximum reflection loss (RL) reached −19.4 dB at 5.84 GHz. Besides, they were found to be environmentally stable in 2 mol L^{-1} NaOH solution for 2 h and at high temperatures of 500°C in air atmosphere. The developed SiC nanofibers were suggested as candidates for EMI shielding materials in harsh environments [75]. In another study [76], they incorporated graphite into SiC/Si_3N_4 composite nanofibers and investigated the relationship between processing, fiber microstructure, and their electromagnetic wave absorption performance. The annealing atmosphere and temperature were observed to affect the electromagnetic wave absorption capability and effective absorption bandwidth. The nanofibers after annealing at 1300°C in Ar showed a minimum RL of −57.8 dB at 14.6 GHz with EAB of 5.5 GHz. The nanofibers annealed at 1300°C in N_2 exhibited a minimum RL value of −32.3 dB at a thickness of 2.5 mm, and the EAB reached 6.4 GHz over the range of 11.3–17.7 GHz. The superior EMI shielding properties (high reflection loss together with wider EAB) imparted the composite nanofibers the potential to be used as reinforcements in polymer and ceramic matrix composites with EMI shielding properties [76]. Huo et al. [72] prepared heterogeneous SiC/ZrC/SiZrOC hybrid nanofibers with different ZrC contents and analyzed them in terms of electrical conductivity, average diameter, and microwave-absorbing capability. When the ZrC concentration increased from 0 to 10 wt.%, decrease in average nanofiber diameter from 800 nm to 200 nm and increase in electrical conductivity from 0.3448 to 2.5676 S cm^{-1} were observed. The minimum reflection loss was measured as −40.38 dB at 14.1 GHz for the SiC/ZrC/SiZrOC hybrid nanofibers and they were suggested as promising high temperature microwave-absorbing materials [72]. Using a combined process of electrospinning and calcination, Huo et al. [73] produced silicon carbide/carbon (SiC/C) hybrid nanofibers, which possessed a high aspect ratio and a scaly surface.

The EMI shielding properties of the SiC/C composite nanofibers were studied in the range from 2 GHz to 18 GHz. A paraffin matrix, which was reinforced with 30 wt. % composite nanofibers and had a thickness of 3 mm showed satisfactory dielectric loss value and a minimal reflection loss of approximately −36 dB at 6.8 GHz. Moreover, the maximum effective absorption (<−10 dB) bandwidth (EAB) was approximately 4.1 GHz (12.6–16.7 GHz) with a 1.5 mm thickness, which could cover most of the Ku-band [73]. Hou et al. [74] also fabricated flexible and lightweight ZrC/SiC hybrid nanofiber mats. When ZrC was added into SiC electrospinning solution, the viscosity decreased and the conductivity increased as a result of which the average diameter reduced from 2.6 μm to 330 nm, the specific BET surface area (SBET) increased from 51.5 to 131.1 $m^2\ g^{-1}$ and the electrical conductivity increased from 1.5×10^{-6} to 1.3×10^{-1} S cm^{-1}. It is found that 3-layer ZrC/SiC nanofiber mats with a thickness of 1.8 mm showed EMI shielding effectiveness (SET) of 18.9 dB, which could be further improved to 20.1 dB at 600°C [74].

3.7 Thermal insulation materials

Ceramic nanofibers exhibit excellent thermal stability and low thermal conductivity, which make them highly desirable for high-temperature thermal insulation applications [26]. Additionally, polymer phase used in electrospinning of ceramic nanofibers results in the formation of nanosized pores after calcination and helps to improve the insulation properties of ceramic nanofibers [131].

Si et al. [77] electrospun ultra-soft and strong silica nanofibers (SNF) from a sol–gel solution containing NaCl, the incorporation of which significantly enhanced the tensile strength of the SNF membranes from 3.2 to 5.5 MPa. The calcination temperature and the NaCl content in the precursor solution were found to affect the morphology and mechanical properties of the membranes. The membranes exhibited a ultra-softness of 40 mN, relative high tensile strength of 5.5 MPa and an ultra-low thermal conductivity of 0.0058 W m^{-1} K^{-1}, which made them promising candidates for bunker clothing [77]. Dong et al. [79] fabricated fine-grained mullite nanofibers derived from the diphasic mullite sol (polymethyl siloxane and aluminum tri-sec-butoxide) by electrospinning and subsequent pyrolysis at 1500°C. Mullite fibers with 216 nm average diameter and ~ 100 nm grain size were obtained after sintering at 1500°C were suggested as high-temperature thermal insulation materials [79]. Si et al. [78] reported a scalable strategy to create superelastic lamellar-structured ceramic nanofibrous aerogels by combining SiO_2 nanofibers with aluminoborosilicate matrices. The developed nanofibrous aerogels exhibited the integrated properties of flyweight densities of >0.15 mg cm^{-3}, zero Poisson's ratio, rapid recovery from 80% strain, and temperature-invariant superelasticity up to 1100°C, robust fire resistance and thermal insulation performance [78]. Dou et al. [81] designed a hierarchical cellular structured silica nanofibrous aerogel by using electrospun SiO_2 nanofibers (SNFs) and SiO_2 nanoparticle aerogels (SNAs) as the matrix and SiO_2 sol as the high-temperature analogue. They obtained randomly deposited SNFs with the SNAs evenly distributed on the fibrous cell wall. The unique hierarchical cellular structure of the ceramic nanofibrous aerogels exhibited ultralow density of ~0.2 mg cm^{-3}, ultralow thermal conductivity (23.27 mW m^{-1} K^{-1}), negative Poisson's ratio, temperature-invariant superelasticity from −196 to 1100°C, and editable shapes on a large scale. The novel nanofibrous aerogels with their favorable properties were suggested as thermal insulation materials for aerospace, industrial, and even extreme environmental conditions [81]. Using electrospinning technique, Zhang et al. [26] prepared multi-phase SiZrOC nanofiber membranes composed of amorphous SiOC and ZrO_2 nanocrystals to solve the incompatibility between thermal stability and low thermal conductivity of single-phase ceramic nanofibers at high temperatures, which limit their practical use. The fabricated SiZrOC nanofibers exhibited excellent high-temperature stability (~1200°C in Ar) and low thermal conductivity (~0.1392 W m^{-1} K^{-1} at 1000°C in N_2). The unique combination of amorphous SiOC and ZrO_2 nanocrystals offered a novel strategy to produce high-performance thermal insulation materials. While the multi-phase interfaces and the ZrO_2 nanocrystals created thermal transfer barriers to reduce the heat transfer, the SiOC phase effectively suppressed radiative heat transfer [26]. Zhang et al. [82] produced ultra-strong, superelastic, and high temperature resistant, lamellar multiarc structured ceramic nanofibrous aerogels by combining $ZrO_2 - Al_2O_3$ nanofibers with $Al(H_2PO_4)_3$ matrices. The resulting $ZrO_2 - Al_2O_3$ nanofibrous aerogels displayed a recovery of 90%, compression strength of more than 1100 kPa, temperature-invariant superelasticity, and high fatigue resistance, thermal insulation performance with low thermal conductivity (0.0322 W m^{-1} K^{-1}), and temperature resistance up to 1300°C [82]. Peng et al. [80] fabricated yttria-stabilized zirconia mixed silica (YSZ/SiO_2) nanofibers by the electrospinning method. The use temperature of the nanofibers increased by nearly 300°C and their maximum strength reached 5.9 ± 0.8 MPa. The YSZ/SiO_2 nanofibers, showing good thermal insulation performance and hydrophobic properties, were suggested for use as high-temperature insulation materials [80].

3.8 Other applications

CNFs have been intensely investigated and applied in many different applications since their first production in 2002. There is still an ever-growing interest in the field with emerging applications that influence different aspects of life.

Apart from the above discussed applications, ceramic nanofibers are also utilized in separation of nuclear waste and recycling of nuclear fuels. Liu et al. [84] electrospun TiO_2 nanofibers with high porosity and functionalized the membranes with silver nanoparticles and nanocrystal metal organic frameworks to capture gases of interest in order to recycle nuclear and industrial waste. These ceramic nanofiber materials showed high porosity, loading capacity, permeability, stability in extreme conditions and were effective in recycling of nuclear waste back into the fuel cycle [84].

Ramaseshan and Ramakrishna [85] produced zinc titanate ceramic nanofibers by electrospinning technology using polyvinylpyrrolidone as a binder for use as catalysts for the detoxification of chemical warfare agents, which are known to react with metal oxides (Mg, Al, Fe, Ti, Zn, Cr, Cu, Mn, etc.) and detoxify them into nontoxic harmless by-products. The zinc titanate nanofibers were found to be effective in decontamination of agents, such as dimethyl methyl phosphonate and chloroethyl ethyl sulfide. The extent of detoxification was measured and the products of reaction of zinc titanate against the simulants were found to be relatively harmless. The nanofibers obtained were suggested to replace conventional-activated carbon by electrospun ceramic nanofibers for face masks and protective clothing [85].

Abdo et al. [86] fabricated titanium dioxide (TiO_2) and carbon nanofibers by electrospinning technique followed by calcination process and utilized these nanofibers to reinforce magnesium matrix composites. When 5 wt.% TiO_2 nanofibers were added the ultimate compressive strength increased to 281 MPa, which was about 12.4% higher than the pure Mg. As a result of the addition of carbon nanofibers into magnesium matrix composites, hardness increased to 64.4% with slight sacrifice in the mechanical properties [86].

CNFs were also utilized in development of highly sensitive and reliable capacitive pressure sensors. Using flexible TiO_2 ceramic nanofibrous networks, Fu et al. [87] developed a capacitive pressure sensor and studied the capacitance-to-pressure sensitivity. The ceramic pressure sensors, which exhibited high sensitivity ($\approx$4.4 k Pa^{-1}), fast response speed (<16 ms), ultralow limit of detection (<0.8 Pa), low fatigue over 50000 loading/unloading cycles, high temperature stability, were suggested for use as real-time health monitoring and motion detection [87].

4. Conclusions

The development of ceramic nanofibers has attracted a significant interest over the recent years. Among different production techniques such as magnetron sputtering, solution blowing, laser spinning, chemical vapor deposition, template synthesis, phase separation, hydrothermal treatment; electrospinning enabled the production of uniform CNFs with high surface areas, very small diameters, extremely long length, and small pore size. Ceramic nanofibers became indispensable materials in many applications. In this chapter, electrospun ceramic nanofibers are reviewed with an emphasis on their applications in tissue engineering, gas sensors, water remediation, batteries, catalyst supports/catalysts, electromagnetic interference shielding and thermal insulation materials. Although there has been a lot of progress since their first production, there is still a lot to be explored regarding their production, and properties, and there is a great potential for their uses in many different fields.

Funding

The author(s) received no financial support for the research, authorship, and/or publication of this article.

Conflict of interest

The author(s) declared no potential conflicts of interest with respect to the research, authorship, and/or publication of this article.

Author details

Nuray Kizildag
Integrated Manufacturing Technologies Research and Application Center, Sabanci University, Istanbul, Turkey

*Address all correspondence to: nuray.kizildag@sabanciuniv.edu

References

[1] Wang L, Yu Y, Chen PC, Zhang DW, Chen CH. Electrospinning synthesis of C/Fe_3O_4 composite nanofibers and their application for high performance lithium-ion batteries. J Power Sources. 2008;183(2):717-723.

[2] Liu L, Lyu J, Mo J, Yan H, Xu L, Peng P, et al. Comprehensively-upgraded polymer electrolytes by multifunctional aramid nanofibers for stable all-solid-state Li-ion batteries. Nano Energy. 2020;69:104398.

[3] Cho C-J, Chang Y-S, Lin Y-Z, Jiang D-H, Chen W-H, Lin W-Y, et al. Green electrospun nanofiber membranes filter prepared from novel biomass thermoplastic copolyester: Morphologies and filtration properties. Journal of the Taiwan Institute of Chemical Engineers. 2020;106:206-14.

[4] Yu X, Li C, Tian H, Yuan L, Xiang A, Li J, et al. Hydrophobic cross-linked zein-based nanofibers with efficient air filtration and improved moisture stability. Chem Eng J. 2020;396:125373.

[5] Kumar V, Mirzaei A, Bonyani M, Kim K-H, Kim HW, Kim SS. Advances in electrospun nanofiber fabrication for polyaniline (PANI)-based chemoresistive sensors for gaseous ammonia. TrAC, Trends Anal Chem. 2020;129:115938.

[6] Li X, Chen S, Zhang X, Li J, Liu H, Han N, et al. Poly-l-Lactic Acid/ Graphene Electrospun Composite Nanofibers for Wearable Sensors. Energy Technology. 2020;8(5):1901252.

[7] Ito Y, Hasuda H, Kamitakahara M, Ohtsuki C, Tanihara M, Kang I-K, et al. A composite of hydroxyapatite with electrospun biodegradable nanofibers as a tissue engineering material. Journal of Bioscience and Bioengineering. 2005;100(1):43-9.

[8] Lin W, Chen M, Qu T, Li J, Man Y. Three-dimensional electrospun nanofibrous scaffolds for bone tissue engineering. Journal of Biomedical Materials Research Part B: Applied Biomaterials. 2020;108(4):1311-21.

[9] Khajavi R, Abbasipour M, Bahador A. Electrospun biodegradable nanofibers scaffolds for bone tissue engineering. J Appl Polym Sci. 2016;133(3).

[10] Liu Y, Zhou S, Gao Y, Zhai Y. Electrospun nanofibers as a wound dressing for treating diabetic foot ulcer. Asian Journal of Pharmaceutical Sciences. 2019;14(2):130-43.

[11] Bakhsheshi-Rad HR, Ismail AF, Aziz M, Akbari M, Hadisi Z, Omidi M, et al. Development of the PVA/CS nanofibers containing silk protein sericin as a wound dressing: In vitro and in vivo assessment. Int J Biol Macromol. 2020;149:513-21.

[12] Sousa MGC, Maximiano MR, Costa RA, Rezende TMB, Franco OL. Nanofibers as drug-delivery systems for infection control in dentistry. Expert Opinion on Drug Delivery. 2020;17(7):919-30.

[13] Celebioglu A, Uyar T. Electrospun formulation of acyclovir/cyclodextrin nanofibers for fast-dissolving antiviral drug delivery. Materials Science and Engineering: C. 2021;118:111514.

[14] Daelemans L, Kizildag N, Van Paepegem W, D'Hooge DR, De Clerck K. Interdiffusing core-shell nanofiber interleaved composites for excellent Mode I and Mode II delamination resistance. Compos Sci Technol. 2019;175:143-50.

[15] Daelemans L, Verschatse O, Heirman L, Van Paepegem W, De Clerck K. Toughening mechanisms

responsible for excellent crack resistance in thermoplastic nanofiber reinforced epoxies through in-situ optical and scanning electron microscopy. Compos Sci Technol. 2021;201:108504.

[16] Barhoum A, Pal K, Rahier H, Uludb H, Kim I, Bechelany M. Nanofibers as new-generation materials: From spinning and nano-spinning fabrication techniques to emerging applications. Applied Materials Today. 2019;17:1-35.

[17] Lupan O, Guérin VM, Ghimpu L, Tiginyanu IM, Pauporté T. Nanofibrous-like ZnO layers deposited by magnetron sputtering and their integration in dye-sensitized solar cells. Chem Phys Lett. 2012;550:125-9.

[18] Huang Z, Kolbasov A, Yuan Y, Cheng M, Xu Y, Rojaee R, et al. Solution Blowing Synthesis of Li-Conductive Ceramic Nanofibers. ACS Applied Materials & Interfaces. 2020;12(14):16200-8.

[19] Quintero F, Mann AB, Pou J, Lusquiños F, Riveiro A. Rapid production of ultralong amorphous ceramic nanofibers by laser spinning. Appl Phys Lett. 2007;90(15):153109.

[20] Honda S-i, Baek Y-G, Ikuno T, Kohara H, Katayama M, Oura K, et al. SiC nanofibers grown by high power microwave plasma chemical vapor deposition. Appl Surf Sci. 2003;212-213:378-82.

[21] Liu S, Shan H, Xia S, Yan J, Yu J, Ding B. Polymer Template Synthesis of Flexible SiO_2 Nanofibers to Upgrade Composite Electrolytes. ACS Applied Materials & Interfaces. 2020;12(28):31439-47.

[22] Zhang Z-J, Zhao J, Qiao Z-J, Wang J-M, Sun S-H, Fu W-X, et al. Nonsolvent-induced phase separation-derived TiO_2 nanotube arrays/porous Ti electrode as high-energy-density anode for lithium-ion batteries. Rare Metals. 2020.

[23] Chan JX, Wong JF, Hassan A, Shrivastava NK, Mohamad Z, Othman N. Green hydrothermal synthesis of high aspect ratio wollastonite nanofibers: Effects of reaction medium, temperature and time. Ceram Int. 2020;46(14):22624-34.

[24] Wen Z, Song X, Chen D, Fan T, Liu Y, Cai Q. Electrospinning preparation and microstructure characterization of homogeneous diphasic mullite ceramic nanofibers. Ceram Int. 2020;46(8, Part B):12172-9.

[25] Asadi-Pakdel K, Mehdinavaz Aghdam R, Shahedi Asl M, Faghihi Sani MA. Synthesis and morphology optimization of electrospun SiBNC nanofibers. Ceram Int. 2020;46(5):6052-9.

[26] Zhang X, Wang B, Wu N, Han C, Wu C, Wang Y. Flexible and thermal-stable SiZrOC nanofiber membranes with low thermal conductivity at high-temperature. J Eur Ceram Soc. 2020;40(5):1877-85.

[27] Kizildag N, Ucar N. Electrospinning Functional Polyacrylonitrile Nanofibers with Polyaniline, Carbon Nanotubes, and Silver Nitrate as Additives. In: Sajjad Haider, Haider A, editors. Electrospinning - Material, Techniques, and Biomedical Applications: IntechOpen; 2016. p. 25-43.

[28] Kizildag N, Ucar N, Onen A, Karacan I. Polyacrylonitrile/polyaniline composite nano/microfiber webs produced by different dopants and solvents. Journal of Industrial Textiles. 2016;46(3):787-808.

[29] Eren O, Ucar N, Onen A, Kizildag N, Karacan I. Synergistic effect of polyaniline, nanosilver, and carbon nanotube mixtures on the structure and properties of polyacrylonitrile

composite nanofiber. J Compos Mater. 2016;50(15):2073-86.

[30] Ucar N, Demirsoy N, Onen A, Karacan I, Kizildag N, Eren O, et al. The effect of reduction methods and stabilizer (PVP) on the properties of polyacrylonitrile (PAN) composite nanofibers in the presence of nanosilver. Journal of Materials Science. 2015;50(4):1855-64.

[31] Pascariu P, Homocianu M. ZnO-based ceramic nanofibers: Preparation, properties and applications. Ceram Int. 2019;45(9):11158-73.

[32] Liu Y, Ding Y, Zhang L, Gao P-X, Lei Y. CeO_2 nanofibers for in situ O_2 and CO sensing in harsh environments. RSC Advances. 2012;2(12):5193-8.

[33] Han T, Ma SY, Xu XL, Xu XH, Pei ST, Tie Y, et al. Rough $SmFeO_3$ nanofibers as an optimization ethylene glycol gas sensor prepared by electrospinning. Mater Lett. 2020;268:127575.

[34] Ma L, Ma SY, Shen XF, Wang TT, Jiang XH, Chen Q, et al. $PrFeO_3$ hollow nanofibers as a highly efficient gas sensor for acetone detection. Sensors and Actuators B: Chemical. 2018;255:2546-54.

[35] Dai X, Shivkumar S. Electrospinning of hydroxyapatite fibrous mats. Mater Lett. 2007;61(13):2735-8.

[36] Wu Y, Hench LL, Du J, Choy K-L, Guo J. Preparation of hydroxyapatite fibers by electrospinning technique. J Am Ceram Soc. 2004;87(10):1988-91.

[37] Keskin S, Uslu İ, Tunç T, Öztürk M, Aytimur A. Preparation and characterization of neodymia doped PVA/Zr-Ce oxide nanocrystalline composites via electrospinning technique. Mater Manuf Processes. 2011;26(11):1346-51.

[38] Starbova K, Petrov D, Starbov N, Lovchinov V. Synthesis of supported fibrous nanoceramics via electrospinning. Ceram Int. 2012;38(6):4645-51.

[39] Zhang P, Jiao X, Chen D. Fabrication of electrospun Al_2O_3 fibers with CaO–SiO_2 additive. Mater Lett. 2013;91:23-6.

[40] Shi Q, Vitchuli N, Nowak J, Caldwell JM, Breidt F, Bourham M, et al. Durable antibacterial Ag/polyacrylonitrile (Ag/PAN) hybrid nanofibers prepared by atmospheric plasma treatment and electrospinning. Eur Polym J. 2011;47(7):1402-9.

[41] Reneker DH, Yarin AL. Electrospinning jets and polymer nanofibers. Polymer. 2008;49(10):2387-425.

[42] Du Z, Zhou X, Ye P, Zeng X, Gan CL. Shape-memory actuation in aligned zirconia nanofibers for artificial muscle applications at elevated temperatures. ACS Applied Nano Materials. 2020;3(3):2156-66.

[43] Franco PQ, João CFC, Silva JC, Borges JP. Electrospun hydroxyapatite fibers from a simple sol–gel system. Mater Lett. 2012;67(1):233-6.

[44] Kim H-W, Kim H-E. Nanofiber generation of hydroxyapatite and fluor-hydroxyapatite bioceramics. Journal of Biomedical Materials Research Part B: Applied Biomaterials. 2006;77B(2):323-8.

[45] Wang X, Gittens RA, Song R, Tannenbaum R, Olivares-Navarrete R, Schwartz Z, et al. Effects of structural properties of electrospun TiO_2 nanofiber meshes on their osteogenic potential. Acta Biomaterialia. 2012;8(2):878-85.

[46] Aly IHM, Abed Alrahim Mohammed L, Al-Meer S, Elsaid K, Barakat NAM. Preparation and

characterization of wollastonite/titanium oxide nanofiber bioceramic composite as a future implant material. Ceram Int. 2016;42(10):11525-34.

[47] Nagarajan S, Belaid H, Pochat-Bohatier C, Teyssier C, Iatsunskyi I, Coy E, et al. Design of boron nitride/gelatin electrospun nanofibers for bone tissue engineering. ACS Applied Materials & Interfaces. 2017;9(39):33695-706.

[48] Teli MD, Nadathur GT. Reversible colourimetric sensing of volatile phase by dye doped electrospun silica based nanofibers. Journal of Environmental Chemical Engineering. 2020;8(4):103920.

[49] Tong Y, Zhang Y, Jiang B, He J, Zheng X, Liang Q. Effect of lanthanides on acetone sensing properties of $LnFeO_3$ nanofibers (Ln = La, Nd, and Sm). IEEE Sens J. 2017;17(8):2404-10.

[50] Nalbandian MJ, Zhang M, Sanchez J, Choa Y-H, Nam J, Cwiertny DM, et al. Synthesis and optimization of Fe_2O_3 nanofibers for chromate adsorption from contaminated water sources. Chemosphere. 2016;144:975-81.

[51] Huh JY, Lee J, Bukhari SZA, Ha J-H, Song I-H. Development of TiO_2-coated YSZ/silica nanofiber membranes with excellent photocatalytic degradation ability for water purification. Scientific Reports. 2020;10(1):17811.

[52] Nalbandian MJ, Greenstein KE, Shuai D, Zhang M, Choa Y-H, Parkin GF, et al. Tailored synthesis of photoactive TiO_2 nanofibers and Au/TiO_2 nanofiber composites: Structure and reactivity optimization for water treatment applications. Environmental Science & Technology. 2015;49(3):1654-63.

[53] Malwal D, Gopinath P. Efficient adsorption and antibacterial properties of electrospun CuO-ZnO composite nanofibers for water remediation. J Hazard Mater. 2017;321:611-21.

[54] Pascariu P, Airinei A, Olaru N, Olaru L, Nica V. Photocatalytic degradation of Rhodamine B dye using $ZnO–SnO_2$ electrospun ceramic nanofibers. Ceram Int. 2016;42(6):6775-81.

[55] Pant B, Ojha GP, Kim H-Y, Park M, Park S-J. Fly-ash-incorporated electrospun zinc oxide nanofibers: Potential material for environmental remediation. Environ Pollut. 2019;245:163-72.

[56] Allabashi R, Arkas M, Hörmann G, Tsiourvas D. Removal of some organic pollutants in water employing ceramic membranes impregnated with cross-linked silylated dendritic and cyclodextrin polymers. Water Res. 2007;41(2):476-86.

[57] Zaidi SDA, Wang C, Shao Q, Gao J, Zhu S, Yuan H, et al. Polymer-free electrospun separator film comprising silica nanofibers and alumina nanoparticles for Li-ion full cell. Journal of Energy Chemistry. 2020;42:217-26.

[58] Cui J, Zhou Z, Jia M, Chen X, Shi C, Zhao N, et al. Solid polymer electrolytes with flexible framework of SiO_2 nanofibers for highly safe solid lithium batteries. Polymers. 2020;12(6):1324.

[59] Jing P, Liu M, Wang P, Yang J, Tang M, He C, et al. Flexible nonwoven ZrO_2 ceramic membrane as an electrochemically stable and flame-resistant separator for high-power rechargeable batteries. Chem Eng J. 2020;388:124259.

[60] Gangaja B, Chandrasekharan S, Vadukumpully S, Nair SV, Santhanagopalan D. Surface chemical analysis of CuO nanofiber composite

electrodes at different stages of lithiation/delithiation. J Power Sources. 2017;340:356-64.

[61] Hu R, Zhang H, Bu Y, Zhang H, Zhao B, Yang C. Porous Co_3O_4 nanofibers surface-modified by reduced graphene oxide as a durable, high-rate anode for lithium ion battery. Electrochim Acta. 2017;228:241-50.

[62] Yang H, Bright J, Chen B, Zheng P, Gao X, Liu B, et al. Chemical interaction and enhanced interfacial ion transport in a ceramic nanofiber–polymer composite electrolyte for all-solid-state lithium metal batteries. Journal of Materials Chemistry A. 2020;8(15):7261-72.

[63] Zhang W, Wang X, Zhang Q, Wang L, Xu Z, Li Y, et al. $Li_7La_3Zr_2O_{12}$ Ceramic nanofiber-incorporated solid polymer electrolytes for flexible lithium batteries. ACS Applied Energy Materials. 2020;3(6):5238-46.

[64] Dai Y, Chai Y, Sun Y, Fu W, Wang X, Gu Q, et al. New versatile Pt supports composed of graphene sheets decorated by Fe_2O_3 nanorods and N-dopants with high activity based on improved metal/support interactions. Journal of Materials Chemistry A. 2015;3(1):125-30.

[65] Dai Y, Formo E, Li H, Xue J, Xia Y. Surface-functionalized electrospun titania nanofibers for the scavenging and recycling of precious metal ions. ChemSusChem. 2016;9(20):2912-6.

[66] Formo E, Peng Z, Lee E, Lu X, Yang H, Xia Y. Direct oxidation of methanol on Pt nanostructures supported on electrospun nanofibers of anatase. The Journal of Physical Chemistry C. 2008;112(27):9970-5.

[67] Formo E, Yavuz MS, Lee EP, Lane L, Xia Y. Functionalization of electrospun ceramic nanofibre membranes with noble-metal nanostructures for catalytic applications. J Mater Chem. 2009;19(23):3878-82.

[68] Fu W, Dai Y, Li JPH, Liu Z, Yang Y, Sun Y, et al. Unusual hollow Al_2O_3 nanofibers with loofah-like skins: Intriguing catalyst supports for thermal stabilization of Pt nanocrystals. ACS Applied Materials & Interfaces. 2017;9(25):21258-66.

[69] Hosseini SR, Ghasemi S, Kamali-Rousta M. Preparation of CuO/NiO composite nanofibers by electrospinning and their application for electro-catalytic oxidation of hydrazine. J Power Sources. 2017;343:467-76.

[70] Liu S, Tian J, Yin K, Li Z, Meng X, Zhu M, et al. Constructing fibril-in-tube structures in ultrathin CeO_2-based nanofibers as the ideal support for stabilizing Pt nanoparticles. Materials Today Chemistry. 2020;17:100333.

[71] Wang L, Yang G, Peng S, Wang J, Ji D, Yan W, et al. Fabrication of $MgTiO_3$ nanofibers by electrospinning and their photocatalytic water splitting activity. Int J Hydrogen Energy. 2017;42(41):25882-90.

[72] Huo Y, Zhao K, Miao P, Kong J, Xu Z, Wang K, et al. Microwave absorption performance of SiC/ZrC/SiZrOC hybrid nanofibers with enhanced high-temperature oxidation resistance. ACS Sustainable Chemistry & Engineering. 2020;8(28):10490-501.

[73] Huo Y, Zhao K, Xu Z, Tang Y. Electrospinning synthesis of SiC/Carbon hybrid nanofibers with satisfactory electromagnetic wave absorption performance. J Alloys Compd. 2020;815:152458.

[74] Hou Y, Cheng L, Zhang Y, Du X, Zhao Y, Yang Z. High temperature electromagnetic interference shielding of lightweight and flexible ZrC/SiC nanofiber mats. Chem Eng J. 2021;404:126521.

[75] Wang P, Cheng L, Zhang Y, Wu H, Hou Y, Yuan W, et al. Flexible, hydrophobic SiC ceramic nanofibers used as high frequency electromagnetic wave absorbers. Ceram Int. 2017;43(10):7424-35.

[76] Wang P, Cheng L, Zhang Y, Zhang L. Flexible SiC/Si_3N_4 composite nanofibers with in situ embedded graphite for highly efficient electromagnetic wave absorption. ACS Applied Materials & Interfaces. 2017;9(34):28844-58.

[77] Si Y, Mao X, Zheng H, Yu J, Ding B. Silica nanofibrous membranes with ultra-softness and enhanced tensile strength for thermal insulation. RSC Advances. 2015;5(8):6027-32.

[78] Si Y, Wang X, Dou L, Yu J, Ding B. Ultralight and fire-resistant ceramic nanofibrous aerogels with temperature-invariant superelasticity. Science Advances. 2018;4(4):eaas8925.

[79] Dong X, Liu J, Li X, Zhang X, Xue Y, Liu J, et al. Electrospun mullite nanofibers derived from diphasic mullite sol. J Am Ceram Soc. 2017;100(8):3425-33.

[80] Peng Y, Xie Y, Wang L, Liu L, Zhu S, Ma D, et al. High-temperature flexible, strength and hydrophobic YSZ/SiO_2 nanofibrous membranes with excellent thermal insulation. J Eur Ceram Soc. 2020.

[81] Dou L, Zhang X, Cheng X, Ma Z, Wang X, Si Y, et al. Hierarchical cellular structured ceramic nanofibrous aerogels with temperature-invariant superelasticity for thermal insulation. ACS Appl Mater Interfaces. 2019;11(32):29056-64.

[82] Zhang X, Wang F, Dou L, Cheng X, Si Y, Yu J, et al. Ultrastrong, superelastic, and lamellar multiarch structured ZrO_2–Al_2O_3 nanofibrous aerogels with high-temperature resistance over 1300 °C. ACS Nano. 2020.

[83] Malwal D, Gopinath P. Fabrication and characterization of poly(ethylene oxide) templated nickel oxide nanofibers for dye degradation. Environmental Science: Nano. 2015;2(1):78-85.

[84] Liu H, Bell N, Cipiti BB, Lewis TG, Sava DF, Nenoff TM. Functionalized ultra-porous titania nanofiber membranes as nuclear waste separation and sequestration scaffolds for nuclear fuels recycle. Albuquerque, New Mexico: Sandia National Laboratories; 2012. Report No.: SAND2012-8025.

[85] Ramaseshan R, Ramakrishna S. Zinc titanate nanofibers for the detoxification of chemical warfare simulants. J Am Ceram Soc. 2007;90(6):1836-42.

[86] Abdo HS, Khalil KA, El-Rayes MM, Marzouk WW, Hashem A-FM, Abdel-Jaber GT. Ceramic nanofibers versus carbon nanofibers as a reinforcement for magnesium metal matrix to improve the mechanical properties. Journal of King Saud University - Engineering Sciences. 2020;32(5):346-50.

[87] Fu M, Zhang J, Jin Y, Zhao Y, Huang S, Guo CF. A highly sensitive, reliable, and high-temperature-resistant flexible pressure sensor based on ceramic nanofibers. Advanced Science. 2020;7(17):2000258.

[88] Zhang J, Lu H, Lu H, Li G, Gao J, Yang Z, et al. Porous bimetallic Mo-W oxide nanofibers fabricated by electrospinning with enhanced acetone sensing performances. J Alloys Compd. 2019;779:531-42.

[89] Wu J, Xing X, Zhu Z, Zheng L, Chen J, Wang C, et al. Electrospun hollow CuO modified V_2O_5 nano-string of pearls with improved acetone sensitivity. Chem Phys Lett. 2019;727:19-24.

[90] Jeong YJ, Koo W-T, Jang J-S, Kim D-H, Kim M-H, Kim I-D. Nanoscale PtO_2 catalysts-loaded SnO_2 multichannel nanofibers toward highly sensitive acetone sensor. ACS Applied Materials & Interfaces. 2018;10(2):2016-25.

[91] Cheng Y, He Y, Li S, Wang Y, Zhao Y, Li Y, et al. Ultra-sensitive and selective acetone gas sensor with fast response at low temperature based on Cu-doped α-Fe_2O_3 porous nanotubes. Journal of Materials Science: Materials in Electronics. 2018;29(13):11178-86.

[92] Jeong YJ, Koo W-T, Jang J-S, Kim D-H, Cho H-J, Kim I-D. Chitosan-templated Pt nanocatalyst loaded mesoporous SnO_2 nanofibers: a superior chemiresistor toward acetone molecules. Nanoscale. 2018;10(28):13713-21.

[93] Kou X, Xie N, Chen F, Wang T, Guo L, Wang C, et al. Superior acetone gas sensor based on electrospun SnO_2 nanofibers by Rh doping. Sensors and Actuators B: Chemical. 2018;256:861-9.

[94] Yang HM, Ma SY, Yang GJ, Chen Q, Zeng QZ, Ge Q, et al. Synthesis of La_2O_3 doped Zn_2SnO_4 hollow fibers by electrospinning method and application in detecting of acetone. Appl Surf Sci. 2017;425:585-93.

[95] Jiang Z, Yin M, Wang C. Facile synthesis of Ca^{2+}/Au co-doped SnO_2 nanofibers and their application in acetone sensor. Mater Lett. 2017;194:209-12.

[96] Zhang J, Lu H, Yan C, Yang Z, Zhu G, Gao J, et al. Fabrication of conductive graphene oxide-WO_3 composite nanofibers by electrospinning and their enhanced acetone gas sensing properties. Sensors and Actuators B: Chemical. 2018;264:128-38.

[97] Koo W-T, Jang J-S, Choi S-J, Cho H-J, Kim I-D. Metal–organic framework templated catalysts: Dual sensitization of PdO–ZnO composite on hollow SnO_2 nanotubes for selective acetone sensors. ACS Applied Materials & Interfaces. 2017;9(21):18069-77.

[98] Pascali CD, Signore MA, Taurino A, Francioso L, Macagnano A, Avossa J, et al. Investigation of the gas-sensing performance of electrospun TiO_2 nanofiber-based sensors for ethanol sensing. IEEE Sens J. 2018;18(18):7365-74.

[99] Li F, Gao X, Wang R, Zhang T. Design of WO_3-SnO_2 core-shell nanofibers and their enhanced gas sensing performance based on different work function. Appl Surf Sci. 2018;442:30-7.

[100] Chen D, Yi J. One-pot electrospinning and gas-sensing properties of $LaMnO_3$ perovskite/SnO_2 heterojunction nanofibers. J Nanopart Res. 2018;20(3):65.

[101] Huang B, Zhang Z, Zhao C, Cairang L, Bai J, Zhang Y, et al. Enhanced gas-sensing performance of ZnO@In_2O_3 core@shell nanofibers prepared by coaxial electrospinning. Sensors and Actuators B: Chemical. 2018;255:2248-57.

[102] Gao X, Li F, Wang R, Zhang T. A formaldehyde sensor: Significant role of p-n heterojunction in gas-sensitive core-shell nanofibers. Sensors and Actuators B: Chemical. 2018;258:1230-41.

[103] Andre RS, Kwak D, Dong Q, Zhong W, Correa DS, Mattoso LHC, et al. Sensitive and selective NH_3 monitoring at room temperature using ZnO ceramic nanofibers decorated with poly(styrene sulfonate). Sensors (Basel). 2018;18(4):1058.

[104] Pang Z, Nie Q, Lv P, Yu J, Huang F, Wei Q. Design of flexible PANI-coated CuO-TiO_2-SiO_2 heterostructure nanofibers with high ammonia sensing

response values. Nanotechnology. 2017;28(22):225501.

[105] Pang Z, Nie Q, Wei A, Yang J, Huang F, Wei Q. Effect of In_2O_3 nanofiber structure on the ammonia sensing performances of In_2O_3/PANI composite nanofibers. Journal of Materials Science. 2017;52(2):686-95.

[106] Sankar Ganesh R, Durgadevi E, Navaneethan M, Patil VL, Ponnusamy S, Muthamizhchelvan C, et al. Low temperature ammonia gas sensor based on Mn-doped ZnO nanoparticle decorated microspheres. J Alloys Compd. 2017;721:182-90.

[107] Das M, Sarkar D. One-pot synthesis of zinc oxide - polyaniline nanocomposite for fabrication of efficient room temperature ammonia gas sensor. Ceram Int. 2017;43(14):11123-31.

[108] Xu L, Dong B, Wang Y, Bai X, Chen J, Liu Q, et al. Porous In_2O_3:RE (RE = Gd, Tb, Dy, Ho, Er, Tm, Yb) Nanotubes: Electrospinning Preparation and Room Gas-Sensing Properties. The Journal of Physical Chemistry C. 2010;114(19):9089-95.

[109] Bai S, Fu H, Zhao Y, Tian K, Luo R, Li D, et al. On the construction of hollow nanofibers of ZnO-SnO_2 heterojunctions to enhance the NO_2 sensing properties. Sensors and Actuators B: Chemical. 2018;266:692-702.

[110] Jaroenapibal P, Boonma P, Saksilaporn N, Horprathum M, Amornkitbamrung V, Triroj N. Improved NO_2 sensing performance of electrospun WO_3 nanofibers with silver doping. Sensors and Actuators B: Chemical. 2018;255:1831-40.

[111] Ponnuvelu DV, Abdulla S, Pullithadathil B. Novel Electro-Spun Nanograined ZnO/Au Heterojunction Nanofibers and Their Ultrasensitive NO_2 Gas Sensing Properties. ChemistrySelect. 2018;3(25):7156-63.

[112] Wang C, Ma S, Sun A, Qin R, Yang F, Li X, et al. Characterization of electrospun Pr-doped ZnO nanostructure for acetic acid sensor. Sensors and Actuators B: Chemical. 2014;193:326-33.

[113] Katoch A, Sun G-J, Choi S-W, Byun J-H, Kim SS. Competitive influence of grain size and crystallinity on gas sensing performances of ZnO nanofibers. Sensors and Actuators B: Chemical. 2013;185:411-6.

[114] Wei S, Yu Y, Zhou M. CO gas sensing of Pd-doped ZnO nanofibers synthesized by electrospinning method. Mater Lett. 2010;64(21):2284-6.

[115] Galstyan V, Comini E, Baratto C, Faglia G, Sberveglieri G. Nanostructured ZnO chemical gas sensors. Ceram Int. 2015;41(10, Part B):14239-44.

[116] Koo W-T, Choi S-J, Kim S-J, Jang J-S, Tuller HL, Kim I-D. Heterogeneous sensitization of metal–organic framework driven metal@metal oxide complex catalysts on an oxide nanofiber scaffold toward superior gas sensors. J Am Chem Soc. 2016;138(40):13431-7.

[117] Li W, Ma S, Li Y, Yang G, Mao Y, Luo J, et al. Enhanced ethanol sensing performance of hollow ZnO–SnO_2 core–shell nanofibers. Sensors and Actuators B: Chemical. 2015;211:392-402.

[118] Lee C-S, Kim I-D, Lee J-H. Selective and sensitive detection of trimethylamine using ZnO–In_2O_3 composite nanofibers. Sensors and Actuators B: Chemical. 2013;181:463-70.

[119] Deng J, Yu B, Lou Z, Wang L, Wang R, Zhang T. Facile synthesis and enhanced ethanol sensing

properties of the brush-like ZnO–TiO_2 heterojunctions nanofibers. Sensors and Actuators B: Chemical. 2013;184:21-6.

[120] Lou Z, Deng J, Wang L, Wang R, Fei T, Zhang T. A class of hierarchical nanostructures: ZnO surface-functionalized TiO_2 with enhanced sensing properties. RSC Advances. 2013;3(9):3131-6.

[121] Wan GX, Ma SY, Li XB, Li FM, Bian HQ, Zhang LP, et al. Synthesis and acetone sensing properties of Ce-doped ZnO nanofibers. Mater Lett. 2014;114:103-6.

[122] Tang W, Wang J, Yao P, Li X. Hollow hierarchical SnO_2-ZnO composite nanofibers with heterostructure based on electrospinning method for detecting methanol. Sensors and Actuators B: Chemical. 2014;192:543-9.

[123] Katoch A, Choi S-W, Sun G-J, Kim HW, Kim SS. Mechanism and prominent enhancement of sensing ability to reducing gases in p/n core–shell nanofiber. Nanotechnology. 2014;25(17):175501.

[124] Song X, Wang Z, Liu Y, Wang C, Li L. A highly sensitive ethanol sensor based on mesoporous ZnO–SnO_2 nanofibers. Nanotechnology. 2009;20(7):075501.

[125] Sun Ye, Zhang D, Chang H, Zhang Y. Fabrication of palladium–zinc oxide–reduced graphene oxide hybrid for hydrogen gas detection at low working temperature. Journal of Materials Science: Materials in Electronics. 2017;28(2):1667-73.

[126] Kim D-H, Jang J-S, Koo W-T, Kim ID. Graphene oxide templating: facile synthesis of morphology engineered crumpled SnO_2 nanofibers for superior chemiresistors. Journal of Materials Chemistry A. 2018;6(28):13825-34.

[127] Abideen ZU, Katoch A, Kim J-H, Kwon YJ, Kim HW, Kim SS. Excellent gas detection of ZnO nanofibers by loading with reduced graphene oxide nanosheets. Sensors and Actuators B: Chemical. 2015;221:1499-507.

[128] Li B, Cao H, Shao J, Li G, Qu M, Yin G. Co_3O_4@graphene composites as anode materials for high-performance lithium ion batteries. Inorg Chem. 2011;50(5):1628-32.

[129] Song MJ, Kim IT, Kim YB, Shin MW. Self-standing, binder-free electrospun Co_3O_4/carbon nanofiber composites for non-aqueous Li-air batteries. Electrochim Acta. 2015;182:289-96.

[130] La Monaca A, Paolella A, Guerfi A, Rosei F, Zaghib K. Electrospun ceramic nanofibers as 1D solid electrolytes for lithium batteries. Electrochem Commun. 2019;104:106483.

[131] Panda PK. Ceramic Nanofibers by Electrospinning Technique—A Review. Transactions of the Indian Ceramic Society. 2007;66(2):65-76.

Chapter 5

Electrospun Nanofibers: Characteristic Agents and Their Applications

Lingayya Hiremath, O. Sruti, B.M. Aishwarya, N.G. Kala and E. Keshamma

Abstract

This study aimed to introduce antibacterial nanofibers, produced by electrospinning as a novel technique in constructing nanostructured materials. The large size and less bioavailability due to impenetrable (or partial/improper penetration) membrane has resulted in production of nanofibers. These nano sized Fibers were successful in delivering the active ingredients and served the purpose of using plants for its cause. Some of the active ingredients include antimicrobial compounds that are incorporated into various products to prevent unwanted microbial growth. As higher bioavailability is one of the most crucial parameters when it comes to medical solutions, electro spun nanofibers are highly preferred. This method is preferable for organic polymers as they have high flexibility, high specific surface area and surface functionalization. Electrospinning technology has been used for the fabrication and assembly of nanofibers into membranes, which have extended the range of potential applications in the biomedical, environmental protection, nanosensor, electronic/optical, protective clothing fields and various other fields.

Keywords: electrospinning, nanofiber, antibacterial agent, metallic nanoparticle, tissue engineering

1. Introduction

Nanofibers are fabricated thin threads arising from physical processes using synthetic chemicals. Nano, when used as a prefix, restricts the diameter of the thin fibres to nano range i.e., 20–400 nm [1]. Nano being a common term these days has proved it worth in various fields, well known being drug delivery and other biological ones. The high surface to volume ratio serves as the driving factor in most cases for its application. This is also the reason fibres when made in nano-range perform better (form highly porous mesh) and have been universally used [1].

Our purpose here is to highlight the technique, electrospinning, which is used to make nano-fibres with antimicrobial properties. It is a popular technique in tissue engineering that uses polymer solutions and strong electric fields to produce nanofibers as close as the natural extracellular matrix (especially in tissue engineering) [1, 2]. When it comes to industries, scalability is a major issue and electrospinning technique is favoured due to this reason and it has a simple setup [3]. Synthetic fibres are used more often than natural ones. There is no standard size of a nanofiber when it comes to biological usage. This is due to the stability issue and hence the porosity, morphology and shape are all variables and are adjusted to create the best possible product [2]. These products are also affected by the technique used for the production (here electrospinning- electric field, flow rate etc.). Environment and solution used to be the other two affecting parameters. There are a lot of correlations to be taken care of before finalising the nanofiber structure.

As of now, we have an idea of what electrospun nanofibers are, however why these fibres are important is still a mystery to solve. With some basic knowledge of biology in earlier classes, we have concluded that staying away or preventing contact with microorganisms can reduce a lot of biological stress in our body. In short, using antimicrobials is a good option when it is available. Thus, Electro spun nanofibers when incorporate antimicrobial properties, become a great deal of interest even to common population e.g., electro spun nanofibers with essential oils (to prevent the side effect from synthetic compounds) [4], Electro spun nanofibers of zein and PVA have been proposed as carriers and stabilisers of epigallocatechin-gallate (EGCG) [5]. There are few studies on the incorporation and release profile of a drug loaded in biodegradable electro spun nanofibrous membranes, based on the drug-polymer interactions, on top of its ability to hinder bacterial growth. A polymer blend composed of/poly (butylene adipate-co-terephthalate) (PBAT)/ poly (lactic acid) (PLA), loaded with different. Cloxacillin (CLOX) contents were fruitfully produced by using electrospinning technique (as shown in the following **Figure 1**) [6]. The confirmation for the encapsulation of drug was done using characterisation techniques like Fourier transform infrared spectroscopy (FTIR), thermogravimetric analysis (TGA) and differential scanning calorimetry (DSC). The effect was measured by the pH when the drug (20% of CLOX) was released (antibacterial activity).

Along with antimicrobials, these nanofibers can possess anti-inflammatory and antioxidant properties as well. Such properties tend to degrade in terms of effect when not incorporated properly but with a controlled release, this degradation can be avoided [5]. A detailed description of this technique, usage and the flaws will be discussed in the later part of this chapter.

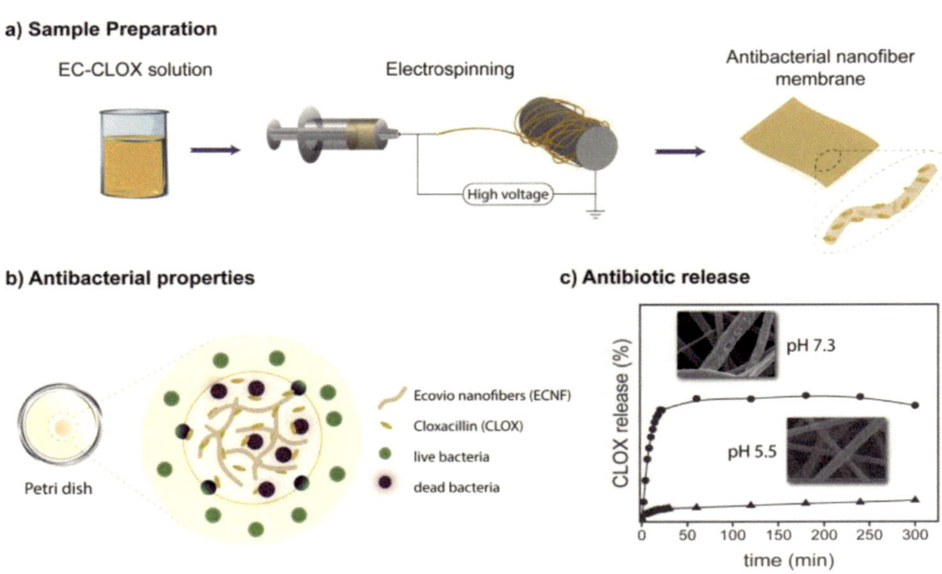

Figure 1.
Antibacterial activity and effect of pH on its release (electrospun nanofibers containing cloxacillin) [6].

2. Nanofiber's production

Fibres can be natural - from animals or plants, or synthetic – man made. When fibres come from plants, cellulose is the polymer made from the sugar glucose that makes those fibres well-built. Natural fibres from plants include cotton, jute, hemp, sisal, and flax. Silk is formed from the cocoons of silkworms. Wool is soft hair that is cut from animals like sheep, goats, alpacas, llamas, and even rabbits. Both silk and wool are protein. Cashmere is an extra-soft fur from goats. Mohair is wool from angora goats. Angora rabbits give us angora fibres [7].

A lot of different polymers can be made into fibres. Fibres are formed when polymer chains are all lined up in the same direction [7]. Metallic fibres, Carbon fibres, fibreglass, mineral fibres, and polymer fibres are all subtypes of synthetic fibres [8].

Diverse types of natural and synthetic types of polymers are used to make nanofibers henceforth they exhibit unlike properties and applications. Instances of manmade polymers include poly lactic acid, polyurethane, polycaprolactone, polyethylene-co-vinyl acetate plus poly 3-hydroxybutyrate-co-3-hydroxyvalerate. Natural polymers comprise cellulose, collagen, gelatine, keratin, silk fibroin, then alginate and chitosan. Polymer chains are linked via covalent bonds [9].

There are many different methods to make nanofibers, including bicomponent extrusion, drawing, electrospinning, thermal-induced phase separation, self-assembly template synthesis and centrifugal spinning [9, 10]. Electrospinning is an extensively used method of producing nanofibers.

2.1 Electrospinning mechanism

When the electrostatic force is applied on solutions or melts, electro-spinning produces fibres with diameters ranging from micrometre to nano-meter scale. A general electrospinning setup consists of three primary components: a syringe with a metallic needle, a high voltage power supply (usually in the kV range), and a grounded collector. When we consider a typical electrospinning course, high power is applied on melts/solutions. Consequently, suspended droplets are formed. A suspended droplet will collapse into a conical droplet as the electrostatic repulsion

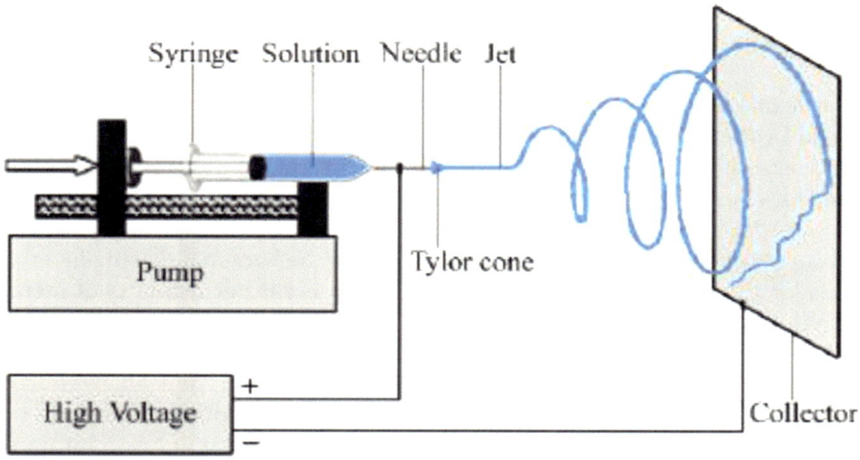

Figure 2.
Electrospinning setup [11].

starts to overcome the surface tension of the fluid. A fine, charged jet of polymer solution is ejected from the tip of the needle as the electrostatic force overcomes the surface tension of the conical droplet. The action between the electric field and surface tension of the fluid outstretches the jet stream and then it encounters a whipping motion and that results in the evaporation of the solvent. This led to the jet stream to be stretched out continuously as a thin and long filament. Subsequently this filament will harden and will be eventually settled onto a grounded collector, and finally results in the development of a uniform fibre (**Figure 2**) [11].

2.2 Electrospun nanofibers with antimicrobial properties

The electro spun nanofibers built-in with antibacterial agents have been fabricated for antimicrobial applications. The electro spun nanofibers exhibit enhanced antimicrobial performance compared to conventional antimicrobial materials. They play significant roles in wound-dressing materials, filtration, tissue scaffolds, protective textiles, and biomedical devices [12].

The electro spun nanofibers with antimicrobial properties fabrication methods are grouped into two categories. Antimicrobial nanofibers can be obtained by one step process or by the following two steps. In one stage process, the suspensions with a mixture of antimicrobial agents and polymer undergo electrospinning. The formulation of this homogeneous mixture is censorious to make up a smooth and continuous nanofiber. The properties of electrospinning solutions are affected by antimicrobial agents. Such vital characteristics that play a significant role in the process and resultant are conductivities and viscosities.

Whereas in the two steps method include, producing an initial electro spun polymeric nanofibers and then post-functionalizing nanofibers with antimicrobial materials. Multiple functionalization approaches have been managed to link the antimicrobial agents onto surfaces of electro spun nanofibers by using various chemical and physical methods [13].

Antimicrobial electro spun nanofibers built-in with different antimicrobial agents: including metallic nanoparticles (silver, zinc, titanium, copper, and cobalt), carbon nanomaterials, antibiotics, and antimicrobial biopolymers.

3. Antimicrobial agents

3.1 Volatile oils

Volatile oils well known as Essential oils are plant derived concentrated hydrophobic and volatile compounds. They are a combination of different compounds such as carvacrol, eugenol, and cymene derived from aromatic plants. The best examples of essential oils are terpenoids and hydrophobic phenolic compounds [14, 15]. The hydrophobic nature of essential oils decides their activity mechanism against microbes. These essential oils break up into the bacterial plasma membrane lipid bilayer and then disrupt its structure. This alters the permeability of membrane to ions and other cellular contents. Consequently, the proton pump collapses and results in cell death [16].

Sadri and his team prepared PEO nanofibers/electro spun chitosan, to which they linked two distinct types of thyme essential oils into this nanofiber. They used broadleaf and narrow leaf thyme essential oils to their study. The nanofibers/chitosan along with the thyme oils were trialled against *P. aeruginosa* and *Staphylococcus aureus*. After 24 hrs, the inhibition of narrow life was reported as 8 and 15 mm were as in case of broad life it was 10 and 19 mm for *P. aeruginosa* and *S. aureus*, respectively.

Accordingly, the broad leaf resulted in more antibacterial activity than narrow leaf spices in the presence of above-mentioned bacteria's [17].

3.2 Herbal bioactive components

There are many studies that prove the potent antibacterial property of plants-derived herbal bioactive components against a wide range of food borne pathogens. The widely researched bioactive components with antimicrobial properties are gingerol, allicin, shikonin, asiaticoside, and curcumin etc. Curcumin (Cur) which is derived from the rhizome of *Curcuma longa* L. is well-known for its valuable properties, including anti-inflammatory, antioxidant, and anticancer features [16, 18–20].

3.3 Silver

Amongst metallic nanoparticles, silver nanoparticles are the most studied and have been demonstrated to be the most effective antimicrobial agents. Ag is a known biocidal agent that is effective against a range of types of fungi, bacteria, and viruses; on the other hand, it is non-hazardous to human cells. The simplest and most frequently used method for combining Ag nanoparticles with electro spun nanofibers is the suspension of Ag nanoparticles directly into the electrospinning polymer solutions [21, 22].

A research team formulated cellulose acetate nanofibers with the use of electrospinning methods. Cellulose acetate nanofibers were transformed into cellulose nanofibers using alkaline hydrolysis. In addition to this, silver nanoparticles were added to the cellulose nanofiber. Developed antibacterial silver cellulose nanofiber activity was examined against *E. coli* and *S. aureus* grown on Lysogeny broth [LB] medium. After 18 hrs of contacting 1% silver nanoparticles, the inhibition zone was spotted with 16- and 14.4-mm diameter against *E. coli* and *S. aureus*, respectively. Besides, it was also proved that antibacterial activities of the Ag nanofibers were directly influenced by the rising concentration of Ag nanoparticle contents [23].

3.4 Zinc and copper

ZnO appears to restrain the growth of strongly resistant bacteria. There are some reports about the significant antibacterial activity of ZnO, which is credited to the production of reactive oxygen species [ROS], causing the production of oxide substances.

Since olden times, Copper has been used for manufacturing utensils as it is a powerful natural biocidal metal. When bacteria encounter copper, there will be cell wall deformation which causes the death of bacteria. To deal with bacteria, many researchers have developed a method where a polymeric matrix was supported with copper by electro-spinning [24]. Ahire and his research team used electrospinning of Poly-D and PEO, L-lactide method to combine Cu nanoparticles into nanofibers. Due to the presence of copper nanofibers, *S. aureus* and *P. aeruginosa* were reduced by 50% and 40%, respectively after two days [25].

3.5 Antibacterial drugs

For wound health, filtration, and active packaging systems antimicrobial nanofibers incorporated with antibacterial drugs have become one of the promising nano-scale materials. A vast range of antibacterial drugs such as peptides and antibiotics have been formulated physically or chemically within electro spun nanofibers or on their surfaces. The polymer degradation, release profile, and release pathway of antibacterial drugs from electro spun nanofibers has a linkage with the release mechanism of antibacterial drugs which may be regulated through the composition of polymer and fibre morphology [26].

Antibacterial drugs encapsulated in electro spun nanofibers have been proved to sustain the antibacterial property over a longer time compared to the un-encapsulated form. A sodium alginate electro spun nanofibers loaded with ciprofloxacin was developed by a team and it was tested for its antimicrobial activity against *Staphylococcus aureus*. The minimum inhibitory concentration [MIC] of ciprofloxacin required is found to be 0.125 μg/mL through this study [27]. Similarly, another team worked on developing nisin nanofibers/cellulose acetate which resulted in approx. 99.9% reduction of *S. aureus* [28].

4. Applications of electro spun nanofibers (ESNF)

Electrospinning offers many advantages like control over morphology, porosity and composition using very simple equipment. Due to its different applications in various fields like filtration products, biomedical applications, and tissue engineering to produce artificial blood vessels, non-woven fabrics, fuel cells, fibre mats etc. [29]. Electrospinning technology has been used for the fabrication and assembly of nanofibers into membranes, which have extended the range of potential applications in the biomedical, environmental protection (**Table 1**), nano sensor, electronic/optical, protective clothing fields and various other fields [30].

4.1 Health applications

The ESNF have shown great capacity in the human healthcare applications, for tissue or organ repair and regeneration, as biocompatible and biodegradable medical implant devices, in medical diagnostics and instrumentation, as protective fabrics against environmental and infectious agents in hospitals and as vectors to deliver therapeutics and drugs [31].

For drug delivery or bio separation, nanofibers with strong paramagnetic properties prepared by the coaxial technology, such as Gd_2O_2S, possibly doped with Eu or Dy, were recommended [32].

For controlled delivery of drugs, molecular medicines, body-care supplements and therapeutics nanofibers are used as a promising tool by cosmetics and pharmaceutical industries. To give an example such as DNA which is attached covalently to a patterned array of carbon fibre and inserted into cells by centrifuging these cells onto the array will not affect cell's viability and the expression of genes encoded by

Materials	Solvent system	Materials	Solvent system
Cellulose acetate	Acetone/DMAc	Chitosan	TFA
Chitin	HFIP/PBS	Hyaluronic acid	DMF/water
Silk fibroin	Formic acid	Fibrinogen	—
Gelatin	TFE/HFIP	Elastin	Water
Collagen	HFIP	Soy protein	HFIP
Wheat protein	HFIP	Whey protein	Acidic aqueous solution

Table 1.
Natural biopolymer electrospun nanofibers [30].

the inserted DNA. This could build a way for the development of a 'smart' polymeric drug delivery system [31]. After alignment, stacking, mechanical properties, diameter, porosity, and biodegradability optimisation nanofiber-based scaffolds have been explored to enhance the repair or regeneration of various types of tissues, including heart, blood vessel, nerves, skin, musculoskeletal system, and tissue interfaces [33].

4.2 Wound dressing

The naturally extracted bioactive agents using electrospinning technique have been majorly promoted for the development of advanced level of dressings which paves way for rapid and efficient wound repair. Electrospun scaffolds consists of several advantages over the traditional dressings for the treatment of chronic as well as acute wounds, high absorption of exudates from the site of wound, efficient exchange of gases and nutrients for cell's proliferation, protection of the injured tissue, and the possibility to release functional molecules [34].

The distinctive features of ESNF scaffolds such as their inter-fibre and intra-fibre pores and high surface area stimulate the fibroblastic cells response by rapidly initiating cell signalling pathways. Additionally, electrospinning technique can be used because of its application in the fabrication of cosmetic masks which are used for skin cleansing and skin healing. The high surface area of an electrospun skin mask facilitates the flow of additives from and to the skin (**Figure 3**) [30].

Many crude extracts of plants have been successfully encapsulated into electrospun fibres, such as *Centella asiatica,* baicalein, green tea, *Garcinia mangostana*, *Tecomella undulata*, *Aloe vera*, *Grewia mollis*, chamomile, grape seed, *Calendula officinalis*, *Indigofera aspalathoides*, *Azadirachta indica*, *Memecylon edule* and *Myristica andamanica* which has been used for wound healing [34].

ESNF has been effectively explored as a wound healing dressing material. By developing nanofibers to provide topographical and biological cues, the migration

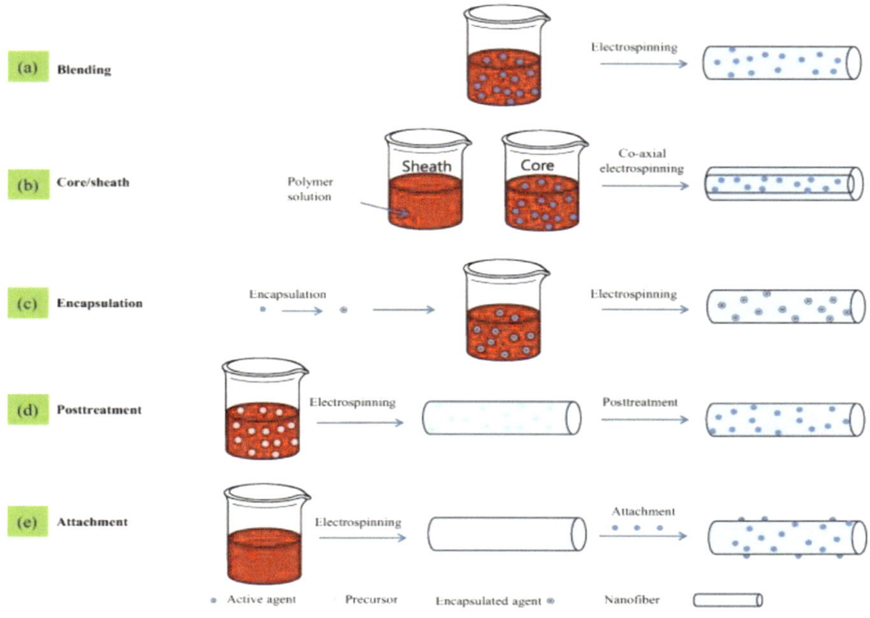

Figure 3.
Various strategies used to prepare suitable wound dressing [31].

and infiltration of repairable cells can be improved. Once the nanofiber-based scaffolds have been optimised accordingly in vitro for the promotion of cell migration and/or delivery of biomolecules they will be subjected for wound healing evaluation in vivo using a mouse, rat, or rabbit model [33].

4.3 Tissue engineering

In tissue regeneration, biocompatible and biodegradable fibrous scaffolds are usually preferred over traditional scaffolds because of their uniqueness and capacity to provide the target cells or tissues with a local environment by imitating the extracellular matrix. Hence, the use of ESNF in tissue engineering is increasing day by day [30].

Osteogenic properties in medicinal plants such as *Cissus quadrangularis* (CQ) and Asian *Panax ginseng* root have been suggested for regeneration of bone. The combined effect of CQ and hydroxyapatite (HA) has been explored by producing PCL-CQ-HA electrospun scaffolds. Proliferation of human foetal osteoblasts (hFOBs) on the composite scaffolds and increased adhesion was observed. Furthermore, increased levels of mineralisation and osteocalcin expression were detected which are fundamental in bone formation [34].

Bio or natural polymers (hyaluronic acid, alginate, collagen, silk protein, fibrinogen, chitosan, starch, and poly (3-hydroxybutyrate-*co*-3-hydroxyvalerate (PHBV)) have been mainly focused by the researchers until recently for tissue engineering, because these polymers showed excellent biocompatibility and biodegradability. However, in recent years, attempts have been made to utilise a wide range of natural and synthetic polymers for the regeneration of new tissues, specifically cartilage tissue, dermal tissue, and bones.

A synthetic polymer poly (lactic acid-co-glycolic acid) (PLGA) is the ideal material for tissue regeneration because of its tuneable and biodegradable nature, easy spinnability, and the presence of multiple focal adhesion points [30].

4.4 Food industry

ESNF is used for the encapsulation of plant extracts with the aim of preserving the integrity and controlling the release of the active ingredients in food processing and packaging. Electrospinning majorly offers the advantage of being a cost-effective manufacturing procedure that operates at room temperature and it is compatible with most edible polymers and materials approved for food contact in these sectors [34].

A hydrophobic prolamin, Zein which can be extracted from corn consists of marvellous film-forming properties with a high thermal resistance. Earlier, zein films were used as edible coatings on tomatoes to delay the colour changes, weight losses, and on nuts to delay rancidity during storage. However, the zein based electrospun mats may provide additional attributes for food packaging [35].

Functional molecules extracted from plants have been exploited for prolonging food shelf-life and avoiding bacteria colonisation in food packaging applications. In one of the modern studies, electrospun mats of β-cyclodextrin (PVA/CEO/β-CD) and PVA containing cinnamon essential oil (CEO) have been developed and tested against *S. aureus* and *E. coli*. The combination of CEO with β-CD enhanced the antibacterial action of this essential oil [34].

A cost competitive plant protein which is a soy protein is partially purified and concentrated from soybeans in various forms, such as soy protein concentrate, defatted soy flour, and soy protein isolate (SPI). Though there is a great amount of interest in developing soy protein as an electrospinning matrix, pure soy protein cannot be electrospunned easily.

5. When are electrospun nanofibers not a good option?

Electrospinning is an impressive technique however, the size of the fibres being nano, is a disadvantage when it comes to control. The limited control of the pore size (Electrospun scaffold) is a diameter dependent which reflects on the cellular infiltration (decrease due to smaller average pore size) [36].

Another reason being the degradation effect (introduced in the latter half of the introduction). The rapid degradation of nanofibrous constructs can adversely affect the ability of the scaffolds to support tissue growth. The structure of the nanofibers plays an important role, especially when it comes to nanoscale, the high surface area to volume ratio serves as the reason for its selection. However, in case of degradation effect, due to this property, the nanofibers are prone to hydrolytic degradation. Hence long-term processes should not employ such scaffolds as before the entire process (observation, selection or any other research studies) is completed, the culture will have no support to grow [36]. Crystallinity in polymers can treat this problem however the size of the fibres (diameter) are still a variable with a high probability of variation (purpose dependent and needs a lot of testing before it can be finally put into use). The poor infiltration of cells into scaffolds is still an issue to deal with, especially when we want to add various properties into nanofibers.

As mentioned in the introduction, electrospinning is an easy to setup and scalable technique. The cost parameter is in our favour whereas the volume imposes some difficulties in terms of production. It is quite difficult to produce a large volume scaffold and if the critical factors do not meet the threshold level the final structure might not be at its best form. This will drastically affect the application part. This will also affect special properties like antimicrobial/inflammatory/oxidant. The release of the drug will be questionable in such cases (**Table 2**).

The drug loading process when it includes a high amount of drug can result in a burst. When we submerge the fibres in aqueous solution (prone to hydrolytic degradation), the antimicrobial properties (e.g., antibiotics) are released in a short duration (might not last till it's required). This issue can be solved by using different set-ups. We have learned that electrospinning technique is an easy to set-up one, here if we want to use such nanofibers for a longer duration, a different set-up is required which introduces more complication [37].

When environmental factors are taken into consideration, most frequently relative humidity is studied. When this parameter is a settable, considering higher RH leads to thinner fibre diameter, an appropriate high RH level could be selected.

Diameter	Fibre composition	Application
438 ± 156 nm	Electrospun, aligned, and randomly oriented PCL	In vitro culture of meniscal fibrocartilage cells and human MSCs
519 ± 127 nm		
430 ± 170 nm	Electrospun, aligned, and randomly oriented PLLA	In vitro culture of human tendon stem cells
450 ± 110 nm		
657 ± 183 nm	Electrospun, aligned PU	In vitro culture of human ligament fibroblasts
300–900 nm	Electrospun PLGA nanofibers on top of microfibers	In vitro culture of porcine MSCs

Table 2.
Variations in polymer nanofibers size in ligament and tendon tissue engineering [36].

Since high RH levels may lead to beads, too high RH levels cannot be selected. Whereas if an ambient RH cannot be controlled. Then RH is considered a disturbance of the electrospinning process and the required jet diameter, responding to the desired fibre diameter, should change with an ambient RH. The system's velocity should be adjusted along with the flow rate. This is done to obtain a required jet diameter (the application of electric field in the polymer solution will cause the droplets to take a conical shape) and control the fibre's diameter. This is where production will become an issue. Such adjustments are difficult and are highly variable. Hence, relative humidity should be studied to decide what type of parameter it is (settable or disturbance) to decide the production rate. The operating regime should be selected to achieve the desired fibre diameter while maximising production rate [38].

Along with relative humidity, temperature should also be taken into consideration. The average diameter of nanofibers produced by electrospinning changes significantly through variation of temperature and humidity. At a relatively higher temperature the solvent evaporation rate will increase and the viscosity of the precursor solution to be electrospun will decrease, and as a result thinner nanofiber would be obtained [39].

6. Limitations

Though there are many reports on the successful presentation of electrospinning as a useful platform technique for the fabrication of nanofibers from a variety of materials, several issues are yet to be explained. Electrospinning process simulation models need to be optimised by considering all the liquid properties for electrospinning and all the processing parameters for better elucidation of the phenomenology of electrified jets. If it is successful, one should be able to analyse the behaviour of the electrified jet for the deterministic fabrication of electrospun nanofibers with well-controlled size, structure, and morphology [33].

Even though there is increase in the usage of natural biopolymers in the electrospinning technique for food packaging has developed a massive growing interest in the recent years but due to lesser flexibility of these materials ultimately leads to difficulty of processing in traditional equipments, and most of them are hydrophilic materials which means that they lack necessary mechanical properties and good barrier properties to moisture and oxygen [35].

7. Conclusion

As discussed above, electrospinning is one of the most efficient techniques used for the synthesis of nanomaterials [30]. It enables the incorporation of unique properties including large surface area, small size, and high activity, which are expected to develop advanced packaging systems for fulfilling consumers' needs. However, nano or micro sized components may lead to environmental pollution or even health risks due to their migration into food and drinks, whereas our knowledge regarding the potential threats from the used nanomaterials is still relatively lacking [35]. Research and experimentation by various organisations along with academic individuals have stated that the bioactive molecules that were naturally derived were better incorporated into polymeric nanofibers and also improved the membrane and scaffolds manufacturing using such electrospun nanofibers [34].

Although ESN includes antimicrobial and loading other similar agents, the spectrum of this range is limited to some curing agents. The further investigation is to be focused on broadening this spectrum resulting into a diversified product with composite materials [33]. Currently, the control on the deposition deposition, porosity, inter-linkage and intra-linkages made nanofiber accessible in almost all the fields- food industry, wound healing management etc. [30]. Environmental field applications with surface functionalized nanofibers are facing a few challenges that need to be tackled which include capacity reduction and kinetic slowness after surface modifications. The level of research to uplift the current properties for targeted action is not up to the mark, hence needs further investigation by the agricultural and food industries for a real time response. Apart from this, combining the nanofibers with microfluidic systems is still challenging because it requires nanofibers with the well-controlled diameter and orientation, as well as the reproducibility to place them at specific positions and with the right orientation [33].

Author details

Lingayya Hiremath[1]*, O. Sruti[1], B.M. Aishwarya[1], N.G. Kala[1] and E. Keshamma[2]

1 Department of Biotechnology, R.V. College of Engineering, Autonomous Institution Affiliated to Visvesvaraya Technological University, Belagavi, Bengaluru, Karnataka, India

2 Department of Biotechnology, Maharani's Science College for Women, Bangalore, Karnataka, India

*Address all correspondence to: lingayah@rvce.edu.in

References

[1] N. Bölgen, Electrospun materials for bone and tendon/ligament tissue engineering, Editor(s): Tamer Uyar, Erich Kny, Electrospun Materials for Tissue Engineering and Biomedical Applications, Woodhead Publishing, 2017, Pages 233-260, ISBN 9780081010228, doi:10.1016/B978-0-08-101022-8.00004-1

[2] Stojanov Spase, Berlec Ales. Electrospun Nanofibers as Carriers of Microorganisms, Stem Cells, Proteins, and Nucleic Acids in Therapeutic and Other Applications, Frontiers in Bioengineering and Biotechnology, VOL. 8, 2020, Pages 130. Available from: https://www.frontiersin.org/article/10.3389/fbioe.2020.00130 DOI: 10.3389/fbioe.2020.00130

[3] Seeram Ramakrishna et al. Electrospun nanofibers: solving global issues, Materials Today, Volume 9, Issue 3, 2006, Pages 40-50, ISSN 1369-7021, doi: 10.1016/S1369-7021(06)71389-X

[4] Mohammadi MA, Rostami M, Beikzadeh S, Raeisi M, Tabibiazar M and Yousefi M: Electrospun nanofibers as advanced antibacterial platforms: a review of recent studies. Int J Pharm Sci & Res 2019; 10(2): 463-473. doi: 10.13040/IJPSR.0975-8232.10(2).463-73

[5] Wanwei Zhang, Sara Ronca, Elisa Mele, Electrospun nanofibres containing antimicrobial plant extracts, Nanomaterials (Basel) 2017 Feb; 7(2): 42. doi: 10.3390/nano7020042 PMCID: PMC5333027

[6] Rodrigo Schneider et al, Biocompatible electrospun nanofibers containing cloxacillin: Antibacterial activity and effect of pH on the release profile, Reactive and Functional Polymers, Volume 132, 2018, Pages 26-35, ISSN 1381-5148, doi: 10.1016/j.reactfunctpolym.2018.09.001

[7] Kids' Macrogalleria, Fibres; Polymer Science Learning Centre. Available from: https://www.pslc.ws/macrog/kidsmac/fiber.htm

[8] Fibre; From Wikipedia, the free encyclopaedia. Available from: https://en.wikipedia.org/wiki/Fiber

[9] Nanofibre; From Wikipedia, the free encyclopaedia Available from: https://en.wikipedia.org/wiki/Nanofiber#:~:text=Nanofibers%20are%20fibers%20with%20diameters,physical%20properties%20and%20application%20potentials

[10] Almetwally, A.A. & El-Sakhawy, Mohamed & Elshakankery, Mahmoud & Kasem, M.H.. (2017). Technology of nano-fibers: Production techniques and properties - Critical review. Journal of the Textile Association. 78. 5-14. Available from: https://www.researchgate.net/publication/322774945_Technology_of_nano-fibers_Production_techniques_and_properties_-_Critical_review

[11] Xiaomin Shi, Weiping Zhou, Delong Ma, Qian Ma, Denzel Bridges, Ying Ma, Anming Hu, "Electrospinning of Nanofibers and Their Applications for Energy Devices", Journal of Nanomaterials, vol. 2015, Article ID 140716, 2015. doi:10.1155/2015/140716

[12] Ziabari, M., Mottaghitalab, V., & Haghi, A. K. (2009). Application of direct tracking method for measuring electrospun nanofiber diameter. Brazilian Journal of Chemical Engineering, *26*(1), 53-62. doi:10.1590/S0104-66322009000100006

[13] Croisier, F., Sibret, P., Dupont-Gillain, C.C., Genet, M.J., Detrembleur, C., Jerome, C., 2015. Chitosan-coated electrospun nanofibers with antibacterial activity. J. Mater. Chem. B 3, 3508-3517.

[14] Del Nobile M, Conte A, Incoronato A and Panza O: Antimicrobial efficacy and release kinetics of thymol from zein films. Journal of Food Engineering 2008; 89(1): 57-63.

[15] Burt S: Essential oils: their antibacterial properties and potential applications in the foods-a review. International journal of food microbiology 2004; 94(3): 223-253.

[16] Chantarodsakun T, Vongsetskul T, Jangpatarapongsa K, Tuchinda P, Uamsiri S, Bamrungcharoen C, Kumkate S, Opaprakasit P and Tangboriboonrat P: [6]-Gingerol-loaded cellulose acetate electrospun fibers as a topical carrier for controlled release. Polymer bulletin 2014; 71(12): 3163-3176.

[17] Sadri M, Karimi-Nazari E, Hosseini H and Emamgholi A: New Chitosan/Poly (ethylene oxide)/Thyme Nanofiber Prepared by Electrospinning Method for Antimicrobial Wound Dressing. Journal of Nanostructures 2016; 6(4): 322-328.

[18] Ranjbar-Mohammadi M, Rabbani S, Bahrami SH, Joghataei M and Moayer F: Antibacterial performance and in-vivo diabetic wound healing of curcumin loaded gum tragacanth/poly (ε-caprolactone) electrospun nanofibers. Materials Science and Engineering: C 2016; 69: 1183-1191.

[19] Zhu L, Liu X, Du L and Jin Y: Preparation of asiaticoside-loaded coaxially electrospinning nanofibers and their effect on deep partial-thickness burn injury. Biomedicine & Pharmacotherapy 2016; 83: 33-40.

[20] Han J, Chen T-X, Branford-White CJ and Zhu LM: Electrospun shikonin-loaded PCL/PTMC composite fiber mats with potential biomedical applications. International journal of pharmaceutics 2009; 382(1-2): 215-221.

[21] Park SW, Bae HS, Xing ZC, Kwon OH, Huh MW and Kang IK: Preparation and properties of silver-containing nylon 6 nanofibers formed by electrospinning. Journal of Applied Polymer Science 2009; 112(4): 2320-2326.

[22] Yeo SY, Lee HJ and Jeong SH: Preparation of nanocomposite fibers for permanent antibacterial effect. Journal of Materials Science 2003; 38(10): 2143-2147.

[23] Kalwar K, Hu L, Li DL and Shan D: AgNPs incorporated on deacetylated electrospun cellulose nanofibers and their effect on the antimicrobial activity. Polymers for Advanced Technologies 2018; 29(1): 394-400.

[24] Raffi M, Mehrwan S, Bhatti TM, Akhter JI, Hameed A, Yawar W and ul-Hasan MM: Investigations into the antibacterial behavior of copper nanoparticles against *Escherichia coli*. Annals of Microbiology 2010; 60(1): 75-80.

[25] Ahire JJ, Hattingh M, Neveling DP and Dicks LM: Copper-containing anti-biofilm nanofiber scaffolds as a wound dressing material. PloS one 2016; 11(3): e0152755.

[26] Qi R, Guo R, Zheng F, Liu H, Yu J and Shi X: Controlled release and antibacterial activity of antibiotic-loaded electrospun halloysite/poly (lactic-co-glycolic acid) composite nanofibers. Colloids and Surfaces B: Biointerfaces 2013; 110: 148-155.

[27] Liu X, Nielsen LH, Kłodzińska SN, Nielsen HM, Qu H, Christensen LP, Rantanen J and Yang M: Ciprofloxacin-loaded sodium alginate/poly (lactic-co-glycolic acid) electrospun fibrous mats for wound healing. European Journal of Pharmaceutics and Biopharmaceutics 2018; 123: 42-49.

[28] Han D, Sherman S, Filocamo S and Steckl AJ: Long-term antimicrobial effect of nisin released from electrospun

triaxial fiber membranes. Acta biomaterialia 2017; 53: 242-249.

[29] Electrospinning: promising technology for producing nanofibers. Fibre2Fashion. Oct 2008. Available from: https://www.fibre2fashion.com/industry-article/3692/electrospinning-promising-technology-for-producing-nanofiber

[30] Adnan Haider, Sajjad Haider, Inn-Kyu Kang. A comprehensive review summarizing the effect of electrospinning parameters and potential applications of nanofibers in biomedical and biotechnology, Arabian Journal of Chemistry, Volume 11, Issue 8, 2018, Pages 1165-1188, ISSN 1878 5352, doi: 10.1016/j.arabjc.2015.11.015

[31] Seeram Ramakrishna et al. Electrospun nanofibers: solving global issues, Materials Today, Volume 9, Issue 3, 2006, Pages 40-50, ISSN 1369-7021, doi: 10.1016/S1369-7021(06)71389-X

[32] Blachowicz, Tomasz, and Andrea Ehrmann. Most Recent Developments in Electrospun Magnetic Nanofibers: A Review. Journal of Engineered Fibers and Fabrics, (January 2020). doi:10.1177/1558925019900843.

[33] Jiajia Xue, Tong Wu, Yunqian Dai, Younan Xia Chem Rev. Electrospinning and Electrospun Nanofibers: Methods, Materials, and Applications. 119(8): 5298-5415. doi: 10.1021/acs.chemrev.8b00593. PMCID: PMC6589095

[34] Wanwei Zhang, Sara Ronca, Elisa Mele. Electrospun Nanofibres Containing Antimicrobial Plant Extracts. Nanomaterials (Basel) 2017 Feb; 7(2): 42. doi: 10.3390/nano7020042. PMCID: PMC5333027

[35] Zhang C, Li Y, Wang P, Zhang H. Electrospinning of nanofibers: Potentials and perspectives for active food packaging. Compr Rev Food Sci Food Saf. 2020; 19: 479-502. doi: 10.1111/1541-4337.12536

[36] Engineering Rebecca L. Dahlin, F. Kurtis Kasper, Antonios G. Mikos. Polymeric Nanofibers in Tissue, Tissue Eng Part B Rev. 2011 Oct; 17(5): 349-364. doi: 10.1089/ten.teb.2011.0238 PMCID: PMC3179616

[37] Nicoletta Ditaranto, Francesco Basoli, Marcella Trombetta , Nicola Cioffi, and Alberto Rainer. Received: 2 August 2018; Accepted: 7 September 2018; Published: 13 September 2018.

[38] Cai, Y., Gevelber, M. The effect of relative humidity and evaporation rate on electrospinning: fiber diameter and measurement for control implications. J Mater Sci 48, 7812-7826 (2013). doi:10.1007/s10853-013-7544-x

[39] De Vrieze, Sander & Camp, Tamara & Nelvig, A. & Hagström, Bengt & Westbroek, Philippe and Clerck, Karen. (2009). The effect of temperature and humidity on electrospinning. Journal of Materials Science. 44. 1357-1362. 10.1007/s10853-008-3010-6

Chapter 6

An Insight into Biofunctional Curcumin/Gelatin Nanofibers

Nand Jee Kanu, Eva Gupta, Venkateshwara Sutar, Gyanendra Kumar Singh and Umesh Kumar Vates

Abstract

Electrospinning (ESPNG) was used to synthesize ultrathin (UT) and uniform nanofibers (from 5 nm to a few hundred nanometers) of various materials which have biomedical applications (BAs) such as dressing of wounds, drug discharge, and so on and so forth. In the first half of the report, there is an audit on the nanofibers having low diameter so that it could have larger surface area to volume proportion, likewise with that it would have sufficient porosity and improved mechanical properties required for wound healing. Nanofibrous mats (NMs) with high biocompatibility could be utilized during healing of wounds by sustained release of curcumin (Cc) and oxygen. The ESPNG was understood through in-depth numerical investigation in the present report. Furthermore, the process parameters (PMs) were reviewed in depth for their contributions in synthesizing UT - Curcumin/Gelatin (Cc/G) nanofibers (NFs) of optimum diameter. The aim of the discussion was to demonstrate that simply optimizing biofunctional (BF) - Cc/G NFs might not be enough to satisfy experts until they are also given access details about the complete ESPNG method (mathematical mechanism) to improve hold over the synthesis of NMs (suitable for BAs) for the release profile of Cc throughout critical periods of healing process.

Keywords: curcumin/gelatin (Cc/G), nanofibers (NFs), nanofibrous mats (NMs), electrospun (ES), electrospinning (ESPNG), biomedical applications (BAs), methanoic acid (HCOOH)

1. Introduction

1.1 Biofunctional (BF) - curcumin/gelatin (cc/G) nanofibers (NFs)

The impact of new revelations in the field of nanotechnology widespreadly affects the wellbeing sciences (**Table 1**). In biomedical field, the possible parts of NFs applications which resemble drug delivery and tissue science and medicine have been researched in the article. In spite of the fact that electrospinning (ESPNG) was considered as a reasonable system for the polymer nanofibers that were polymeric, biodegradable or non-biodegradable, manufactured or common and so on, which were with uniform distances across ranges from 5 nm to a few hundred nanometers [2–4]. The ESPNG procedure was favored over other regular strategies in published papers for the synthesis of polymer nanofibers [5–8]. The requirements for the biopolymers such as their restricted dissolvability in natural

Few ES - NFs loaded with Cc	Potential application in wound healing/ dressing, so on
Few Cc loaded NFs including (a) polycaprolactone-polyethylene glycol; (b) poly (3-hydroxybutyric acid-co-3-hydroxyvaleric acid) (PHBV); (c) poly(lactic acid) hyperbranched polyglycerol; (d) $\varepsilon-$ polycaprolactone/ polyvinylalcohol; and so on	Potential wound healing application
Few Cc loaded NFs including tragacanth/ poly(ε-caprolactone) NFs	Potential application in dressing of diabetic wound based on *in vivo*
Few Cc loaded NFs including almond gum/ polyvinyl alcohol (PVA) NFs	Therapeutic capacity and bioavailability
Few Cc loaded NFs including Zinc-Cc with coaxial NFs	Orthopedic applications
Few Cc loaded NFs including (a) zinc NFs; (b) cellulose acetate/ polyvinylpyrrolidone NFs; (c) polyurethanes NFs; (d) gelatin (G) NFs; and so on	Antibacterial application
Few Cc loaded NFs including chitosan/ poly (vinyl alcohol) (PVA) NFs; and so on	Sustained drug release

Table 1.
Potential applications of biofunctional (BF) - curcumin (cc) based electrospun (ES) NFs (reprinted with permission from ref. [1]. Copyright 2020 IOP publishing).

solvents due to high particle size, as well as their expensive purifying steps and their suitable polymeric solutions because of their inclination to frame hydrogen bonds, were controlled subsequent to mixing with engineered polymers in any case these restrictions may limit their ESPNG process for nanofibrous mats (NMs) [9].

The nanofibrous mats (NMs) which were prepared from ES collagen nanofibers were utilized for applications of tissue science and medicine [8]. Additionally, *Aloe vera* which is a characteristic polymer also holds potential to be used for tissue science and medicine applications because it has a cancer prevention agents and it is totally not harmful to living tissues [10, 11]. The BAs of some other ES nanofibers were discussed in **Table 1**. Gelatin (G), a polymer made up of proteins and peptides, is not harmful to living tissues. As a result, it was thought to be a fair and healthy option when it came to dressing dangerous injuries, such as diabetic ulcers.

Due to their considerable tensile strength compared to traditional fibers (with diameters ranging over 100 nm), the low profile NFs can serve as a suitable material while healing and act as barriers to protect the wound (**Figures 1** and **2**) [11–14]. Gelatin is also noted for its high water absorption and fluid affinity, making it an ideal option for the moist healing process. In methanoic acid, gelatin (a natural biopolymer that is a denatured form of collagen) is quite soluble. Collagen is a protein found in the extracellular matrix (ECM) of humans and animals, but it is costly due to its production processes [15–25]. The properties of these nanofibers can also be regulated according to requirements by optimizing input process parameters (PMs) such as high potential power supply, solution's resistance to the flow, length between the NFs collector and emitter, and feed rate, according to the authors. Methanoic acid was clearly used as a natural unstable dissolvable in the ESPNG to disintegrate gelatin (G) at room temperature. Recently, the use of G-nanofibers with sufficient tensile strength for fabricating NMs has got a lot of attention for antimicrobial applications [26–30]. In addition to their light weight (LW), effective spinning of minimum diameter nanofibers provides a large surface area of these nanofibers. It was fundamentally required for the purpose of dressing the wounds and for other BAs (**Table 1**). Mindru et al. [31] succeeded in synthesizing NMs of sufficient strength for BAs using methanoic acid. Rather than cytotoxic

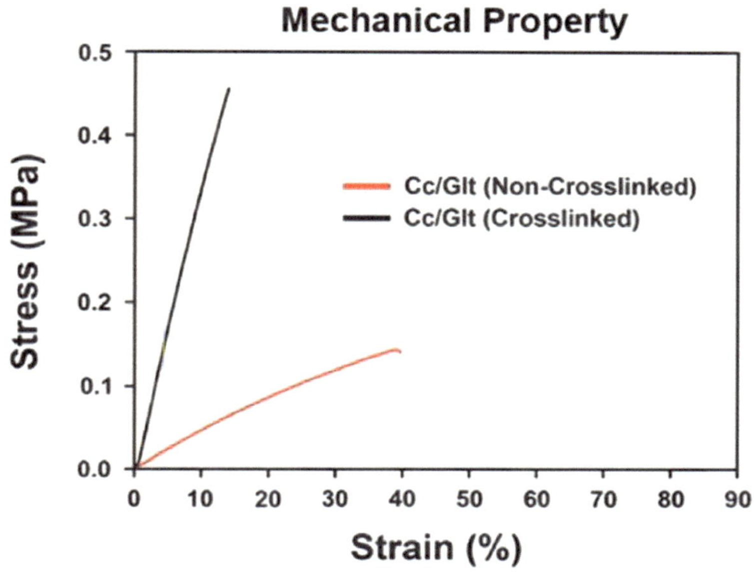

Figure 1.
The BF - ES NFs were crosslinked to improve tensile strength of the NMs (reprinted with permission from ref. 12. Copyright 2017 springer nature).

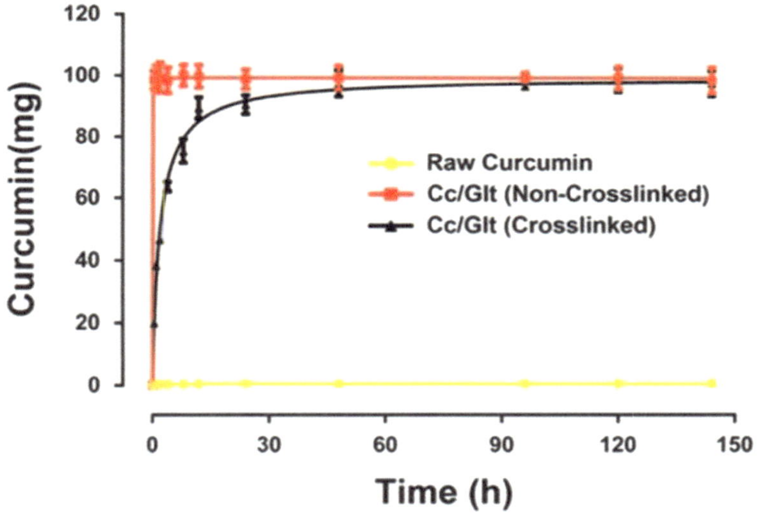

Figure 2.
The event of cc discharge was shown with time (reprinted with permission from ref. 12. Copyright 2017 springer nature). The need to crosslink the ES - NFs was illustrated.

solvents, Maleknia et al. [32] utilized HCOOH/water to get ready solutions for the ESPNG of G-nanofibers which can be utilized for BAs such as dressing of wounds, delivery of pharmaceuticals, and for tissue science and medicine. They were successful in synthesizing G-nanofibers with 197 nm diameter. Chen et al. [33] utilized methanoic acid and ethanol to have the improvement in the volatility of the dissolvable rather than cytotoxic solvents while setting up the dissolvable for preparing ES G-nanofibers. For medication conveyance, the nanofibrous mats broke in a rapid manner in fluid polymeric solutions. Aytac et al. [23] research suggests that ES

G-NFs exemplified with ciprofloxacin/hydroxypropyl-beta-cyclodextrin incorporating complex will break down quicker in water than ES G-nanofibers stacked with ciprofloxacin. Using a dialysis process, Yabing et al. [21] synthesized drug-loaded micelles (poly(ethylene glycol)-block-caprolactone copolymer) and integrated these pharmaceuticals into ES G-NFs. The NMs developed using ES NFs have considerable surface regions and such NFs have a significant contribution in tissue science and medicine. The solvent utilized here was the methanoic acid for ESPNG BF-nanofibers which leads to different BAs such as enzyme immobilization, materials for bone recovery, antifungal and antibacterial exercise in the release of medications, bioactive materials encapsulation during packaging of food and dressing of wounds [34].

The turmeric extracted from *Curcuma longa*, which was regular turmeric (herbaceous plant) and is broadly utilized in Asian countries like India and China, as a bioactive compound with potent anti-inflammatory and antioxidant properties in medicine. Synthetic dimethoxycurcumin has been found to be more effective than natural curcumin (Cc) at destroying cancer cells (which is the leading cause of death in the world) (derived from the plant) [35–45]. Ramrezagudelo et al. [36] incorporated antibiotic doxycycline pharmaceuticals (mitochondrial biogenesis inhibitors that may limit cancer stem cells in the early stages of breast cancer) into ES hybrid poly-caprolactone/gelatin/hydroxyapatite soft NMs and assessed these drug delivery meshes as effective antitumor and antibacterial scaffolds (**Figure 3**). The utilization of methanoic acid as dissolvable for solutes such as Cc and gelatin (G) has been the favored decision in numerous BAs. Researchers have successfully prepared solutions of Cc and dimethoxycurcumin utilizing methanoic acid [35, 46–48]. After 12 hours, higher concentrations of Cc, such as 17 percent Cc loaded poly (ε-caprolactone) (PCL) NFs, should release more Cc at a particular rate than lower concentrations such as 3 percent Cc loaded PCL NFs. (**Figure 2**) [11, 12]. Utilization of PCL-Cc polymeric solutions, BF-ES nanofibers were prepared [11–14, 48].

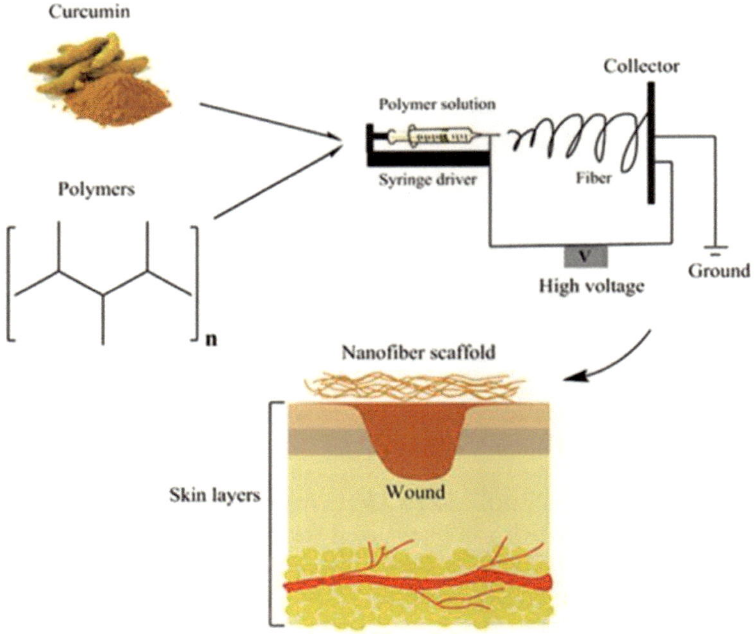

Figure 3.
Preparation of cc loaded ES NFs for sustained release of pharmaceuticals for potential healing process (reprinted with permission from ref. 13. Copyright 2018 John Wiley and Sons).

Xinyi et al. [12] synthesized curcumin/gelatin (Cc/G) nanofibrous mats and studied the arrival of Cc on rodent models (intense injury) by means of an in vitro approach. The healing process was tested by treating rodents utilizing the Cc/G nanofibrous mats (investigations done on the third, seventh, and fifteenth days subsequent to injuring). It inspired us to create Cc-loaded gelatin NFs suitable for the fabrication of NMs for the application of Cc and oxygen to the wound on a long-term basis (during healing) [13]. These NMs will then have antioxidant and anti-inflammatory properties, making them ideal for the healing process [13, 48–66].

1.2 Mechanism behind electrospinning (ESPNG) of cc/G nanofibers (NFs)

The process of electrospinning (ESPNG) utilizes an electric field applied to the emitter and a ground terminal to pull back a thread of polymeric solution out of the opening of the emitter. In the process of ESPNG, the Maxwell electrical pressure was set as per ratio, $\frac{V^2}{d^2}$; where permittivity was $'\varepsilon'$, a high potential power supply was $'V'$ and the electrode spinning gap was shown with $'d'$. The critical high potential power supply (V_c) was $\sqrt{\frac{\gamma d^2}{\varepsilon R}}$, and it must exceeded before any jet could spread out from the needle tip. For, $\gamma = 10^{-2} kg/s^2, d = 10^{-2} m, \varepsilon = 10^{-10} C^2/(Jm)$ $and R = 10^{-4} m$, a high potential power supply around 10 KV was necessary to form a jet of any type. The polymeric solution of the Laplace condition (utilized in the modeling) in the feeble polarization limit depicts the electrostatics in the fluid stage in an axisymmetric indirect support framework (r, θ, ϕ) with the vertices of the Taylor cone at the source which can be shown using general Eq. (1).

$$\psi_1(r,\theta) = A_n r^n P_n(\cos\theta); for \theta_0 \geq \theta \geq 0,$$
$$\psi_g(r,\theta) = B_n r^n P_n(\pi - \theta); for \pi \geq \theta \geq \theta_0 \quad (1)$$

Eq. (1), $V = \frac{4}{3}\pi r^3$, was the drop in the volume of fluid, here the spinning gap r which was from the cone vertex of angle $2\theta_0$ to the emitter tip and the state of the drop was said to be utilizing a Taylor cone, accordingly was described as $r = R(z)$. Later on the z- axis was corresponding to the applied electric field with $z \in [-l, l]$, where l was the length if the semi-long pivot of the drop and limit condition $\theta_0 \geq \theta \geq 0$ represents the boundary condition of the fluid. P_n $[x]$ was the Legendre's function, where A_n and B_n were constants. They suggested a model for ESPNG polymeric nanofibers which relies upon a sink-like flow towards the vertex of the Taylor cone. The course of action of the flow in axisymmetric polar headings $(r, \varepsilon, 0)$ was given utilizing conditions (2) and (3).

$$v_r = \frac{vF(\varepsilon)}{r} \quad (2)$$

$$F(\varepsilon) = b\left\{3\tanh^2\left[\left(\sqrt{\frac{-b}{2}}\right)(\alpha - \varepsilon) + 1.146\right] - 2\right\} \quad (3)$$

In the above Eqs. (2) and (3), velocity of the radial feed, v_r, the kinematic solution's resistance to the flow of feed, v, the wedge/Taylor cone half angle, a, the parameter, b which serves to decides the inertial concentration of stream on the Taylor cone vertex/Taylor cone. With that the mass and charge conservations led to expressions for v and σ in terms of R and E, also the force and E-field conditions were assessed utilizing second-degree differential equations. Inclination of the stream surface (R) was supposed as the highest from the origin of the nozzle and hence the initial result of z was equal to zero. Furthermore, PMs were discussed using the set of Eqs. (4) [67, 68].

$$\begin{aligned} R(0) &= 1 \\ E(0) &= E_0 \\ \tau_{prr} &= 2r_n \frac{R'_0}{{R_0}^3} \\ \tau_{pzz} &= -2T_{prr} \end{aligned} \tag{4}$$

Here Eq. (4), the radius of the jet initially was $'R_0'$ and the formula used for calculating the jet velocity, $\upsilon_0 = \frac{Q}{\pi R_0^2 K}$, where the rate of discharge of the polymeric solution, Q, and the conductivity of the liquid solution, K. Moreover, the electric field (E_0) was calculated using the formula, $E_0 = \frac{I}{\pi R_0^2 K}$ and the surface charge density (σ_0) was calculated using the formula, $\overline{\varepsilon} E_0$, where the dielectric constant of ambient air was $'\overline{\varepsilon}'$ and the constant was $'E_0'$ which was to be used during simulation of the ESPNG. The viscous stress (τ_0) was calculated using the formula, $\tau_0 = \frac{\eta_0 \nu_0}{R_0}$. A Newtonian liquid law of force for the liquid was summed up and for that the shear pressure (τ) was given as $\tau = K \left(\frac{\partial v}{\partial y}\right)^m$ as shown using Eq. (5) [67]. The electric field will overcome the surface tension of the polymer liquid and thereafter through Taylor's cone NFs will be pulled out and ES over the moving cylinder collector.

$$\frac{d(\sigma R)}{dz} \simeq - \left(2R\frac{dR}{dz}\right)/Pe \tag{5}$$

Furthermore, the response of $'E'$ is a function of axial position (z) and it can be shown in Eq. (6) [1, 69, 70].

$$\frac{d(E)}{dz} = \ln \chi \left(\frac{d^2R^2}{dz^2}\right)/Pe \tag{6}$$

The model discussed above so far was found fit for foreseeing the conduct of the PMs of the ESPNG [67].

2. Electrospinning (ESPNG) of cc/G nanofibers (NFs)

In the present investigation, we utilized set-up for ESPNG (**Figure 4(a)**). The prominent four parts that were related to the ESPNG – PMs such as spinning gap between the emitter and drum collector, high potential power supply, rate of feed, and solution's resistance to the flow of a polymeric solution (taken in a 2 ml needle syringe). For the ESPNG process, a high potential power supply has been set across the length of the moving cylindrical drum collector (covered with an aluminum sheet) to pull NFs from Taylor's cone formed at the tip of the syringe's needle. The NFs were stretched up from the polymeric solution containing a polar natural solvent and a polymer solute in a definite proportion. After that, these NFs were collected over the moving cylinder which was turned with a speed of around 1000 rpm so that the NFs with UT - diameters could be synthesized across by

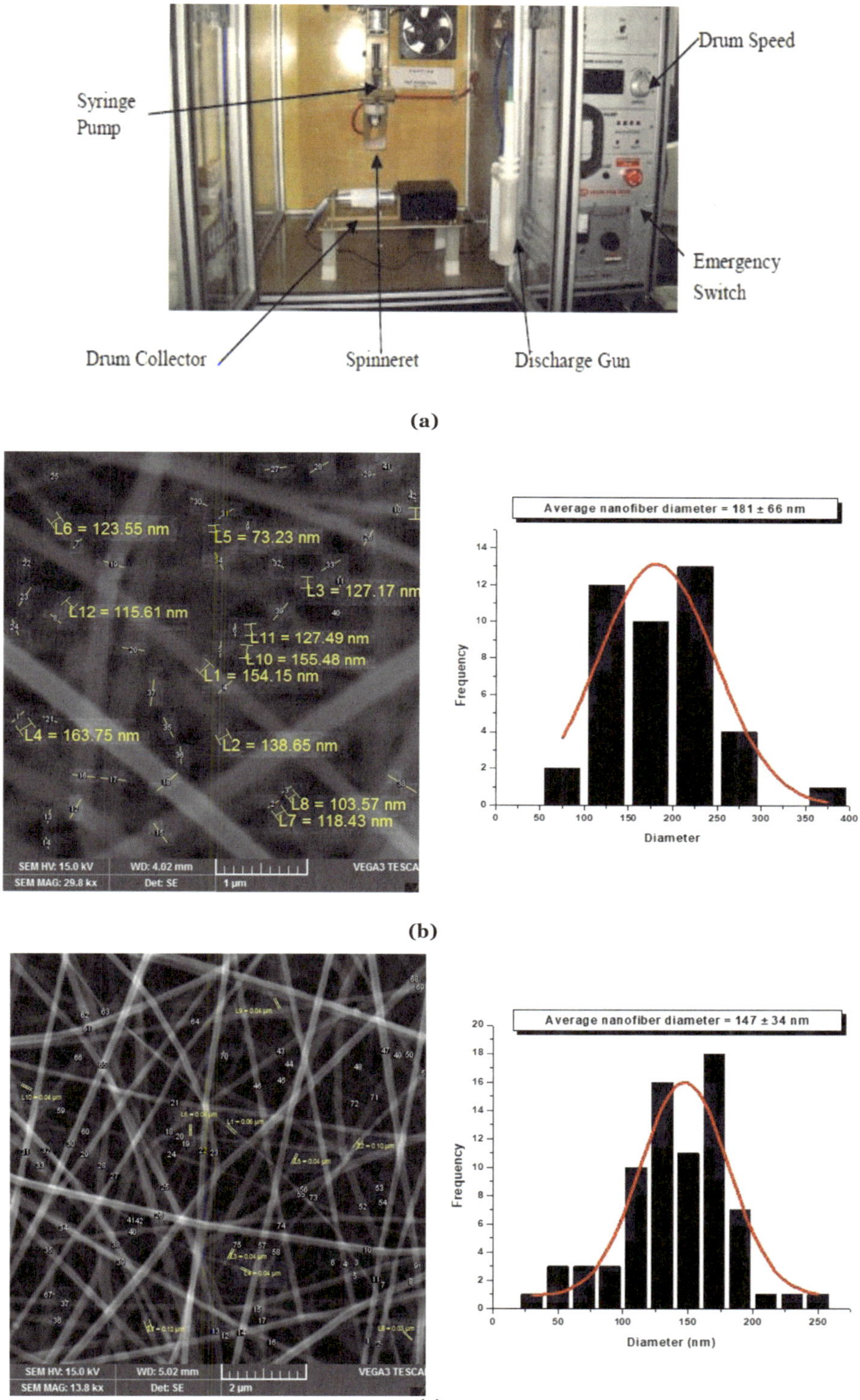

Figure 4.
The (a) ESPNG set up was used to synthesize (b), and (c) the UT - and BF - Cc/G NFs (Reprinted with permission from Ref. [70]. Copyright 2020 IOP Publishing).

extending them and adjusting them directly as well as improving their mechanical properties. The four PMs were the spinning gap between the collector and needle's tip, rate of feed, solution's resistance to the flow, and the high potential power supply were considered [1, 69, 70].

For synthesizing the BF - NFs parameters were considered. The polymeric solution was prepared after blending 1 percent curcumin (Cc) (0.1 g) with 1.5 percent G (0.15 g) in 10 ml of methanoic acid (98 percent concentrated). Other than that, each other polymeric solution was prepared by mixing 1.2 percent Cc (0.12 g) with 2 percent G (0.2 g) in 10 ml of methanoic acid (98 percent concentrated), both at room temperature. The examinations were done at room temperature, in encompassing air which had moisture around 80 percent.

The synthesis of NFs was done by varying the spinning gap between the tip of the needle (10 cm and 15 cm), the rate of feed (0.1 ml h^{-1} and 0.15 ml h − 1), the possible high potential power supply (15 KV and 20 KV), and the solution's resistance to the flow (65 cP and 70 cP, on account of the additional substances obsessions). For 48 hours, the mats were dried at room temperature to completely remove the methanoic acid. The diameters of the NFs were then examined using scanning electron microscopy (SEM) (**Figure 4(b)**, and **(c)**).

3. The electrospun (ES) cc/G nanofibers (NFs): a state-of-the-art review

The diameter (nm) of the NFs was synthesized during the process of electro spinning measures (**Table 2**). The differences in the outcomes (as far as the

Runs	Spinning gap (cm) A	Feed Rate (ml/h) B	High potential power supply (KV) C	Solution's resistance to the flow (cP) D	Mean Diameter (nm)
1	Low - 10	Low - 0.1	Low - 10	Low - 65	*205 ± 22.5*
2	High - 15	Low - 0.1	Low - 10	Low - 65	*181 ± 66*
3	Low - 10	High - 0.15	Low - 10	Low - 65	*270 ± 16*
4	High - 15	High - 0.15	Low - 10	Low - 65	*280 ± 20*
5	Low - 10	Low - 0.1	High - 15	Low - 65	*260 ± 26.5*
6	High - 15	Low - 0.1	High - 15	Low - 65	*254 ± 28*
7	Low - 10	High - 0.15	High - 15	Low - 65	*147 ± 34*
8	High - 15	High - 0.15	High - 15	Low - 65	*286 ± 31*
9	Low - 10	Low - 0.1	Low - 10	High - 70	*287 ± 77*
10	High - 15	Low - 0.1	Low - 10	High - 70	*288 ± 57*
11	Low - 10	High - 0.15	Low - 10	High - 70	*375 ± 96*
12	High - 15	High - 0.15	Low - 10	High - 70	*206 ± 56*
13	Low - 10	Low - 0.1	High - 15	High - 70	*308 ± 74*
14	High - 15	Low - 0.1	High - 15	High - 70	*229.5 ± 60*
15	Low - 10	High - 0.15	High - 15	High - 15	*235 ± 47*
16	High - 15	High - 0.15	High - 15	High - 70	*274 ± 53*
					Total $\sum X = 4085.5$

Table 2.
The effect of PMs on diameters of cc/G NFs (reprinted with permission from ref. 1. Copyright 2020 IOP publishing).

diameters of the NFs) synthesized were as per the following: at a high potential power supply such as 15 KV (at 15 cm distance, 0.1 ml h^{-1} rate of feed and 65 cP solution's resistance to the flow) utilizing a solution having 1.5 percent G, 1 percent Cc in 10 ml of 98 percent concentrated methanoic acid, NFs with diameters around 254 nm (254 $\pm$ 28 nm) which was quite larger than the 181 nm (181 $\pm$ 66 nm) (**Figure 4(b)**) spinning gap across got at 10 KV using similar polymeric solution and keeping PMs at same levels. At a higher rate of feed such as 0.15 ml h^{-1} (at 10 cm distance, 15 KV high potential power supply, and 65 cP solution's resistance to the flow) utilizing a solution having 1.5 percent G, 1 percent cc in 10 ml of 98 percent concentrated methanoic acid, the diameter across of NFs were prepared around 147 nm (147 $\pm$ 34 nm) (**Figure 4(c)**) which was quite smaller than 260 nm (260 $\pm$ 26.5 nm) as the diameter synthesized at 0.1 ml h^{-1} rate of feed utilizing similar polymeric solution and keeping PMs at same levels. At a higher rate of feed such as 0.15 ml h^{-1} (at 15 cm distance, 10 KV power supply, and 70 cP solution's resistance to the flow) utilizing an solution having 2 percent G, 1.2 percent Cc in 10 ml of 98 percent concentrated methanoic acid, the diameter of NFs were prepared around 206 nm (206 $\pm$ 56 nm) which was smaller than 229.5 nm (229.5 $\pm$ 60 nm) as the diameter synthesized at 0.1 ml h^{-1} (at 15 cm distance, 15 KV high potential power supply and 70 cP solution's resistance to the flow) utilizing a similar polymeric solution. For a higher concentration (2 percent G, 1.2 percent Cc in 10 ml of 98 percent concentrated methanoic acid), the solution's resistance to the flow was prepared (utilizing a solution's resistance to the flow - measurement set up) to be 70 cP and afterward the diameter of the NFs increased to 235 nm (235 $\pm$ 47 nm) at 10 cm distance, 0.15 ml h^{-1} rate of feed and 15 KV high potential power supply, from 147 nm (147 $\pm$ 34 nm) (**Figure 4(c)**) (prepared at 1.5 percent G, 1 percent Cc in 10 ml of 98 percent concentrated methanoic acid) at 10 cm distance, 0.15 ml h^{-1} rate of feed, 15 KV high potential power supply and 65 cP solution's resistance to the flow. At a spinning gap between the collector and needle's tip such as 15 cm (0.15 ml h^{-1} rate of feed, 15 KV high potential power supply, and 70 cP solution's resistance to the flow) utilizing an solution having 2 percent G, 1.2 percent Cc in 10 ml of 98 percent concentrated methanoic acid, the diameters of NFs were prepared around 274 nm (274 $\pm$ 53 nm) which was larger than the 235 nm (235 $\pm$ 47 nm) diameter obtained for 10 cm spinning gap utilizing similar polymeric solution and keeping PMs at same levels [1, 69, 70].

3.1 Design of experiments

The 2^k factorial design algorithm was run to test the basic variables or PMs such as the gap between collector and needle's tip, rate of feed, high potential power supply, and solution's resistance to the flow (each changed at two unique levels such as low (−) and high (+)) [71–73]. Accordingly, the total number of runs or trials were 2^4 i.e., 16. Each of the 16 examples was inspected under scanning electron microscopy (SEM) for characterization of diameters in *nm* (as listed in **Table 2**). A few samples of the Cc/G NFs analyzed under SEM were shown in **Figure 4(b)**, and **(c)**). The UT – spongy NMs were synthesized under all process conditions [1, 69, 70].

3.2 Analysis of variance

Analysis was performed to find the effects of PMs on the diameter of BF nanofibers. Correction factor, CF (to calculate the sum of squares of PMs) was evaluated using relationship (7).

3.2.1 Correction factor (CF)

For diameter (nm), the correction factor (CF) was calculated as

$$CF = \frac{(\Sigma X)^2}{n} = \frac{(4085.5)^2}{16} \cong 1043207 \tag{7}$$

Where the gross total of observed diameters ΣX and the number of iterations n, was 16.

The effect of the factors can be assessed using Eq. (8).

$$\frac{[\Sigma Y_{low}]^2}{n} + \frac{[\Sigma Y_{high}]^2}{n} - CF \tag{8}$$

Where Y is an input variable such as the spinning gap (A), Y_{high} and Y_{low} represents the aggregate of all mean diameters synthesized at low (−)and high (+) levels, individually, for the specific info variable with each whole assumed control over the high and low estimations of different factors. The mean diameters for the low (−) and high (+) levels of PMs were taken from **Table 2** [1, 69, 70].

1. Sum of squares, spinning gap variable (cm),SS_A

$$\frac{[\Sigma A_{low}]^2}{n} + \frac{[\Sigma A_{high}]}{n} - CF$$
$$= \frac{[205 + 270 + 260 + 147 + 287 + 375 + 308 + 235]^2}{8}$$
$$+ \frac{[181 + 280 + 254 + 286 + 288 + 206 + 229.5 + 274]^2}{8} - 1043207$$
$$= 489.5$$

The sum of the square of any interaction was assessed using Eq. (9).

$$\frac{[\Sigma AB_{low}]^2}{n} + \frac{[\Sigma AB_{high}]^2}{n} - CF \tag{9}$$

For any interaction such as AB, the SS_{AB} was evaluated as follows:

2. Sum of squares for interaction AB, SS_{AB}

$$\frac{[\Sigma AB_{low}]^2}{n} + \frac{[\Sigma = AB_{high}]^2}{n} - CF$$
$$= \frac{[181 + 270 + 254 + 147 + 288 + 375 + 229.5 + 235]^2}{8} +$$
$$\frac{[205 + 280 + 260 + 286 + 2870206 + 308 + 274]^2}{8} - 1043207$$
$$= 1000$$

Out of all interactions, the SS_{ABC} was recorded for its *highest result* around *9925*. The errors were added together and the ratio, MS_{error} was calculated using Eq. (10).

$$MS_{error} = SS_{error}/V_{error} = 21 \tag{10}$$

Where, representation of the number of errors was done by V_{error}, and in our case it was one. Now, ratio which was calculated, using the F-distribution Table, which was F esteemed for 95 percent degree of certainty as 7.71 and further inferred that the diameter relies on factors: (a) - Interaction between spinning gap (cm), (ml h^{-1}) rate of feed and (KV) high potential power supply, (b) - Interaction between spinning gap (cm) and solution's resistance to the flow (cP), (c) D-Solution's resistance to the flow (cP), (d) - Interaction between rate of feed (ml h^{-1}) and high potential power supply (KV), (e) - Connection between spinning gap (cm) and high potential power supply (KV), (f) - Interaction between rate of feed (ml h^{-1}), high potential power supply (KV) and solution's resistance to the flow (cP), (g) - Interaction between spinning gap (cm), (ml h^{-1}) rate of feed, (KV) high potential power supply and solution's resistance to the flow (cP), (h) - Interaction between spinning gap (cm) and rate of feed (ml h^{-1}), (I) - Interaction between high potential power supply (KV) and solution's resistance to the flow (cP), (j) - Interaction between rate of feed (ml h^{-1}) and solution's resistance to the flow (cP), (k) - High potential power supply (KV), (l) A - spinning gap (cm), and (m) - Rate of feed (ml h^{-1}).

3.3 Regression analysis

Every process parameter here has two levels such as low (−) and high (+) levels and a degree of freedom (DOF), in this way we utilized a common regression model to compute the minimum diameter of NFs based on the effects of interactions such as $\beta_1, \beta_2, \beta_3, \beta_4, \beta_5, \beta_6, \beta_7, \beta_8, \beta_9$ and β_{10} (in terms of contributions of interactions between ABC-Interaction between spinning gap (cm), rate of discharge of polymeric solution (ml h^{-1}) and high potential power supply (KV), AD -Interaction between spinning gap (cm) and solution's resistance to the flow (cP), BC - Interaction between rate of discharge of polymeric solution (ml h^{-1}) and high potential power supply (KV), AC -Interaction between spinning gap (cm) and high potential power supply (KV), BCD - Interaction between rate of discharge of polymeric solution (ml h^{-1}), high potential power supply (KV) and solution's resistance to the flow (cP), $ABCD$ - Interaction between spinning gap (cm), rate of discharge of polymeric solution (ml h^{-1}), high potential power supply (KV) and solution's resistance to the flow (cP), AB -Interaction between spinning gap (cm) and rate of discharge of polymeric solution (ml h^{-1}), CD - Interaction between high potential power supply (KV) and solution's resistance to the flow (cP), BD-Interaction between rate of discharge of polymeric solution (ml h^{-1}) and solution's resistance to the flow (cP), respectively) as well as the main effects such as $\beta_3, \beta_{11}, \beta_{12,}$ and β_{13} (in terms of contributions of D - Solution's resistance to the flow (cP), C - High potential power supply (KV), A -Spinning gap (cm), and B - rate of discharge of polymeric solution (ml h^{-1}), respectively), in Eq. (11).

$$Y(T_n) = \beta_0 + \beta_1 T_1 + \beta_2 T_2 + \beta_3 T_3 + \ \dots \ + \beta_n T_n + \eta \tag{11}$$

Where,

$$\beta_0 = \sum\nolimits_{i=1}^{N} \frac{Y_i}{N} = \frac{4085.5}{16} = 255.344$$

Furthermore, the influence of each process parameter, P, was computed using the relationship, $Y_P = \bar{Y}_{P+} - \bar{Y}_{p-}$, where $\bar{Y}_{P+}$ and $\bar{Y}_{p-}$ stand for the sum of all mean diameters prepared at low (−) and high (+) levels, respectively, for the particular input variable. The results of the corresponding mean diameters for the low (−) and high (+) levels of the particular process parameter were taken from **Table 2**.

Therefore, the percentage contribution for ABC = (9925/41257.5) × 100 = 24.

$\beta_1 = \frac{1}{2} \times$ The influence of the interaction, ABC= 25.

The general form of the regression equation was formulated and shown in Eq. (12). Using Eq. (12), the minimum diameter of curcumin/gelatin (Cc/G) NFs (nm) for sustained release of Cc could be evaluated after determination of the coefficients (such as $\beta_1, \beta_2, \beta_3, \beta_4, \beta_5, \beta_6, \beta_7, \beta_8, \beta_9$ and β_{10}) of the interaction effect (such as $X_{ABC}, X_{AD}, X_{BC}, X_{AC}, X_{BCD}, X_{ABCD}, X_{AB}, X_{CD}$ and X_{BD}) as well as the coefficients (such as $\beta_3, \beta_{11}, \beta_{12,}$ and β_{13}) of the basic PMs (such as $X_D, X_C, X_{A,}$ and X_B).

$$\begin{aligned} \text{Diameter (nm)} = {} & 255.344 + 25X_{ABC} - 20.5X_{AD} + 20X_D - 17.75X_{BC} + 17.25X_{AC} + 13X_{BCD} \\ & +11X_{ABCD} + 8X_{AB} - 7.5V_{CD} - 6.5X_{BD} - 6X_C - 5.5X_A + 3.75X_B \end{aligned} \tag{12}$$

The above model Eq. (12) was valid for the boundary conditions such as (a) $10 \le X_A \le 15$ (cm), (b) $0.10 \le X_B \le 0.15$ (ml h^{-1}), (c) $10 \le X_c \le 15$ (KV), (d)$65 \le X_D \le 70$ (cP).

The mean diameters (nm) of Cc/G NFs were varied with respect to PMs as shown in **Figure 5(a)**. It was observed that (a) with an increase in spinning gap (cm), and high potential power supply (KV), the mean diameters (nm) of BF - NFs were reduced; and (b) with the increase in the rate of feed (ml h^{-1}), and the solution's resistance to the flow (cP), the mean diameters (nm) of the NFs were increased. The influence of ABC - Interaction between spinning gap (cm), rate of feed (ml h^{-1}) and high potential power supply (KV), AD - Interaction between spinning gap (cm) and solution's resistance to the flow (cP), D-Solution's resistance to the flow (cP), BC - Interaction between rate of feed (ml h^{-1}) and high potential power supply (KV), and AC - Interaction between spinning gap (cm) and high potential power supply (KV) were quite significant.

The contour plots (2D plots) of the mean diameters of NFs with respect to basic PMs were shown in **Figure 5(b)** and **(d)** [1]. **Figure 5(c)** and **(e)** [1] show the fitted model's predicted 3D response surface plots of the mean diameters (nm) of Cc/G NFs formed. The contributions of ABC-Interaction between spinning gap (cm), feed rate (mL/h), and power supply (KV), AD-Interaction between spinning gap (cm) and solution's resistance to the flow (cP), D-Solution's resistance to the flow (cP), BC-Interaction between feed rate (mL/h) and power supply (KV), and AC-Interaction between spinning gap (cm) and power supply (KV) had significant effects of 24 percent, 15.5 percent, 12 percent, and 11.5 percent, respectively, over the preparation of Cc/G NFs with minimum diameters. **Figure 5(f)** shows the optimized parameter settings. Shifting the red lines to find the optimum results of PMs within the range may be used to estimate the effects of important PMs on the mean diameter (nm) of Cc/G NFs. The composite desirability, D, in our case is 0.8129, which is similar to 1. The current response results are represented by the horizontal blue line (**Figure 5(f)**).

The mean diameter of UT - Cc/G NFs was predicted to be 189.6563 nm using a configured setting of 1.5 percent G and 1 percent Cc in 10 mL of 98 percent concentrated methanoic acid, with an electrospining unit with a power supply of 10 KV, a spinning gap from the emitter to collector drum of 15 cm, a feed rate of 0.1 mL/h, a solution's resistance to the flow of 65 cP, and a drum collector speed of 1000 rpm. The SEM image of Cc/G NFs with an mean diameter of 181 nm (181 ± 66 nm) synthesized under similar conditions using the same solution was shown in **Figure 4(b)**. As a result, the approximate diameter (nm) of Cc/G NFs in

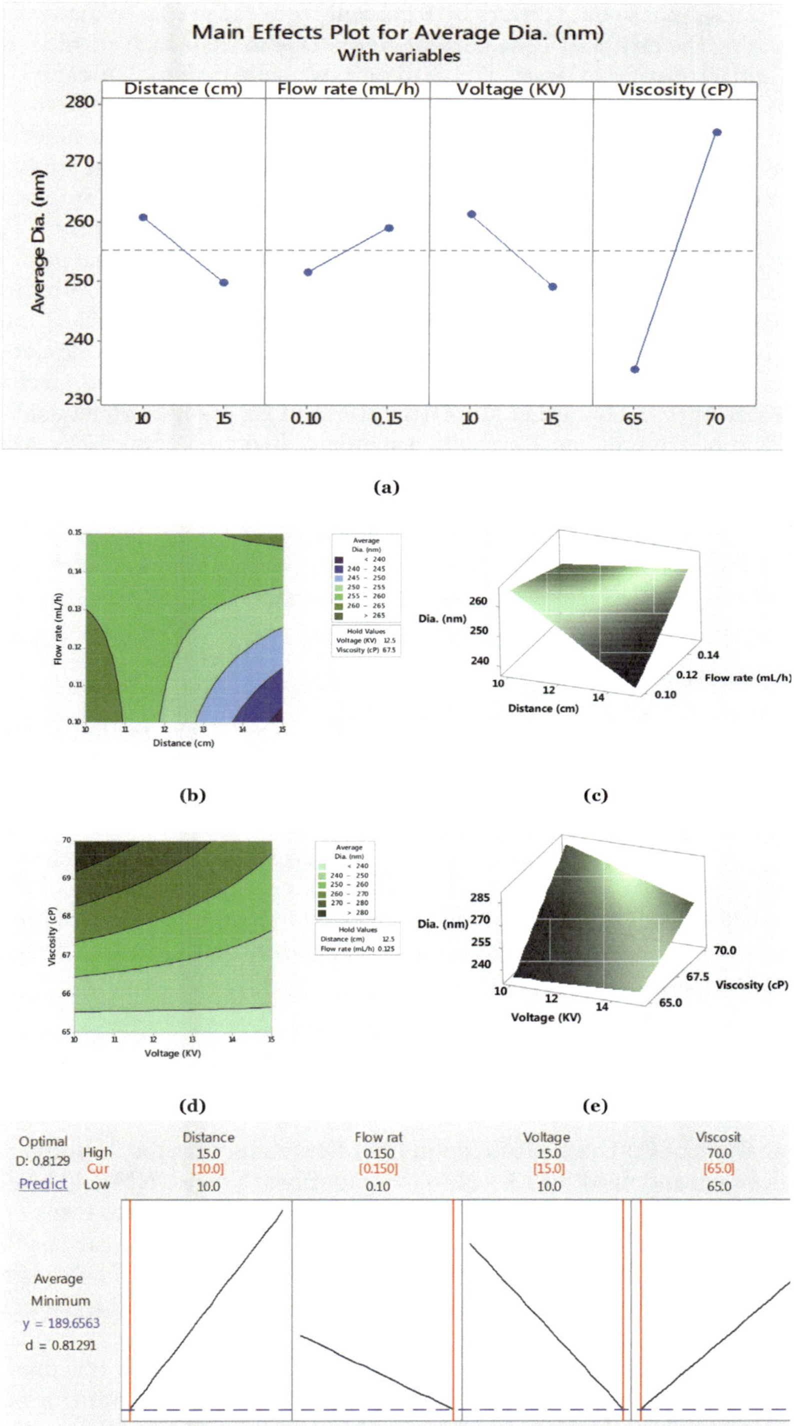

(f)

Figure 5.
Electrospining PMs optimization (reprinted with permission from ref. 1. Copyright 2020 IOP publishing). (a) Mean diameters of NFs versus PMs. (b), (d) two dimensional contour plots for mean diameter of NFs with respect to PMs. (c), (e) three dimensional plots for mean diameter of NFs with respect to PMs. (f) the study of optimized setting of PMs for synthesis of UT - NFs.

the optimization phase only differs by 8 percent from the prepared diameter, demonstrating the efficacy of the current study. Due to their high surface area to volume ratio in relation to length and diameter, we believe these LWs and UT - NFs with sufficient film porosity could be used in the healing process.

The optimum conditions for synthesizing the minimum mean diameter (181 ± 66 nm) of UT - Cc/G NFs were achieved (**Figure 4(b)**) in the study, which could be ideal for dressing diabetic chronic ulcers due to its specific properties such as LW, not harmful to living tissues, water absorbent, and fluid affinity.

Using the optimized environment of a polymeric solution, Sharjeel et al. [72] were effective in ESPNG, novel and hybrid polymeric nanofibrous mesh for dressing burn wounds after integrating gabapentin (a neuropathic pain killer) into polyethylene NFs and acetaminophen (a class of analgesics) into sodium alginate NFs (mixed in 80:20 blend proportion). In the healing process, the hybrid mechanism may be a safe option. Sharjeel et al. [73] synthesized ES - polyethylene oxide and chitosan NFs of 116 nm diameter (with a standard deviation of only 21 nm) using the response surface methodology with acetic acid and water (50:50, v/v) as the solvent (each dissolved separately in acetic acid and water solution in a 5 percent weight-to-volume ratio) (the ratio of polyethylene oxide and chitosan in the polymeric solution was 80:20).

4. Future researches

It is still a challenge to investigate the use of curcumin (Cc) loaded nanofibers (NFs) for efficient drug release during different stages of the healing process. Specific polymers for ES Cc NFs must be chosen based on the types of pharmaceutical to be released and the different stages of the healing process. That being said, the application of cytotoxic chemicals in drug delivery, especially during skin treatment, can negatively impact recent research findings. Current reviews of Cc in NFs have revealed a new area of research for the development of possible biomaterials for use in bone tissue science and medicine, diabetic chronic ulcer treatment, cancer treatment, and other applications [74–77].

5. Conclusion

Our analysis of Cc-based electrospun (ES) NFs underlines the importance, such as the relevance and need for BF - NFs and nanofibrous mats (NMs) in healing process, cancer care, tissue science and medicine, and other BAs, to inspire researchers interested in working in this cutting-edge area to solve various BAs with BF - NFs. To ease in the synthesis of UT - Cc/G NFs, the ESPNG mechanism (mathematical investigation of the process) was analyzed in detail in the first paper of the article.

The mechanism behind ESPNG was explored in this study as it was used to prepare curcumin/gelatin (Cc/G) NFs that could be used in the healing process. Gelatin (G) was chosen for the fiber system because it is not harmful to living tissues, as well as being water absorbent (fluid affinity), allowing for a moist healing process in the future. Since gelatin is commercially available at a low cost, it was an obvious option for the current study. The LW-UT and spongy NFs with mean diameter of 147 nm (147 ± 34 nm) were successfully synthesized using ESPNG at a higher power supply, such as 15 KV (at 10 cm distance, 0.15 mL/h feed rate, 65 cP solution's resistance to the flow, and drum collector speed of 1000 rpm) with a solution containing 1.5 percent G and 1 percent Cc in 10 mL of 98 percent concentrated methanoic acid (**Figure 4(c)** and **Table 2**).

We came to the following conclusions after deciding the relative effects of the different ESPNG influences: (a) the effects of ABC-Spinning gap (cm), feed rate (mL/h), and higher potential power supply interaction (KV), AD-Interaction between spinning gap (cm) and solution's resistance to the flow (cP), D-Solution's resistance to the flow (cP), BC-Interaction between feed rate (mL/h) and high potential power supply (KV), and AC-Interaction between spinning gap (cm) and high potential power supply (KV) are 24 percent, 16 percent, 15.5 percent, 12 percent, and 11.5 percent, respectively, over the preparation of the Cc/G NFs' minimum diameters; (b) BCD-Feed rate (mL/h), high potential power supply (KV), and solution's resistance to the flow interaction (cP), ABCD-Spinning gap (cm), feed rate (mL/h), high potential power supply (KV), and solution's resistance to the flow interaction (cP), AB-Interaction between feed rate (mL/h) and spinning gap (cm), CD-High potential power supply (KV) and solution's resistance to the flow interaction (cP), C-High potential power supply (KV), A-Spinning gap (cm), and B-Feed rate (mL/h), BD-Interaction between feed rate (mL/h) and solution's resistance to the flow (cP) have a major influence on the preparation of Cc/G NFs with a minimum diameter; and (c) the diameter (nm) is affected by the ACD-Interaction between spinning gap (cm), high potential power supply (KV), and solution's resistance to the flow (cP) by just 0.05 percent, which is not important.

The 2^k factorial design of the experiment was used to investigate the effects of all four important PMs on the diameter of the NFs empirically. The MINITAB 17 programme was used to generate the results to investigate the difference in NFs' diameters as a function of input parameters. The differences in NFs diameters with respect to the critical PMs that were observed included (a) a higher spinning gap yielded lower diameters, (b) a higher potential power supply yielded lower diameters, (c) the diameter of the NFs increased with an increase in feed rate, and (d) the diameters of the NFs increased with an increase in solution's resistance to the flow.

Using the optimized setting of a solution containing 1.5 percent G and 1 percent Cc in 10 mL of 98 percent concentrated methanoic acid, and the electrospining machine with a high potential power supply of 15 KV, a spinning gap from the emitter to collector drum of 15 cm, a feed rate of 0.1 mL/h, solution's resistance to the flow of 65 cP, and a drum collector speed of 1000 rpm, the optimum condition for the production of UT - Cc/G NFs with an 189.6563 nm mean diameter was calculated. The approximate mean diameter (nm) of Cc/G NFs in the optimization phase differs by just 8 percent from the prepared mean diameter, i.e., 181 nm (181 ± 66 nm), demonstrating the efficacy of the current study.

Such UT - NFs with sufficient film porosity are not harmful to living tissues in nature, and it was suggested that they could be used in dressing problematic wounds, such as diabetic chronic ulcers, because they have unique properties, such as a high surface area to volume ratio and light weight, that allow for sustained Cc release during healing. The research paper that has been presented thus far is unique in that it covers (a) the entire ESPNG process (numerical investigations of the mechanism) to improve control over the preparation of UT - NFs, and (b) the applications of NMs (incorporating BF - NFs) that are currently in use. Eventually, the ESPNG PMs were optimized (to obtain UT - NFs) to prepare NMs for BAs such as the healing process (through sustained release of Cc during crucial hours of healing).

Author details

Nand Jee Kanu[1,2*], Eva Gupta[3,4], Venkateshwara Sutar[2], Gyanendra Kumar Singh[5] and Umesh Kumar Vates[3]

1 S. V. National Institute of Technology, Surat, India

2 JSPM NTC, Pune, India

3 Amity University, Noida, India

4 AGCE, Satara, India

5 Adama Science and Technology University, Adama, Ethiopia

*Address all correspondence to: nandssm@gmail.com

References

[1] Kanu NJ, Gupta E, Vates UK and Singh GK. Electrospinning process parameters optimization for biofunctional Curcumin/Gelatin nanofibers. Materials Research Express. 2020. (doi:10.1088/2053-1591/ab7f60)

[2] Khajavi R, Abbasipour M. Electrospinning as a versatile method for fabricating core–shell, hollow and porous nanofibers. Scientia Iranica F. 2012;19: 2029–2034.

[3] Skinner JL, Andriolo JM, Murphy JP and Ross BM. Electrospinning for nano- to mesoscale photonic structures. Nanophotonics. 2017; 6: 765–787.

[4] Huang ZX, Wu JW, Wong S-C, Qu JP and Srivatsan TS. The technique of electrospinning for manufacturing core–shell nanofibers. Materials and Manufacturing Processes. 2017. (https://doi.org/10.1080/10426914.2017.1303144)

[5] Ramakrishna S, Fujihara K, Teo WE, Yong T, ZuweiM and Ramakrishna R. Electrospun nanofibers: solving global issues. Materials Today. 2006;9:40–50.

[6] Pengcheng L, Yanbo L, Jinbo and Yao A. Review on the research status of massive production of nanofibers via electrospinning technology. Proceedings of the International Conference on Information Technology and Scientific Management. Scientific Research. 2010; 326–328.

[7] Reneker DH and Chun I. Nanometer diameter fibers of polymer, produced by electrospinning. Nanotechnology. 1996; 7:216–223.

[8] Jamil AM, Gary EW, DavidGS and Gary LB. Electrospinning of collagen nanofibers. Biomacromolecules. 2002;3: 232–238.

[9] Kriegel C, Arrechi A, Kit K, McClements DJ and JochenW. Fabrication, functionalization, and application of electrospun biopolymer nanofibers. Critical Reviews in Food Science and Nutrition. 2008; 48: 775–797.

[10] Shekh R, PrincetonCand Narayan B. Aloe vera for tissue engineering applications J. Funct. Biomater. 2017; 8. DOI:10.3390/jfb8010006.

[11] Merrell JG, McLaughlin SW, Tie L, LaurencinCT, Alex FC and Lakshmi SN. Curcumin loaded poly (ε-Caprolactone) nanofibers: Diabetic wound dressing with antioxidant and anti-inflammatory properties. Clin Exp Pharmacol Physiol. 2009; 36:1149–1156.

[12] Xinyi D, Juan L, Huaiyuan Z, Johannes W, Ursula H, Stefanie S, Carina P, Yi S, Machens HG and Arndt FS. Nano-formulated curcumin accelerates acute wound healing through Dkk-1-mediated fibroblast mobilization and MCP-1-mediated anti-inflammation. NPG Asia Materials. 2017; 9: e368.

[13] Narges F, Majid D, Jebraeel M and Azadeh S. Curcumin nanofibers for the purpose of wound healing. J. Cell. Physiol. 2018; 1–18.

[14] Mouthuy PA, Škoc MS, Gašparović AČ, Milković L, Carr AJ and Zarkovic N. Investigating the use of curcumin-loaded electrospun filaments for soft tissue repair applications. International Journal of Nanomedicine. 2017; 12:3977–3991.

[15] Chang SK, Doo HB, Kyung DG, Ki HL, In C.U. and Young HP. Characterization of gelatin nanofiber prepared from gelatin-formic acid solution. Polymer. 2005; 46: 5094–5102.

[16] Deng L, Kang X, Liu Y, Feng F and Hui Z. Characterization of gelatin/zein films fabricated by electrospinning versus solvent casting. Food Hydrocoll. 2018;74:324–332.

[17] Sadaf SG, Sima H and Hosein N. Fabrication and characterization of chitosan/gelatin/thermoplastic polyurethane blend nanofibers. Journal of Textiles and Fibrous Materials. 2018;1: 1–8.

[18] Ditpon K, Masaharu H, Mayuko O, Satoshi U, Masashi I, Tetsuya F and Hiroshi T. Preparation and characterization of electrospun gelatin nanofibers for use as nonaqueous electrolyte in electric double-layer capacitor. Journal of Nanotechnology. 2019. (https://doi.org/10.1155/2019/2501039)

[19] Lin L, Gu Y and Cui H. Novel electrospun gelatin-glycerin-ε-poly-lysine nanofibers for controlling *Listeria monocytogenes* on beef. Food Packaging and Shelf Life. 2018; 18: 21–30.

[20] Bazbouz MB, Liang H and Tronci G. A UV-cured nanofibrous membrane of vinylbenzylated gelatin-poly (ε-caprolactone) dimethacrylate co-network by scalable free surface electrospinning. Materials Science and Engineering C. 2018;91: 541–555.

[21] Yabing W, Haoxuan L, Yanhuizhi F, Peilin J, Jiansheng S and Chen H. Dual micelles-loaded gelatin nanofibers and their application in lipopolysaccharide-induced periodontal disease. International Journal of Nanomedicine. 2019; 14: 963–976.

[22] Jang HJ, Kim YM, Yoo BY and Seo YK. Wound-healing effects of human dermal components with gelatin dressing .J. Biomater. Appl. 2018; 32: 716–724.

[23] Aytac Z, Ipek S, Erol I, Durgun E and Tamer U. Fast-dissolving electrospun gelatin nanofibers encapsulating ciprofloxacin/ cyclodextrin inclusion complex. Colloids Surf. B Biointerfaces. 2019; 178:129–136.

[24] Piran M, Shiri M, Soufi ZM, Esmaeili E, Soufi ZM, Vazifeh SN, Mahboudi H, Daneshpazhouh H, Dehghani N and Hosseinzadeh S. Electrospun triple-layered PLLA/gelatin. PRGF/PLLA scaffold induces fibroblast migration. J. Cell. Biochem. 2019;1–13.

[25] Novickij V, Švedienė J, Paškevičius A, Markovskaja S, Girkontaitė I, Zinkevičienė A, Lastauskienė E and Novickij J. Pulsed electric field-assisted sensitization of multidrug-resistant *Candida albicans* to antifungal drugs. Future Microbiol.2018; 13: 535–546.

[26] Christina K, Alessandra A, Kevin K, McClements DJ and Weiss J. Fabrication, functionalization, and application of electrospun biopolymer nanofibers. Critical Reviews in Food Science and Nutrition. 2008; 48: 775–797.

[27] Liu X, Nielsen LH, Klodzinska SN, Nielsen HM, Qu H, Christensen LP, Rantanen J and Yang M.Ciprofloxacin-loaded sodium alginate/poly (Lactic-Co-Glycolic Acid) electrospun fibrous mats for wound healing. Eur. J. Pharm. Biopharm. 2018; 123: 42–49.

[28] Chouhan D, Janani G, Chakraborty B, Nandi S K and Mandal BB. Functionalized PVA-silk blended nanofibrous mats promote diabetic wound healing via regulation of extracellular matrix and tissue remodeling. J. Tissue Eng. Regen. Med. 2018;12: e1559–e1570.

[29] Gizaw M, Thompson J, Faglie A and Lee SY. Electrospun fibers as a dressing material for drug and biological agent delivery in wound healing applications. Bioengineering. 2018;5:1–33.

[30] Tonda-Turo C, Ruini F, Ceresa C, Gentile P, Varela P, Ferreira AM, Fracchia L and Ciardelli G. Nanostructured scaffold with biomimetic and antibacterial properties for wound healing produced by 'green electrospinning approach Colloids Surf. B. 2018;1: 1–34.

[31] Mindru TB, Mindru IB, Malutana T and Turab V. Electrospinning of high concentration gelatin solutions. Journal of Optoelectronics and advanced Materials. 2007; 9: 3633–3638.

[32] Maleknia L and Majdi ZR. Electrospinning of gelatin nanofiber for biomedical application. Oriental Journal of Chemistry. 2014; 30: 2043–2048.

[33] Chen HC, Jao WC and Yang MC. Characterization of gelatin nanofibers electrospun using ethanol/formic acid/ water as a solvent. Polymer Advanced Technologies. 2008;20:98–103.

[34] Vitalij N, Eglė L, Gediminas S, Irutė G, Auksė Z, Jurgita Š, Algimantas P, Svetlana M and Jurij N. Low concentrations of acetic and formic acids enhance the inactivation of *Staphylococcus aureus* and *Pseudomonas aeruginosa* with pulsed electric fields. BMC Microbiology. 2019;19. (https:// doi.org/10.1186/s12866-019-1447-1.)

[35] Constantin T, Konstantinos D, Zacharias DS, Sophia H, Han Z, Liu ZL, Wyche JH and Pantazis P. Metabolism and anticancer activity of the curcumin analogue, dimethoxycurcumin. Clin Cancer Res. 2007;13:1269–1277.

[36] Ramírezagudelo R, Scheuermann K, Galagarcía A, Monteiro A, Pinzón-García AD, Cortés ME and Sinisterra RD. Hybrid nanofibers based on poly-caprolactone/gelatin/ hydroxyapatite nanoparticles-loaded doxycycline: effective anti-tumoral and antibacterial activity. Materials Science and Engineering C. 2018;83: 25–34.

[37] Wong K, Ngai S, Lee L, Goh B, Chan KG and Chuah LH. Curcumin nanoformulations for colorectal cancer: a review Front.Pharmacol. 2018;10: 152.

[38] Chen Y, Du Q, Guo Q, Huang J, Liu L, Shen X and Peng J. A W/O emulsion mediated film dispersion method for curcumin encapsulated pH-sensitive liposomes in the colon tumor treatment. Drug Dev. Ind. Pharm. 2018; 1–10.

[39] Ferri C, West K, Otero K and Kim YH. Effectiveness of curcumin for treating cancer during chemotherapy. Altern Complement Ther. 2018; 24: 13–18.

[40] Javadi S, Rostamizadeh K, Hejazi J, Parsa M and Fathi M. Curcumin mediated down-regulation of alphaV beta3 integrin and upregulation of pyruvate dehydrogenase kinase 4 (PDK4) in Erlotinib resistant SW480 colon cancer cells. Phytother Res. 2018; 32: 355–364.

[41] Marjaneh RM, Rahmani F, Hassanian SM, Rezaei N, Hashemzehi M, Bahrami A, Ariakia F, Fiuji H, Sahebkar A, Avan A, Khazaei M. Phytosomal curcumin inhibits tumor growth in colitis-associated colorectal cancer. J Cell Physiol. 2018; 233: 6785–6798.

[42] Sesarman A, Tefas L, Sylvester B, Licarete E, Rauca V, Luput L, Patras L, Banciu M and Porfire A. Anti-angiogenic and antiinflammatory effects of long-circulating liposomes co-encapsulating curcumin and doxorubicin on C26 murine colon cancer cells. Pharmacol Rep. 2018; 70: 331–339.

[43] Enas A, Clive JR, Rozita R, Kah HY and Eng KS. Pharmacokinetic and anti-colon cancer properties of curcumin-containing chitosan-pectinate composite nanoparticles. Journal of Biomaterials Science Polymer Edition. 2018. (https:// doi.org/10.1080/09205063.2018.1541500)

[44] Yasmine A, Kaoru O, Ahmed A, Hirokazu S, Kawai T, Lim CT, Kumar V, Okaya S, Kato K, Hiyama E, Yanagida T, Masujima T, Shimizu Y, Honda K. Live single cell mass spectrometry reveals cancer-specific metabolic profiles of

circulating tumor cells. Cancer Sci. 2019;110: 697–706.

[45] Rafael CC, Laura C, Gloria P, Amelia D, Juan ML-R, Consolación M and Jose P. Electrospun nanofibers: recent applications in drug delivery and cancer therapy. Nanomaterials. 2019; 9: 656. (https://dx.doi.org/10.3390%2Fnano9040656)

[46] Feifei W, Zhaoyang S, Jing Y and Lan X. Preparation, characterization and properties of porous PLA/PEG/ Curcumin composite nanofibers for antibacterial application Nanomaterials. 2019;9: 508. (https://dx.doi.org/ 10.3390%2Fnano9040508)

[47] Wang J and Windbergs M. Influence of polymer composition and drug loading procedure on dual drug release from Plga:Peg electrospun fibers. Eur. J. Pharm. Sci. 2018; 124: 71–79.

[48] Negut I, Grumezescu V and Grumezescu AM. Treatment strategies for infected wounds Molecules. 2018; 2329: 1–23.

[49] Hoang MS, Doan NH and Huynh DP. Fabrication of curcumin loaded nano polycaprolactone/chitosan nonwoven fabric via electrospinning technique. Journal of Science and Technology. 2017; 55: 99–108.

[50] Abdollahi E, Momtazi AA, Johnston TP and Sahebkar A. Therapeutic effects of curcumin in inflammatory and immune mediated diseases: a nature-made jack-of-all-trades? Journal of Cellular Physiology. 2018; 233: 830–848.

[51] Dai J, Gu L, Su Y, Wang Q, Zhao Y, Chen X, Deng H, Li W, Wang G and Li K. Inhibition of curcumin on influenza A virus infection and influenzal pneumonia via oxidative stress, TLR2/4, p38/JNKMAPK and NF-κB pathways. International Immunopharmacology. 2018; 54: 177–187.

[52] Panahi Y, Khalili N, Sahebi E, Namazi S, Simental-Mendía LE, Majeed M and Sahebkar A. Effects of curcuminoids plus piperine on glycemic, hepatic and inflammatory biomarkers in patients with type 2 diabetes mellitus: A randomized double-blind placebo controlled trial. Drug Research. 2018;68: 403–409.

[53] Zhao Y, Liu JG, ChenWM and Yu AX. Efficacy of thermosensitive chitosan/β-glycerophosphate hydrogel loaded with βcyclodextrin-curcumin for the treatment of cutaneous wound infection in rats Experimental and Therapeutic Medicine. 2018;15: 1304–1313.

[54] Mamidi N, Romo IL, Gutiérrez HML, Barrera EVand Alex EZ. Development of forcespun fiber-aligned scaffolds from gelatin–zein composites for potential use in tissue engineering and drug release. MRS Communications. 2018; 1–8.

[55] Mamidi N, Romo IL, Barrera EV and Alex EZ. High throughput fabrication of curcumin embedded gelatin-polylactic acid forcespun fiber-aligned scaffolds for the controlled release of curcumin. MRS Communications. 2018; 1–9.

[56] Ravikumar R, Ganesh M, Senthil V, Ramesh YV, Jakki SL and Eun YC. Tetrahydro curcumin loaded PCL-PEG electrospun transdermal nanofiber patch: Preparation, characterization, and in vitro diffusion evaluations. J. Drug Del. Sci. Tech. 2018; 44: 342–348.

[57] Weilan Y, Dingsheng W, Dake C, Juntao Z, Gan H, Jiang F, Liu X, Lao B, Yu W, Guan Y and Zhong G. Simultaneous determination of curcumin, tetrahydrocurcumin, quercetin, and paeoniflorin by UHPLC-MS/MS in rat plasma and its application to a pharmacokinetic study. Journal of Pharmaceutical and Biomedical Analysis. 2019; 172:58–66.

[58] Wang Y and Xu L. Preparation and characterization of porous core-shell

fibers for slow release of tea polyphenols. Polymers. 2018; 10: 144.

[59] Lu H, Qiu Y, Wang Q, Li G and Qufu W. Nanocomposites prepared by electrohydrodynamics and their drug release properties. Materials Science and Engineering C. 2018;91: 26–35.

[60] Zhang L, Wang Z, Xiao Y, Liu P, Shige W, Yili Z, Mingwu S and Xiangyang S. Electrospun PEGylated PLGA nanofibers for drug encapsulation and release. Mater. Sci. Eng.C. 2018;91: 255–262.

[61] Mostafalu P, Tamayol A, Rahimi R, Ochoa M, Akbar K, Gita K, Iman KY, Sara B, Mehmet RD, Babak Z, Sameer RS, Ali K. Smart Bandage for Monitoring and Treatment of Chronic Wounds. Small. 2018; 14: 1703509.

[62] Haley RM and Von RHA. Localized and targeted delivery of Nsaids for treatment of inflammation: a review. Exp. Biol. Med. 2018.

[63] Cheng H, Yang X, Che X, Yang M and Guangxi Z. Biomedical application and controlled drug release of electrospun fibrous materials. Materials Science and Engineering C. 2018;90: 750–763.

[64] Hao S, Zhang Y, Meng J, Liu J, Wen T, GuNand XuH. Integration of a superparamagnetic scaffold and magnetic field to enhance the wound-healing phenotype of fibroblasts. ACS Appl. Mater. Interfaces. 2018; 10 22913–22923.

[65] Abudula T, Gzara L, Simonetti G, Alshahrie A, Salah N, Morganti P, Chianese A, Fallahi A, Tamayol A, Bencherif SA, Memic A. The Effect of Poly (Glycerol Sebacate) Incorporation within Hybrid Chitin–Lignin Sol–Gel Nanofibrous Scaffolds. Materials. 2018; 11: 451.

[66] Novickij V, Zinkevičienė A, Perminaitė E, Čėsna R, Eglė L, Algimantas P, Jurgita Š, Svetlana M, Jurij N and Irutė G. Noninvasive nanosecond electroporation for biocontrol of surface infections: an in vivo study. Sci. Rep. 2018;8: 14516.

[67] Mi Y, Xu J, Tang X, Bian C, Hongliang L, Qiyu Y and Junying T. Scaling relationship of in vivo muscle contraction strength of rabbits exposed to high-frequency nanosecond pulse bursts Technol. Cancer Res. Treat. 2018; 17.

[68] Yeo YL and Friend RJ. Electrospinning carbon nanotube polymer composite nanofibers. J. Exp. Nanosci. 2006; 1 : 177–209.

[69] Rafiei S, Maghsoodlou S, Saberi M, Lotfi S, Motaghttalab V, Noroozi B and Haghi AK. New Horizons in modeling and simulation of electrospun nanofibers: a detailed review. Cellulose Chem. Technol. 2014; 48: 401–424.

[70] Mir RM, Mehdi R and Farhad S. Investigation of effect of electrospinning parameters on morphology of polyacrylonitrile/ polymethyl methacrylate nanofibers: a box-behnken based study Journal of Macromolecular Science, Part B: Physics. 2015. (https://doi.org/10.1080/00222348.2015.1042628)

[71] Vince B and Xuejun W. Effect of electrospinning parameters on the nanofiber diameter and length. Materials Science and Engineering C. 2009;29: 663–668.

[72] Montgomeryv DC. Design and Analysis of Experiments. (New York: Wiley) Student Edition. 2013. 8th Edition.

[73] Ross PJ. Taguchi techniques for quality engineering. (New York: McGraw Hill Book Company). 1989.

[74] Sharjeel A, Tanveer H, Ahsan N, Abdul Z and Nabyl K. A novel double-

layered polymeric nanofiber-based dressing with controlled drug delivery for pain management in burn wounds Polym. Bull. 2019. (https://doi.org/10.1007/s00289-019-02727-w)

[75] Sharjeel A, Tanveer H, Ahsan N, Abdul Z, Seeram R, Misbah H and Nabyl K. Enhanced antibacterial activity of PEO-chitosan nanofibers with potential application in burn infection management. Int. J. Biol. Macromol. 2019; 135: 1222–1236.

[76] Gang G, ShaoZhi F, LiangXue Z, Hang L, Min F, Feng L, ZhiYong Q and YuQuan W. Preparation of curcumin loaded poly(ε- caprolactone)-poly (ethylene glycol)-poly(ε-caprolactone) nanofibers and their in vitro antitumor activity against Glioma 9L cells. Nanoscale. 2011; 3: 3825.

[77] Sharjeel A, Tanveer H, Zulfiqar A R and Ahsan N. Current applications of electrospun polymeric nanofibers in cancer therapy Materials Science and Engineering C. 2019; 97: 966–977.

Chapter 7

Electrochemical Behavior of Cellulose Nanofibrils Functionalized with Dicyanovinyl Groups

Robson V. Pereira, Thais E. Gallina, Marcelo A. Pereira-da-Silva, Kênia S. Freitas and Aparecido J. de Menezes

Abstract

Cellulose is considered one of the most important renewable sources of biopolymers on Earth. It has attracted widespread attention due to its physical–chemical characteristics, such as biocompatibility, low toxicity, biodegradability, low density, high strength, stability in organic solvents, in addition to having hydroxyl groups, which enable its chemical modification. In this study, cellulose nanofibrils (CNFs) were functionalized with dicyanovinyl groups through nucleophilic vinylic substitution (SNV) and used as electrocatalyst in electrochemical of carbon dioxide (CO_2) reduction. Results indicate that introducing dicyanovinyl groups into the structure of nanocellulose increases electrocatalytic activity as compared to that of pure nanocellulose, shifting the onset potential of the electrochemical CO_2 reduction reaction to more positive values as compared to those for the reaction with argon. The atomic force microscopy (AFM) images show no changes in the morphology of CNFs after chemical modification.

Keywords: cellulose nanofibrils, nucleophilic vinylic substitution, electrochemical CO_2 reduction

1. Introduction

The most abundant biopolymer on Earth, cellulose displays wide chemical variability due to the functionalization capability of its hydroxyl groups, via chemical and physical reactions. Moreover, cellulose exhibits diverse morphologies, found in the hierarchical constructions that constitute plants. Cellulose is mainly found in plant cell walls, but it can also be found in other living beings, such as bacteria, fungi, and even in some species of sea mammals [1, 2].

From the standpoint of chemical structure, this biopolymer belongs to the carbohydrate group, classified as a linear polysaccharide consisting of β-D-glycopyranose units joined by β-1,4 glycosidic bonds. As a result, cellulose exhibits a high molecular weight ranging from about 50,000 to 2.5 million g/mol, depending on its source [3]. The repetitive unit of cellulose, known as cellobiosis, is a cellulose dimer (**Figure 1**).

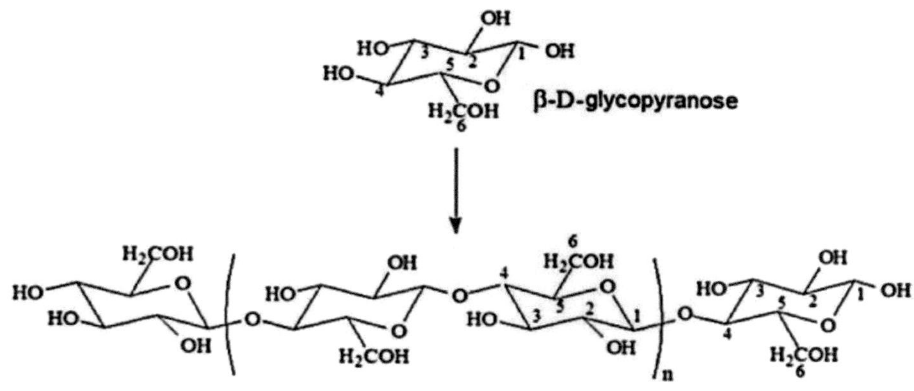

Figure 1.
Molecular structure of cellulose.

Each of these individual chains clusters into larger units, called fibrils or microfibrils, which in turn clump together and form cellulose fibers. This organization may present amorphous regions, in which fibers exhibit an undefined arrangement, or highly organized segments with fibers arranged parallel to each other. Notwithstanding other arrangements, crystalline and amorphous stretches constitute the most common fiber configurations in the polymeric structure of cellulose [1].

The degree of polymerization (DP) of this semi-crystalline biopolymer varies according to the raw material used to obtain it and the method used for its extraction. For instance, wood pulp has a DP between 10,000 and 15,000 glycosidic units, while values for cellulose of bacterial origin range from 2,000 to 6,000 [4].

At nanoscale, cellulose can be obtained by chemical or physical methods or both. Cellulose nanofibrils are usually obtained by physical methods, e.g., high shear rate mechanical treatment, whereas cellulose nanocrystals are usually obtained by chemical methods, e.g., acid hydrolysis. The difference between mechanically-produced nanocellulose and chemically-produced cellulose is that the former can reach a length of up to 2 μm, while the latter, in addition to being crystalline, exhibits a length of the order of 150 nm [5, 6].

Besides its physical–chemical properties, such as low cost, biodegradability, renewability, low toxicity, and stability in organic solvents, nanocellulose exhibits a high aspect ratio and a high specific surface area. These properties combined promote its use in nanocomposites [7–9], hydrogels and aerogels [10, 11], biomedical products [12, 13], pharmaceuticals [14], environmental applications [15], and electrochemistry. In the latter case, it is used mainly in sensors [16], transistors, and solar cells [17, 18].

The introduction of strong electron withdrawing groups such as malononitrile groups in dyes and polymers has been reported in the literature [19, 20], and the presence of these groups leads to Intramolecular Charge Transfer (ICT), increasing the electron density in dicyano groups.

Electrochemical CO_2 reduction is not only an effective way of lowering CO_2 concentrations in the atmosphere, but it is also advantageous. CO_2 can be effectively reduced to renewable fuels, such as ethanol, methane, and methanol, which can contribute to meeting today's growing demand for renewable energy sources [21, 22]. So, some studies on CO_2 reduction catalysis have used conducting polymers as polyethylenimine (PEI) [23], and polyaniline (Pan) [24] that causes an effect of reducing catalytic overpotential and increasing current density and efficiency, besides increased of selectivity for CO_2 reduction was observed for Cobalt phthalocyanine with poly-4-vinyl pyridine polymers (CoPc – P_4VP) [25].

In this study, cellulose nanofibrils were functionalized with dicyanovinyl groups, from use ethoxymethylene-malononitrile (EMMN) as chemical modifier, for use in the electroreduction CO_2, whose excessive presence in the environment can cause serious problems, such as the greenhouse effect and, consequently, climate change [26, 27].

2. Method

2.1 Nanocellulose functionalization

In a flask, 0.5 g (3.1 mmol) by mass of an aqueous dispersion of 3% w/v CNFs (SuzanoPapel & Celulose) was placed under agitation. At that point, sodium hydroxide (NaOH) solution (0.1 M; Vetec; 97%) was added by means of a pipette (dropwise) until pH 10 was reached. The mixture was left under agitation for 30 minutes. Afterwards, EMMN (1.14 g; 9.3 mmol; Sigma Aldrich; 98%) was added to the mixture and left to react at room temperature (**Figure 2**), varying the reaction time and keeping the stoichiometry at 1:3 molar ratio (nanocellulose:malononitrile). The effect of stoichiometry and temperature on the reaction efficiency was evaluated for the best experimental condition.

After the programmed reaction time, the reaction medium was placed in a sintered glass funnel (no. 4) and rinsed with acetone (Vetec; 99.5%), ethanol (Vetec; 99.5%), methanol (Vetec; 99.8%), and distilled water until neutral pH was reached. The sample was then placed in an amber glass bottle and stored in a refrigerator.

2.2 Atomic force microscopy

AFM was conducted on a Dimension ICON microscope (Bruker). The sample was prepared by dripping 5 microliters of a solution containing the CNFs on a mica surface. The mica was cleaved twice right before applying the solution dropwise onto the surface and left to dry for 1 hour at room temperature. To prevent CNFs from dragging, intermittent contact mode with a rectangular silicon probe was used (cantilever spring constant = 40 N/m; oscillation frequency = 330 kHz).

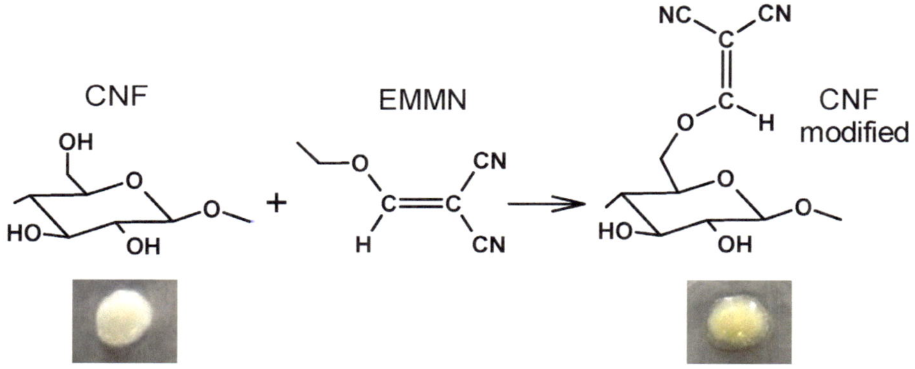

Figure 2.
CNF functionalization with EMMN.

2.3 Elementary analysis

The content of carbon (C), hydrogen (H), and nitrogen (N) in samples of pure and modified nanocellulose was determined by elementary analysis with a Perkin Elmer model 2400 instrument.

2.4 Thermal analysis

The thermogravimetric analysis (TG) of pure and modified nanocellulose was performed on a TG-DSC (Netzsch, model STA 409; PC – Luxx), with 50 mL min-1 nitrogen flow, 25–720 °C analysis interval, and 10 °C min^{-1} heating rate.

2.5 Electrochemical analysis

All electrochemical measurements were performed in a conventional three-electrode electrochemical cell. A platinum plate and Ag/AgCl were used as counter electrode and reference electrode, respectively. The working electrode comprised an ultrathin layer of catalyst (pure and modified nanocellulose) under the pyrolytic graphite layer (0.070 cm and 23.0 mm diameter) of a rotating disk electrode (RDE).

The 1% w/v aqueous suspension of pure and modified nanocellulose was prepared by ultrasonic dispersion in methanol. A 10 μL aliquot of this suspension was pipetted onto the surface of the pyrolytic graphite substrate. Then, the solvent was left to evaporate in a desiccator. Later, a 10 μL aliquot of Nafion® solution was pipetted onto the catalytic layer in order to attach the polymer layer to that of pyrolytic graphite.

The electrochemical behavior of pure and modified nanocellulose was monitored by means of cyclic voltammetry and polarization curves. The potentials applied to the electrodes during the assays were controlled by an Autolabpotentiostat/galvanostat. The electrolyte was saturated with pure argon (Ar) and CO_2 depending on the assay. Polarization curves were obtained using an RDE with potential ranging from −2.0 to 1.0 V vs. Ag/AgCl and a scanning rate of 5 mV s-1.

3. Results and discussion

3.1 Atomic force microscopy

AFM images, **Figure 3**, reveal that the samples are organized as bundles of nanofibrils. In some places, individual nanofibrils can be found on the mica surface, which enabled measuring their diameter. **Figure 3a** and **c** show the relief images of the nanofibrils before and after chemical surface modification whereas **Figure 3b** and **d** show 3D AFM images. This analysis indicates that the average diameters of original nanofibrils and modified nanofibrils are 6.6 ± 1.6 nm and 5.6 ± 1.1 nm, respectively.

Overall, AFM results suggest that chemical modification has not changed the morphology of the nanofibrils, which exhibit diameters in the order of 6.0 nm.

3.2 Elementary analysis

Table 1 shows CHN results at several reaction times for the functionalization of CNFs with EMMN. This reaction occurs through nucleophilic vinylic substitution (SNV) of the ethoxy group with the malononitrile group in the polymer chain [28].

As shown in **Table 1**, the 8-hour reaction (Reaction 3) yielded the highest nitrogen content and is, therefore, the most effective in functionalizing and

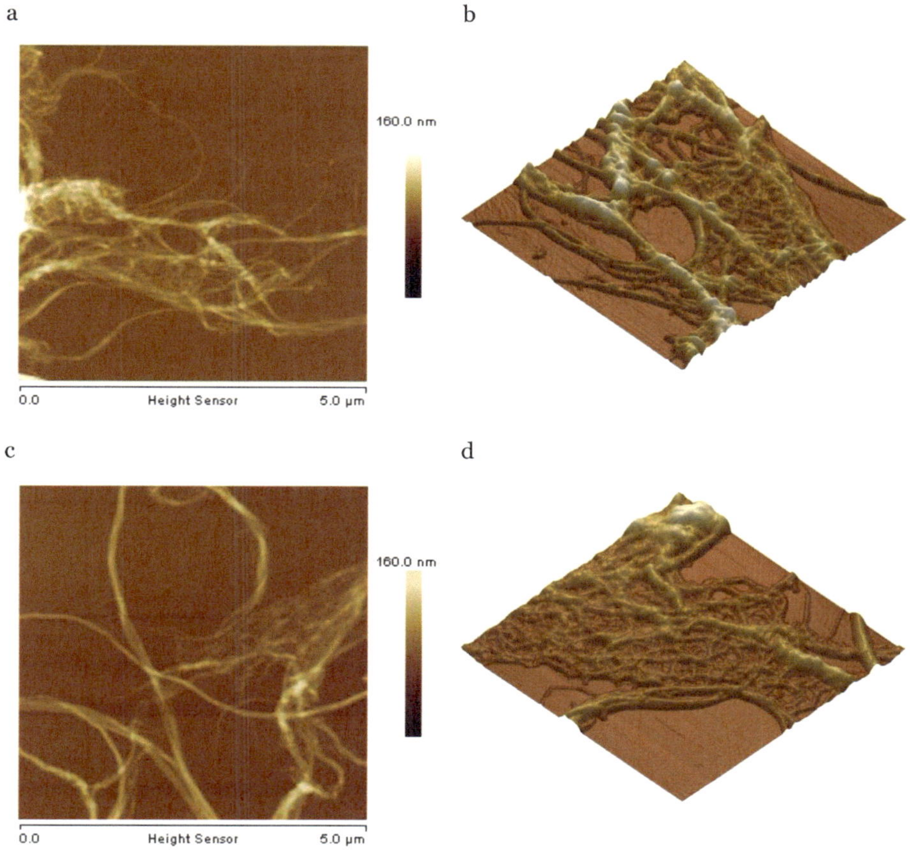

Figure 3.
AFM for pure and modified CNFs in relief (a and c) and 3D (b and d - 2 microns × 2 microns × 15 nm), respectively.

Sample	Time (h)	C (%)	H (%)	N (%)
Pure nanocellulose	—	40.55	6.13	0.017
Reaction 1	2	40.66	5.69	0.280
Reaction 2	4	41.23	6.07	0.810
Reaction 3	8	38.86	5.50	0.920
Reaction 4	24	39.39	6.13	0.760

Table 1.
Elementary analysis results for several times of reaction between CNFs and EMMN at 1:3 molar ratio and room temperature.

incorporating the malononitrile (dicyanovinyl) group into the nanocellulose chain. Increasing the reaction time to 24 h led to a decrease in nitrogen content, probably due to compound degradation.

For the best experimental condition (Reaction 3), the effect of stoichiometry and temperature on reaction yield was investigated. In this case, the same conditions used in Reaction 3 were used, with 1:2 stoichiometry at room temperature and, subsequently, 1:3 stoichiometry at 70 °C. Both assays exhibited a decrease in nitrogen content.

A more accurate way to measure reaction yield is by estimating the degree of substitution (DS), which can be obtained from Eq. (1):

$$DS = \frac{M_{glu} \cdot \%N}{100M_N - M_{mal} \cdot \%N} \tag{1}$$

Where DS is the degree of substitution, Mglu the molar mass of the glucose monomer (162 g/mol), MN the molar mass of the nitrogen atom, Mmal the molar mass of the malononitrile group introduced into the cellulose (77 g), and %N the nitrogen content determined by elementary analysis.

By means of Eq. (1), Reaction 3 exhibits the highest DS value: 0.12. Despite not being very high, this value is close to those reported in the literature for reactions in which amino groups are introduced into the nanocellulose chain [29, 30]. For instance, the nitrogen content found in the functionalization reaction of cellulose nanocrystals with propargylamine was 0.79% [29]. Another study involving nanocellulose amination with 2-hydroxy-3-chloro-propylamine yielded a nitrogen content of 0.9% and a degree of substitution of 0.11 [30].

3.3 Thermal analysis

Figure 4 shows the thermogravimetric analysis for pure and modified CNFs as well as the derivatives of the thermogravimetric curves (dTG). The thermal behavior of both materials exhibits a single decomposition event between 305 °C and 390 °C, with pure CNFs exhibiting greater loss of mass at the end of the process (**Figure 4**). The peak corresponding to maximum mass loss for the functionalized CNFs occurs at a temperature approximately 20 °C lower (Tmax = 349.05 °C) than that for pure CNFs (Tmax = 369.34 °C), as shown in **Figure 3**. Similarly, the beginning of the decomposition process for the modified CNFs also occurs at a temperature approximately 20 °C lower (Tinitial = 306.21 °C) than that for pure CNFs (Tinitial = 327.87 °C).

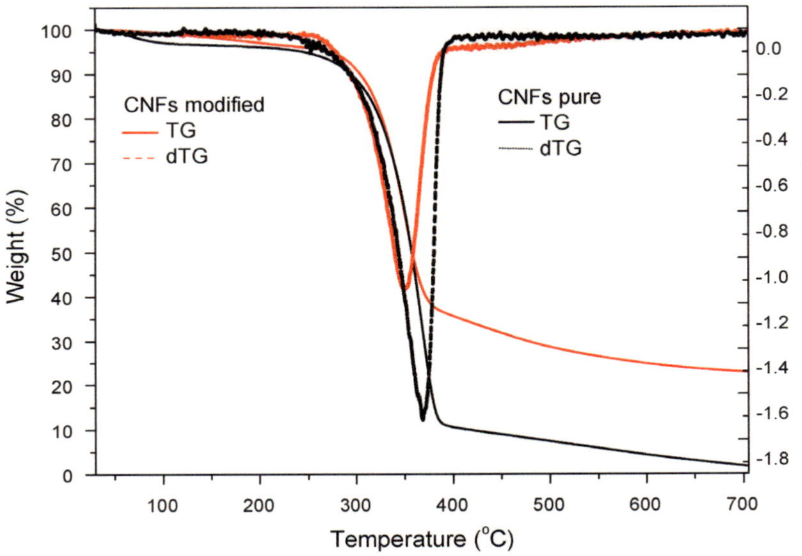

Figure 4.
TG and dTG curves for CNFs pure and modified.

The drop in the decomposition temperature of the modified CNFs may be due to a decrease in crystallinity when the dicyanovinyl group was introduced. There are reports in the literature of cellulose exhibiting lower thermal resistance when carbamate groups are introduced, which indicates a decrease in thermal stability of the functionalized cellulose as compared to that of pure cellulose [31].

3.4 Electrochemical analysis

Figure 5 shows cyclic voltammetry profiles for the CNFs electrocatalysts with and without mode modification at 5 mv/s scanning rate and applied potential ranging from −1.5 to 1.5 (vs Ag/AgCl). Conductivity of the electrocatalyst increases when cyan groups are introduced, as shown by the increase in area and current density.

This increase in conductivity has a positive effect concerning the use of the electrocatalyst as cathode in CO_2 reduction.

CO_2 conversion, whether thermal or electrochemical, is associated with high energy consumption due to CO_2 being a very stable molecule. In the case of electrochemical reduction of CO_2, the source of energy is electricity. It is possible to reduce CO_2 completely by applying a higher potential. However, an appropriate catalyst can significantly reduce energy consumption and increase end-product selectivity.

Figure 6 shows the polarization curves for pure and modified CNFs in an atmosphere of Ar and CO_2. It is possible to observe that CNFs modification with cyan groups increases current density of the CO_2 reduction reaction, which implies a higher CO_2 conversion rate in the products. It also points to the onset potential for CO_2 reduction shifting to more positive values as compared to those observed for pure CNFs. This may be attributed to adsorption/desorption of reaction intermediates in the polymer interface due to the presence of the cyan group.

The use of CNFs modified by the dicyan group has improved the catalytic efficiency of the electrocatalyst, thereby promoting CO_2 reduction, probably due to higher availability of active sites in its fibrillar structure, especially cyan groups on the surface.

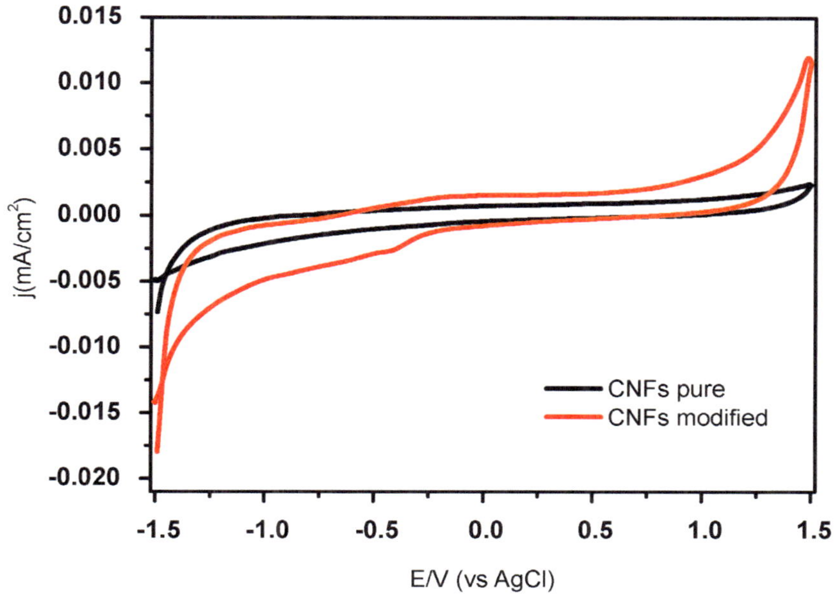

Figure 5.
Cyclic voltammetry for CNFs pure and modified at a scanning rate of 5 mV s⁻¹; electrolyte: K2SO4 0.5 Mol/L saturated with Ar at 25 °C. currents normalized by the geometric area of the electrode.

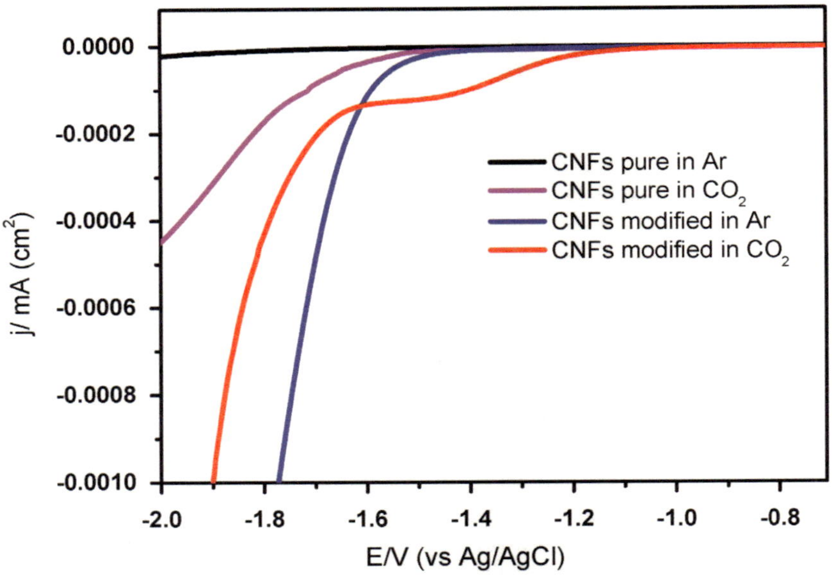

Figure 6.
Steady-state polarization curves for electrodes containing CNFs pure and modified at a scanning rate of 5 mV s^{-1}, K_2SO_4 0.5 Mol/L saturated with Ar and CO_2 at 25 °C. currents normalized by the geometric area of the electrode.

4. Conclusion

AFM results indicate no significant changes in the morphology of modified cellulose nanofibrils. The best experimental conditions for chemical functionalization is 1:3 molar ratio at room temperature. As reported in other studies on chemically modified cellulose, the degradation temperature for modified cellulose nanofibrils is lower than that for pure cellulose.

From the electrochemical perspective, introducing dicyanovinyl groups into the polymeric chain of cellulose nanofibrils has led to significantly higher electrocatalytic activity in CO_2 reduction as compared to that for pure cellulose nanofibrils, shifting the onset potential to more positive values as compared to that observed for the reaction with Ar. Moreover, the CO_2 molecule exhibits affinity for the cellulose polymer and the polymeric layer may play an inhibitory role in water reduction.

Acknowledgements

The authors thanks the Program of Post Graduation in Science of Materials - PPGCM of the Universidade Federal de São Carlos - UFSCar, campus Sorocaba for financial support.

Conflict of interest

The authors declare no conflict of interest.

Author details

Robson V. Pereira[1], Thais E. Gallina[1], Marcelo A. Pereira-da-Silva[2], Kênia S. Freitas[1] and Aparecido J. de Menezes[3]*

1 Grupo de Eletroquímica e Polímeros Naturais (GEPN), Universidade Federal do Rio de Janeiro - UFRJ, Campus Macaé, Macaé - Rio de Janeiro, Brasil

2 Instituto de Física de São Carlos, IFSC, Universidade de São Paulo - USP, São Carlos, São Paulo, Brasil

3 Grupo de Polímeros de Fontes Renováveis - GPFR, Universidade Federal de São Carlos - UFSCar, Campus Sorocaba, Sorocaba, São Paulo, Brasil

*Address all correspondence to: jrmenezes@ufscar.br

References

[1] Habibi Y. Lucia L. A. Rojas O. J. Cellulose Nanocrystals: Chemistry, Self-Assembly, and Applications. Chem. Rev. 2010;110:3479-3500. DOI: 10.1021/cr900339w

[2] Alavi M. Modifications of microcrystalline cellulose (MCC), nanofibrillated cellulose (NFC), and nanocrystalline cellulose (NCC) for antimicrobial and wound healing applications. e-Polymers. 2019;19:103-119. DOI: 10.1515/epoly-2019-0013

[3] Fengel D. Wegener G. Wood: chemistry, ultrastructure and reactions. Berlin: Walter de Gruyter; 1984. 613 p.

[4] Blanco A. et al. Nanocellulose for Industrial Use: Cellulose Nanofibers (CNF), Cellulose Nanocrystals (CNC), and Bacterial Cellulose (BC). In: Hussain C. M. Handbook of Nanomaterials for Industrial Applications. 2018. p. 74-126. DOI: 10.1016/b978-0-12-813351-4.00005-5

[5] Abitbol T. et al. Nanocelulose, a tiny fiber with huge applications, Curr. Opin. In Biotech. 2016;39:76-78. DOI: 10.1016/j.copbio.2016.01.002

[6] Thompson L. et al. Cellulose nanocrystals: Production, functionalization and advanced applications. Rev. Adv. Mater. Sci. 2019;58: 1-16. DOI: 10.1515/rams-2019-0001

[7] Dufresne A. Cellulose nanomaterial reinforced polymer nanocomposites. Current Opinion in Colloid & Interface Science, 2017;29:1-8. DOI: 10.1016/j.cocis.2017.01.004

[8] Gan P. G. et al. Thermal properties of nanocellulose-reinforced composites: A review, J. Appl. Polym. Sci. 2020;137:48544-48558. DOI: 10.1002/app.48544

[9] Mondal S. Review on Nanocellulose Polymer Nanocomposites. Polymer-Plastics Techn. and Eng. 2018;57:1377-1391. DOI: 10.1080/03602559.2017.1381253

[10] France K. J. Hoare T. Cranston, E. D., Review of hydrogels and aerogels containing nanocellulose, Chem. Mater. 2017;29:4609-4631. DOI: 10.1021/acs.chemmater.7b00531

[11] Long L. Y. Weng Y. X. Wang Y.Z. Cellulose aerogels: Synthesis, applications and prospects. Polymers. 2018;10:1-28. DOI: 10.3390/polym10060623

[12] Patel D.K. Duttab S. D. Lim K. T. Nanocellulose-based polymer hybrids and their emerging applications in biomedical engineering and water purification. RSC. Adv. 2019;9:19143-19162. DOI: 10.1039/C9RA03261D

[13] Ngwabebhoh F. A. Yildiz U. Nature-derived fibrous nanomaterial toward biomedicine and environmental remediation: Today's state and future prospects. J. Appl. Polym. Sci. 2019;136:47878. DOI: 10.1002/app.47878

[14] Akhlaghi S. P. Berry R. C. Tam K. C., Surface modification of cellulose nanocrystal with chitosan oligosaccharide for drug delivery applications, Cellulose. 2013;20:1747-1764. DOI: 10.1007/s10570-013-9954-y

[15] Liu L. et al. Adsorption removal of dyes from single and binary solutions using a celulose-based bioadsorbent. ACS Sustainable Chem. Eng. 2015;3:432-442. DOI: 10.1021/sc500848m

[16] Zhang Y. et al. Electrochemical chiral sensor based on cellulose nanocrystals and multiwall carbono nanotubes for discrimination of tryptophan enantiomers. Cellulose. 2018:25:3861-3871. DOI: 10.1007/s10570-018-1816-1

[17] Cheng Q. et al. Construction of transparente celulose-based nanocomposite papers and potential application in flexible solar cells, ACS Sustainable Chem. Eng. 2018;6:8040-8047. DOI: 10.1021/acssuschemeng.8b01599

[18] Hsu H. H. Zhong W. Membranes for Free-Standing Supercapacitors Nanocellulose-Based Conductive: A Review. Membranes. 2019;9:1-21. DOI: 10.3390/membranes9060074

[19] Pereira R. V. Ferreira A. P. G. Gehlen M. H. Fluorescent probes with malononitrile side group in methyl methacrylate copolymers. J. of Photochem. and Photobiol. A, Chem. 2008,198:69-74. DOI: 10.1016/j.jphotochem.2008.02.017

[20] Pereira R. V. Gehlen M. H. Polymerization and Conformational Transition of Poly(methacrylic Acid) Probed by Electronic Spectroscopy of Aminoacridines. Macromolecules 2007,40:2219-2223. DOI: 10.1021/ma062020c

[21] Karamad M. et al. Mechanistic Pathway in the Electrochemical Reduction of CO2 on RuO2. ACS Catalysis. 2015;5:4075-4081. DOI: 10.1021/cs501542n

[22] Jhong H. M. Ma S. Kenis P. Electrochemical conversion of CO2 to useful chemicals: current status, remaining challenges, and future opportunities. Curr. Op. in Chem. Eng. 2013;2:191-199. DOI: 10.1016/j.coche.2013.03.005

[23] Zhang S. et al. Polyethylenimine-Enhanced Electrocatalytic Reduction of CO2 to Formate at Nitrogen-Doped Carbon Nanomaterials. J. Am. Chem. Soc. 2014;136:7845-7848. DOI: 10.1021/ja5031529

[24] Aydin R. Köleli F. Electrochemical reduction of CO2 on a polyaniline electrode under ambient conditions and at high pressure in metanol. J. Electroanal. Chem. 2002;535:107-112. DOI: 10.1016/S0022-0728(02)01151-8

[25] Liu Y. McCrory C.C.L. Modulating the mechanism of electrocatalytic CO2 reduction by cobalt phthalocyanine through polymer coordination and encapsulation. Nat. Commun. 2019;10:1683 DOI: 10.1038/s41467-019-09626-8

[26] Hu J. P. et al. Sensitivity analysis of greenhouse effect with the concentration changes of greenhouse gases. J. of Eng. Thermophysics. 2012;33:1380-1382.

[27] Goldemberg J. Energia e Desenvolvimento Sustentável. São Paulo: Blucher, 2010; 4: 124-126.

[28] Salon J. et al. Nucleophilic vinylic substitution (SNV) of activated alkoxymethylene derivatives with 6-aminoquinoxaline. Eur. J. Org. Chem. 2005; 4870-4878. DOI: 10.1002/ejoc.200500298

[29] Filpponen I., Sadeghifar H., Argyropoulos, D. S. Photoresponsive cellulose nanocrystals, Nanomater. Nanotechnol. 2011;1:34-43. DOI: 10.5772/50949

[30] Akhlaghi S P. et al. Synthesis of amine functionalized cellulose nanocrystals: optimization and characterization, Carbohyd. Res. 2015;409: 48-55. DOI: 10.1016/j.carres.2015.03.009

[31] Vo L.T. et al. Functionalisation of cellulosic substrates by a facile solventless method of introducing carbamate groups. Carbohyd. Polym. 2010;82:1191-1197. DOI: 10.1016/j.carbpol.2010.06.052

Chapter 8

The Nanocellulose Fibers from Symbiotic Culture of Bacteria and Yeast (SCOBY) Kombucha: Preparation and Characterization

Pingkan Aditiawati, Rudi Dungani, Salsabila Muharam, Aminudin Sulaeman, Sri Hartati, Mustika Dewi and Enih Rosamah

Abstract

Symbiotic Culture of Bacteria and Yeast (SCOBY) is a by-product in the form of cellulose polymers produced by bacteria in the kombucha fermentation process. Until now, SCOBY products still have application limitations. Several world designers have succeeded in making works using fabrics based on SCOBY. The resulting fabric has a flexible texture and is brown like synthetic leather. Fabrics based on SCOBY are also considered cheap and more environmentally friendly with short production time. The use of SCOBY as a fabric base material still has problems, where the fabric produced from SCOBY kombucha, directly through the drying process, has the characteristic of being very easy to absorb water. Another problem is that SCOBY production in the kombucha fermentation process is difficult to achieve a uniform thickness and SCOBY production in a large surface area is also difficult to stabilize. The development of SCOBY into cellulose fibers can be done by first changing the structure of SCOBY into nanocellulose. This nanocellulose production can then be developed into nanocellulose fibers in the form of threads and then spun to become a complete fabric. The production of nanocellulose is carried out using cellulase enzymes. It is known that cellulase enzymes can be obtained through the growth of bacteria or specific fungi. One of the groups of fungi and bacteria commonly used to produce cellulase enzymes are *Trichoderma* and *Bacillus.*

Keywords: kombucha, *Trichoderma*, *Bacillus*, cellulases, nanocellulose

1. Introduction

Wood cellulose is found in the form of cellulose bundles that stick together due to bonds by lignin. Bonding between cellulose and lignin can occur with hemicellulose intermediates. Cellulose can be found in two primary structures, namely amorphous and crystalline structures [1]. In the industrial sector, cellulose is applied in the form of cellulose fibers. Cellulose fibers are known to be used as raw material for making fabrics in the textile industry. Until now, the main production of cellulose fiber still depends on the cultivation of cotton plants. Production of cellulose fibers through this method is known to require a production time of around 4–5 months. Cellulose fibers have several advantages, including the resulting cellulose fibers can have unique properties depending on the type of tree used as the source of cellulose, have high mechanical strength and high flexibility. However, cellulose fibers produced from tree cellulose also have drawbacks, namely requiring long stages and a long time in the production process. Several stages that must be passed when carrying out the cellulose fiber production process, including the kraft cooking, bleaching, delignification, and spinning stages [2].

Based on the prediction data of FAO (2016), the demand for cellulosic fiber production will increase by around 1.5% per year, to reach the demand of 28.3 million tonnes in 2025. This prediction is inversely proportional to the prediction that the stock of cellulosic fiber production in the world will decline. This data is predicted to occur due to an imbalance between the speed of cellulosic fiber production and the speed of demand for clothing in the world.

Currently, researchers are developing alternative production methods to meet the deficit in cellulose production. One of the alternative methods being developed is the production of cellulose from bacteria. This method is considered to be able to help the deficit in cellulose production, due to the production process which requires a relatively shorter time. Production of cellulose from bacteria is known to be carried out by groups of acetic acid bacteria, such as *Acetobacter xylinum* or *Gluconobacter sp*. The production of cellulose fibers from bacterial cellulose also has several advantages such as the purity of cellulose in bacterial cellulose which is higher ~90%, does not contain lignin and hemicellulose and can be produced in various substrates which cause lower production costs [3].

Symbiotic Culture of Bacteria and Yeast (SCOBY) Kombucha is a cellulose product from bacteria that is considered a potential substitute for cotton for fabric raw materials. SCOBY is known to be a byproduct in the kombucha industry, which is currently experiencing limited application development. Currently, several world designers have succeeded in making works using fabrics based on SCOBY. The resulting fabric has a flexible texture and is brown like synthetic leather. Fabrics based on SCOBY are also considered cheap and more environmentally friendly because they are easily degraded by the environment [4].

Until now, the use of SCOBY as a fabric base still has problems, where the fabric produced from SCOBY kombucha, directly through the drying process, has the characteristic of being very easy to absorb water. This characteristic is a drawback for SCOBY based fabrics, because the water bound in SCOBY based fabrics is difficult to dry and can make SCOBY return to its original shape. Another problem with the use of SCOBY directly as a fabric base material, is that the production of SCOBY in the kombucha fermentation process is difficult to achieve uniform thickness and SCOBY production in a large surface area is also difficult to stabilize [4].

This nanocellulose production can then be developed into nanocellulose fibers in the form of threads and then spun to become a complete fabric. This method is expected to make fabrics from SCOBY to have characteristics that are more resistant to water. The more uniform structure of the nanocellulose can also make the nanocellulose fibers stronger and more compact so that they can be developed as fabrics with special needs, such as bullet-proof fabrics in the military field [5].

The manufacture of cellulose can be done in three ways, namely mechanically, acid hydrolysis and the help of cellulase enzymes. It is known that cellulase enzymes can be obtained through the growth of bacteria or specific fungi. One of the groups of fungi and bacteria commonly used to produce cellulase enzymes are *Trichoderma* and *Bacillus*. Both groups of fungi and bacteria are degrading cellulose microbes that commonly have habitats in soil [6].

2. Source of cellulose in nature

Naturally, cellulose fibers are most commonly found in plants. Cellulose fibers have an important role in the formation of cell walls in plants. It is known that most of the layers in the plant cell wall can be formed firmly due to the presence of the cellulose microfibrils (CMF) structure which is bound to each other between the cell wall layers. CMF can consist of 30–100 cellulose nanofibrils macromolecules with a modified 1,4-glycosidic extended chain bond, with a diameter ranging from 10-30 nm. CMF in plants naturally binds to hemicellulose through hydrogen bonds (**Figure 1**). This bond occurs to strengthen the structure of plant cell walls, where hemicellulose is known to act as a stabilizer between lignin and cellulose bonds [7].

Apart from plants, bacteria are also known to produce cellulose fibers well. Cellulose fibers produced from bacteria are known as Bacterial Cellulose (BC). BC is one of the primary metabolites produced by acetic acid bacteria, for example the genus of bacteria and *Acetobacter*. The acetic acid bacteria group is known to form a thick gel consisting of CMF and water, under certain fermentation conditions. The degree of polymerization that BC has is between 2000 and 6000. BC has several advantages over cellulose fibers in plants, including BC has a higher purity level, where BC does not contain hemicellulose and lignin. The characteristics of BC can also be modified into certain characters based on the content of microfibrils and cellulose crystallization, by modifying the fermentation conditions of acetic acid bacteria. The production of BC is known to require a shorter time than the production of cellulose fibers in plants, however BC and cellulose fibers in plants have the same molecular structure [8].

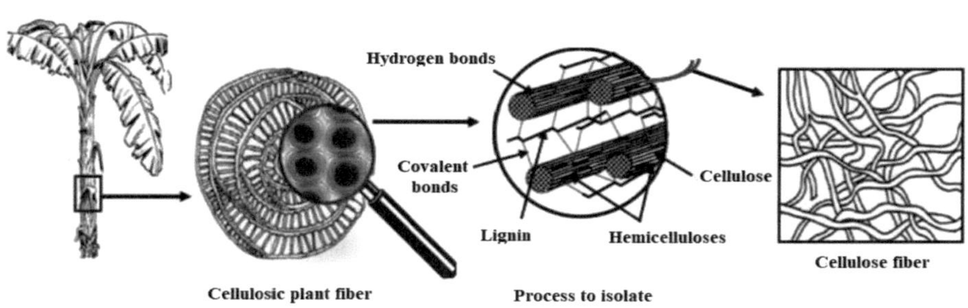

Figure 1.
Structure of cellulose in plants [7].

3. Delignification

The delignification process involves at least 3 types of enzymes, namely: lignin peroxidase, manganese peroxidase, and *lacase* [9]. Lignin peroxidase and manganese peroxidase are enzymes that depend on hydrogen peroxide. In the delignification process, there are at least four mechanisms carried out by the enzyme, namely: breaking ether bonds between monomers; cutting propane side chains; de-methylation; cleans benzene bonds to ketoadipic acid to enter the TCA cycle [10, 11],

Figure 2 shows the delignification process by the lignin peroxidase (LiP) enzyme. This enzyme is an enzyme that depends on the availability of hydrogen peroxide.

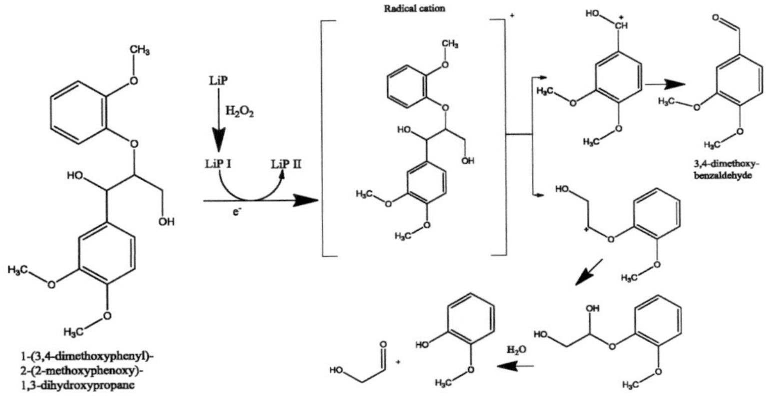

Figure 2.
LiP enzyme delignification mechanism [12].

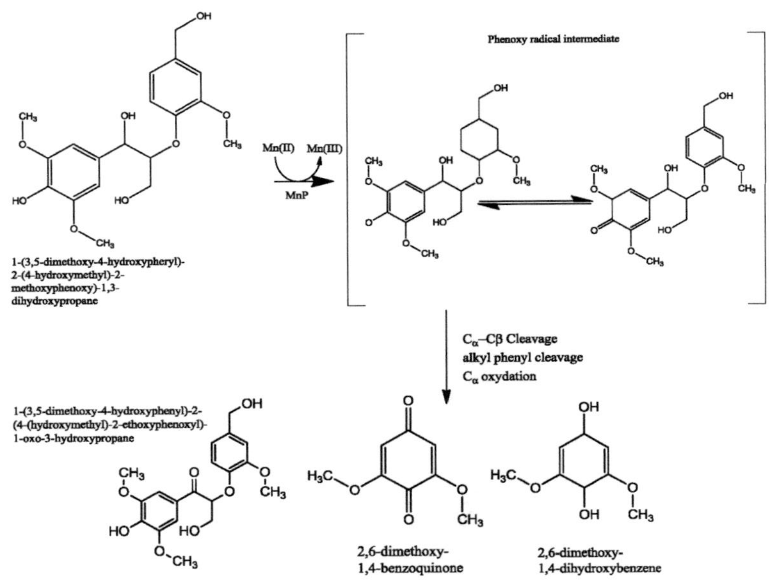

Figure 3.
MnP enzyme delignification mechanism [12].

This enzyme will change the 1- (3,4-dimethoxyphenyl) -2- (2-methoxyphenoxy) -1,3-dihydroxypropane group into a radical cation so that it is less stable and causes the breaking of the bond to become a 3,4-dimethoxy-benzaldehyde compound [12].

Figure 3 shows the working process of the Manganese Peroxidase (MnP) enzyme. This enzyme is an enzyme that depends on the availability of Mn2 + ions. This enzyme works by converting 1- (3,5-dimethoxy-4-hydroxyphenyl) -2- (4- (hydroxymethyl) -2-methoxyphenoxy) -1,3-dihydroxypropane into radical phenolic compounds so that it is unstable which causes its formation compounds such as 2,6-dimethoxy-1,4-dihydroxybenzene.

Figure 4 shows the action of the laccase enzyme. This laccase enzyme directly converts phenolic compounds into simpler compounds without going through intermediates. This process can go through 3 pathways, namely: cleavage of C alpha and beta, addition of alkyl groups, and C alpha oxidation.

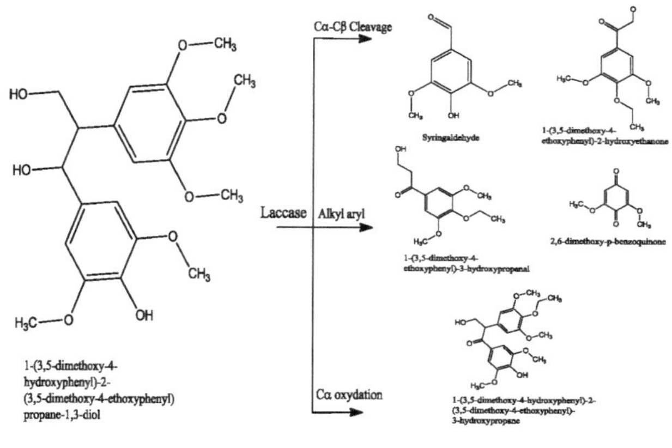

Figure 4.
The mechanism of delignification of the laccase enzyme [12].

4. Cellulose in symbiotic culture of bacteria and yeast (SCOBY)

Symbiotic Culture of Bacteria and Yeast (SCOBY) is a cellulose biopolymer composed of the interaction of acetic acid bacteria and yeast. SCOBY can be formed through the kombucha fermentation process [13]. The cellulose formed in SCOBY has different characteristics with cellulose in plants. Cellulose that is synthesized through bacteria is considered more efficient and effective in the production process because it does not require a long time and a large amount of substrate [14]. Some of the advantages of producing cellulose from bacteria compared to plants include high purity, better mechanical strength, a higher polymerization rate and crystallinity index [15], a higher tensile strength based on a tensile test and a better hydrophobicity ability to water [16].

Bacterial cellulose has a basic structure of microfibrils with a glucan chain arrangement that is bound by hydrogen bonds to form a crystalline domain. Microfibrils in cellulose synthesized by bacteria are known to have a size 100 times smaller than plant cellulose fibers [17]. Electron microscopy observations show that the cellulose produced by bacteria will be synthesized in the form of cellulose fibers. Acetic acid bacteria produce two forms of cellulose, namely cellulose I in the form of a ribbon-like polymer and cellulose II in the form of an amorphous polymer which is more stable. The difference in the synthesis of cellulose I and II is in the process of forming cellulose outside the cytoplasmic membrane (**Figure 5**).

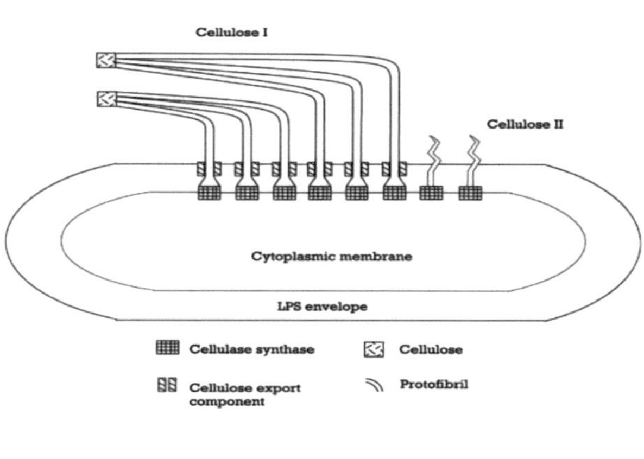

Figure 5.
Synthesis mechanism of cellulose I and cellulose II by Acetobacter xylinum *[16].*

Cellulose I is synthesized to form cellulose complexes that are linked to one another outside the cytoplasmic membrane, while cellulose II is formed to resemble free cellulose fibers outside the cytoplasmic membrane. The structure content of cellulose I and II will affect the tensile strength, polymerization rate and crystallinity index of cellulose fibers. The microfibrils produced by *Acetobacter xylinum* have dimensions of about 3-4 nm in length and 70–80 nm in width [16].

Cellulase is an enzyme that can degrade cellulose by breaking the 1,4-glycosidic bonds in cellulose polymers. Naturally, cellulases can be obtained in nature as metabolites of microbial metabolism, such as bacteria and fungi. Microbes that can produce cellulase enzymes usually have habitats in the soil, where these microbes play a role in the degradation of cellulose in plants. Cellulase is known to be one of the most widely used enzymes in the industrial sector, such as bio-stoning in the textile industry, extraction of fruit and vegetable juices in the food industry, and bleaching in the paper industry [18]. In the manufacture of nanocellulose, cellulase can also be used to degrade the structure of cellulose fibrils into crystalline, so that it can change the size of cellulose to nanocellulose.

4.1 Screening method for cellulase producing bacteria

Screening of bacteria that can produce cellulase can be done using CMC (Carboxymehylcellulose) medium. CMC is a cellulose derivative which is commonly used as a thickener or stabilizer in the industrial field [19]. The composition of the CMC medium includes CMC, yeast extract, MgSO4, NH4H2PO4, and KCl. Bacteria that are thought to be able to produce cellulase are cultivated first on CMC agar medium, under certain conditions. The growth of bacterial colonies on CMC medium can be an early marker of cellulase activity in bacteria. Qualitative confirmation of cellulase activity in bacteria can be done by testing 1% Congo Red and 1 M NaCl. The formation of a clear zone that occurred around the bacterial colony after testing the Congo Red can be used as a qualitative positive result of cellulase activity in bacteria. The formation of this clear zone indicates that bacteria can hydrolyze cellulase contained in the medium to simple sugar (glucose) [18].

4.2 Cellulase catalysis mechanism

The endogilanase randomly acts to cut the 1,4-glycosidic bonds so that the cellulose chain has a new end. Endoglucanases produced by bacteria, fungi, animals and plants have different catalyst modules. In fungi, the endoglucanase produced generally has a catalytic module without carbohydrate-binding module (CBM), while the endogluanase in bacteria is generally supplemented with CBM. CBM is generally located at the N or C terminus in the cellulose structure and functions as a binding site between enzymes and an insoluble substrate, allowing cellulase to break down the crystalline domain regions of cellulose. Most cellulases have an enzyme active site in the form of clefts that allow cellulases to bind and break the cellulose chains to produce glucose, soluble cellodextrins and insoluble cellulose fragments. Some endoglucanases can also act gradually to hydrolyze the crystalline domain of cellulose, which results in the main product being cellobiose or cellodextrin [20].

Exogucanases are known to work specifically at the ends of the cellulose chains and produce the main products in the form of cellobiose and glucose. Exoglucanase can effectively act on the crystalline domain structure of cellulose. Cellobiohydrolase (CBH) is one of the most widely produced exoglucanases. CBH is generally produced by bacteria and fungi, with a variety of different catalyst modules. A recognized significant CBH structure is a tunnel structure formed from two surface loops on the active site of the enzyme. The tunnel-shaped active site of exoglucanase makes the cellulose hydrolysis process unique. In the mechanism of hydrolysis of exoglucanase, the cellulose chain enters the tunnel, where the active side of the enzyme recognizes the end of the cellulose chain and hydrolyzes the 1,4-glycosidic bonds at the end of the cellulose chain. In general, exoglucanase and endoglucanase have an enzyme folding side, the difference in the folding structure of the two is only in the active side of the enzyme [21].

B-glucosidase (BG) is an exogluanase that does not contain CBM and functions to hydrolyze cellubiose and cellodextrin into glucose. BG acts as an enzyme that lowers the level of cellubiose in the substrate which can act as a CBH inhibitor and endogilanase. BG is known to be produced by bacteria, fungi, plants and animals. In aerobic fungi, it is known that BG is produced extracellularly, while in bacteria BG is produced intracellally and is maintained in the cytoplasm. BG has a pocket-shaped enzyme active site, which allows the enzyme to bind to non-reducing glucose units and hydrolyze cellobiose and cellodextrin to glucose [22].

4.3 Factors affecting cellulase performance

The success of the cellulase enzyme to carry out cellulose hydrolysis is influenced by several factors, including the degree of water swelling, the level of crystallinity, and the enzymatic synergistic effect that can occur on cellulase.

4.3.1 Degree of water swelling (DWS)

Water content in cellulose is an important factor that can affect the performance of the cellulase enzyme. Cellulose which has a low DWS level tends to be dry and has a narrow surface area due to shrinkage. The ability of cellulose to swell and shrink is influenced by the nature of the solvent used. Solvents with non-polar characteristics generally find it difficult to swell cellulose structures and increase the surface area, whereas solvents with polar characteristics are known to swell cellulose structures very well. The swollen structure of cellulose has a wider surface area, which allows the cellulase to more easily penetrate the multiple sides of the cellulose [23].

4.3.2 Degree of crystallinity

The degree of crystallinity of cellulose is known to play a role in determining the rate of hydrolysis of cellulase enzymes. This factor is motivated by the data regarding amorphous cellulose which is degraded more quickly to cellobiose, compared to crystalline cellulose. This data is used by researchers as a form of confirmation of cellulase performance, where the increased crystallization data of a cellulose treated with cellulase, indicates good cellulase activity. This theory is believed by looking at the data that the cellulase first hydrolyzes amorphous cellulose and converts it into a crystalline form [23, 24]. SCOBY kombucha is known to consist of 37% crystalline structure and 63% nanofibril structure [22].

Crystallization of cellulose also affects the adsorption rate of cellulase enzymes on cellulose. It is known that cellulose with a higher degree of crystallization has a lower enzyme adsorption rate. The crystalline structure of cellulose generally inhibits its penetration of the hydrolase system, CBM and other enzyme components [23].

4.3.3 Enzyme synergistic effects

The synergistic effect of cellulase is one of the important factors in the hydrolysis of cellulose. This synergistic effect can occur on the performance of endogluanase-endogluanase, endogluanase-exogluxase, exogilanase-exogilanase,

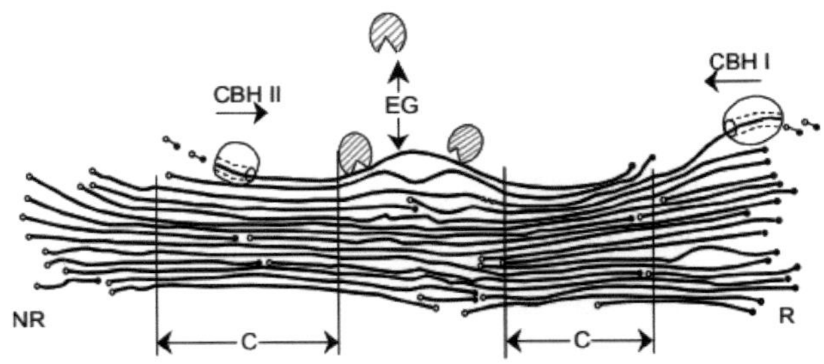

Figure 6.
Cellulase action mechanism; CBH: Cellobiosehydrolase/Exoglukanase; EG: Endoglukanase, NR: Non Reductor; R: Reductor [26].

or endogilanase/exogilanase with CBM simultaneously (**Figure 6**). Several studies have shown that the performance of the endogucanase-exogilanase enzyme can occur synergistically and produce good cellulose hydrolysis products. Other studies have shown that the use of cellulases that have the same cutting-edge (endo-endo-gluanase or exogluanase) can make the two enzymes inhibit each other [25].

5. Characterization of SCOBY nanocellulose fibers

In the enzymatic process of making nanocellulose, it is known that the size of cellulose has succeeded in achieving the characteristics of nanocellulose. However, the production of nanocellulose by enzymatic method has drawbacks, where the breaking of glycosidic bonds and hydrogen bonds in cellulose by cellulase causes a free C structure in cellulose. This free C structure is unstable and tends to form bonds with the surrounding C structures. This condition causes the enzymatic treated nanocellulose to be easily aggregated and has a large size [14]. One way that can be done to avoid this polymerization is the coating process using a buffer, such as CTAB [26] or other compounds, such as chitosan [27]. However, until now there has not been found the right coating process to avoid the polymerization process, because the coating process with CTAB buffer is feared to change the structure of the nanocellulose, while coating using chitosan is feared that it will make the nanocellulose experience an error reading during PSA analysis.

Based on the SEM results in **Figure** 7, it can be seen that the cellulose structure on SCOBY is in the form of microfibrils. The SEM results stated that cellulose has a microfibril structure consisting of amorphous and crystalline structures [1]. The cellulose structure in SCOBY looks more stacked and random. In SEM results, it is known that the cellulose size ranges from 270 nm–740 nm.

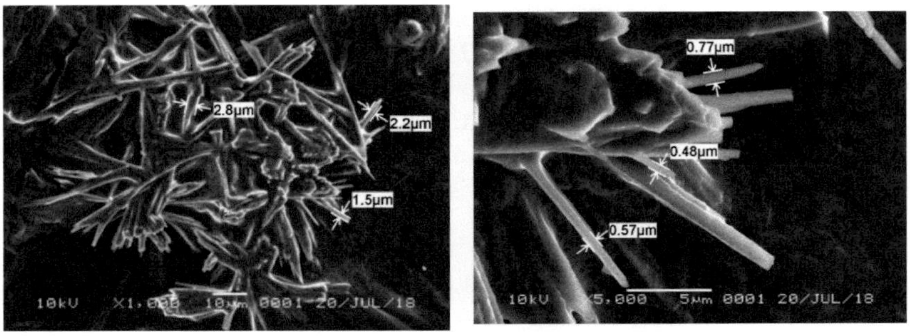

Figure 7.
The cellulose structure is crystalline with single pointed edges on the SEM results.

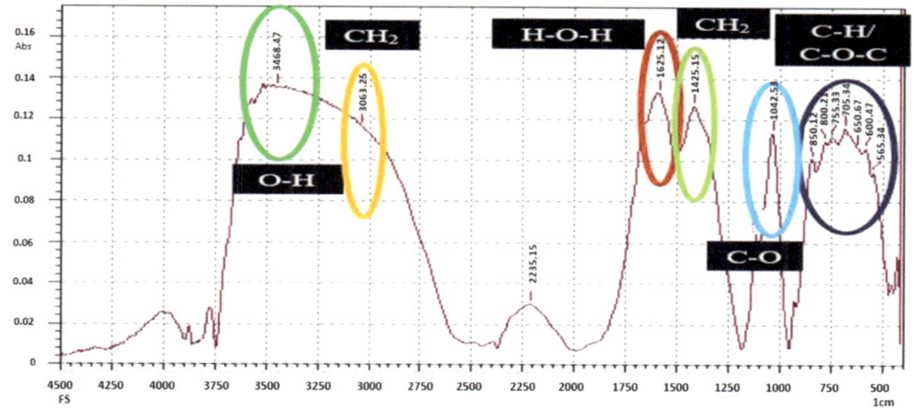

Figure 8.
SCOBY nanocellulose FT-IR absorbance curve with a size of 60 nm.

The structure of cellulose is crystalline with single pointed ends due to the treatment of cellulase enzymes which can break the 1,4-glycosidic bonds in cellulose and remove the amorphous structure of cellulose, so that most of the cellulose structures change to crystalline structures [27]. In SEM results, it is known that the cellulose size ranges from 480 nm–770 nm. This size does not match the PSA results, which state that the nanocellulose particle size of sample 4.2 is 60 nm. This size difference can occur due to re-aggregation between nanocellulose particles. Aggregation can occur starting shortly after the enzymatic process from cellulase is stopped to the drying process of the nanocellulose in the SEM sample preparation process [23]. A fast analysis process is needed to avoid re-aggregation of the nanocellulose, if the nanocellulose is coated.

Cellulose from SCOBY has 5 main functional groups owned by cellulose. The O-H group (3345 cml) is a hydrogen bond that functions to bind the cellulose microfibrils to one another to keep them structured and compact. The CH2 group (2898 cml/1314 cml) is a carboxyl group that can be used to estimate the crystallization rate of cellulose. H-O-H groups (1644–1650 cml) were used to determine the water adsorption rate. The C-O group (1107 cml) is a polyhydroxyl group which can state that SCOBY cellulose is formed from glucose or its derivatives. The C-O-C group (1050–1055 cml) is a glycosidic bond that plays a role in glucose polymer bonds so that it can form cellulose [28].

SCOBY nanocellulose fibers show all the clusters that belong to SCOBY cellulose. These results indicate that the SCOBY nanocellulose fibers were indeed cellulose samples. The difference in FT-IR results between cellulose and nanocellulose SCOBY lies in the absorbance value of the FT-IR results. In general, the FT-IR SCOBY nanocellulose absorbance value was lower than cellulose, especially for the O-H, H-O-H and C-O-C groups. These results indicate that enzymatic treatment on cellulose has succeeded in breaking hydrogen and glycosidic bonds in cellulose so that cellulose can undergo a change in size to 60 nm, which includes the size of nanocellulose. Changes in cellulose size also affect the H-O-H groups in cellulose, where the adsorption power of water on cellulose is smaller [29] (**Figure 8**).

6. Conclusions

The preparation of nanocellulose from SCOBY Kombucha can be done using crude extract of the cellulase enzyme from Bacillus sp. The optimum amount of enzymes used for the manufacture of nanocellulose from SCOBY kombucha is 2: 3 (w/v) SCOBY against crude extract of cellulase enzymes.

SCOBY cellulose fibers, a microbial polysaccharide, has significant potential and suitable for various industrial applications such as food, pharmaceutical, textiles, cosmetic, fashion, and paper. Its material properties has high elasticity, best deformation and comfort properties. SCOBY cellulose fibers forms a strong gel film of crystalline microfibrils and absence of impurities of hemicelluloses and lignin as well as environmentally friendly.

Acknowledgements

The authors would like to thank the Chemical Analytic Laboratory of Faculty of Mathematics and Natural Sciences, Institut Teknologi Bandung. The authors would also like to thank Ministry of Research and Technology/National Research and Innovation Agency (RISTEK/BRIN) Republic Indonesia as a financial. Support.

Author details

Pingkan Aditiawati[1], Rudi Dungani[1*], Salsabila Muharam[1], Aminudin Sulaeman[2], Sri Hartati[3], Mustika Dewi[1] and Enih Rosamah[4]

1 School of Life Science and Technology, Institut Teknologi Bandung, Bandung, Indonesia

2 Faculty of Mathematics and Natural Sciences, Institut Teknologi Bandung, Bandung, Indonesia

3 Faculty of Agriculture, Padjadjaran University, Bandung, Indonesia

4 Faculty of Forestry, Mulawarman University, Samarinda, Indonesia

*Address all correspondence to: dunganir@gmail.com

References

[1] Kargarzadeh H, Loelovich M, Ahmad I, Thomas S, DufresneA. Handbook of nanocellulose and cellulose nanocomposites. 2nd ed. Wiley-VCH Verlag GmBH & Co; 2017. 5p. DOI:10.1002/9783527689972

[2] Anwar JS, Deshmukh NA, Pinjari DV. A critical review of manufacturing processes used in regenerated cellulosic fibres: viscose, cellulose acetate, cuprammonium, LiCl/DMAc, ionic liquids, and NMMO based lyocell. Cellulose. 2019; **26**: 2913-2940. DOI: 10.1007/s10570-019-02318-y

[3] Chan C, Shin J, Jiang S. Development of tailor-shaped bacterial cellulose textile cultivation techniques for zero-waste design. Clothing and Textile Research Journal. 2018;**36:** 33-44. DOI: 10.1177/0887302X17737177

[4] Laurin. S., 2015. Kombucha couture live clothing [Internet]. 2015. Available from: www.kombuchacouture.com [Accessed: 2020-08-28]

[5] Abitbol T, Rivkin A, Cao Y, Nevo Y, Abraham E, Shalom T, Lapidot S, Shoseyov O. Nanocellulose, a tiny fiber with huge applications. Current Opinion in Biotechnology. 2016; **39**:76-88. DOI: 10.1016/j.copbio.2016.01.002

[6] Chawla P, Bajaj I, Survase S, Singhal R. Microbial cellulose: fermentative production and applications. Food Technology Biotechnology. 2009; **47**: 107-124.

[7] Dungani R, Karina M, Subyakto, Sulaeman A, Hermawan D, Hadiyane A. Agricultural waste fibers towards sustainability and advanced utilization: A review. Asian Journal of Plant Sciences. 2016;**15**: 42-55. DOI: 10.3923/ajps.2016.42.55

[8] Fontana JD, Koop HS, Tiboni M, Grzybowski A, Pereira A, Cassandra DK, da Silva MGR, Wielewski LP. New insights on bacterial cellulose. In: Grumezescu AM, Holban AM, editors. Food Biosysthesis. Elsevier; 2017. p. 213-249. DOI: 10.1016/C2016-0-00180-8

[9] Janusz G, Pawlik A, Sulej J, Swiderska-Burek U, Jarosz-Wilkołazka A, Paszczynski A. Lignin degradation: microorganisms, enzymesinvolved, genomes analysis and evolution. FEMS Microbiology Reviews. 2017; 41: 941-962. DOI: 10.1093/femsre/fux049

[10] de Moraes Rocha GJ, Nascimento VM, da Silva VFN. Enzymatic bioremediation of effluent from sugarcane bagasse soda delignification process. Waste Biomass Valorization. 2014; 5: 919-929. DOI: 10.1007/s12649-014-9316-5

[11] Kumar A, Chandra R. Ligninolytic enzymes and its mechanisms for degradation of lignocellulosic waste in environment. Heliyon. 2020; 6: e03170. DOI: 10.1016/j.heliyon.2020.e03170

[12] Oscar S, Sierra R, Alméciga-Díaz CJ. Delignification process of agro-industrial wastes an alternative to obtain fermentable carbohydrates for producing fuel. In: Manzanera M, editor. Alternative fuel. IntechOpen; 2011. p. 111-154. DOI: 10.5772/22381

[13] Dutta H, Paul SK. Kombucha drink: production, quality, and safety aspects. In: Grumezescu AM, Holban AM, editors. Production and management of beverages, vol I. The science of beverages. Elsevier, Amsterdam; 2019. p 259-288

[14] Amarasekara AS, Wang D, Grady TL. A comparison of kombucha SCOBY bacterial cellulose purifcation methods. SN Applied Sciences. 2020; 2: 240: 1-7. DOI: 10.1007/s42452-020-1982-2

[15] Gayathry G, Gopalaswamy G. Production and characterisation of microbial cellulosic fbre from *Acetobacter xylinum*. Indian Journal of Fiber & Textile Research. 2014; 39: 93-96.

[16] Dima S, Panaitescu S, Orban C, Ghiurea M, Doncea S, Fierascu R, Nistor C, Alexandrescu E, Nicolae C, Trica B, Moraru A, Oancea F. Bacterial nanocellulose from side-streams of kombucha beverages production: Preparation and physical-chemical properties. Polymers. 2017; 9: 374-397. DOI: 10.3390/polym9080374

[17] Kovalenko VI. Crystalline cellulose: structure and hydrogen bonds. Russian Chemical Reviews. 2010; 79: 231-238. DOI: 10.1070/RC2010v079n03ABEH004065

[18] Behera B, Sethi B, Mishra R, Dutta S, Thatoi H. Microbial cellulases – diversity & biotechnology with reference to mangrove environment: A review. Journal of Genetic Engineering and Biotechnology. 2017; 25: 197-210. DOI: 10.1016/j.jgeb.2016.12.001

[19] Hollabaugh C, Burt L, Walsh H. Carboxymethylcellulose: Uses and applications. Industrial and Engineering Chemistry. 1954; 37: 943-947. DOI: 10.1021/ie50430a015

[20] Abdul Hamid SB, Islam MM, Das R. Cellulase biocatalysis: key influencing factors and mode of action. Cellulose. 2015; 22: 2157-2182. DOI: 10.1007/s10570-015-0672-5

[21] Attigani A, Sun L, Wang Q, Liu Y, Bai D, Li S, Huang X. The crystal structure of the endoglucanase Cel10, a family 8 glycosyl hydrolase from *Klebsiella pneumoniae*. Acta Crystallographica Section F Structural Biology Communication. 2016; F72: 870-876. DOI: :10.1107/S2053230X16017891

[22] Mahapatra S, Vickram AS, Sridharan TB, Parameswari R, Pathy MS. Screening, production, optimization and characterization of β-glucosidase using microbes from shellfish waste. 3 Biotechnology. 2016; 6: 213-222. DOI:10.1007/s13205-016-0530-7

[23] Karim Z, Afrin S, Husain Q, Danish R. Necessity of enzymatic hydrolysis for production and functionalization of nanocelluloses. Critical Reviews in Biotechnology. 2016; 37: 355-370. DOI: 0.3109/07388551.2016.1163322

[24] Szijarto N, Siika-aho M, Tenkanen M, Alapuranen M, Vehmaanpera J, Reczey K, Viikari L. Hydrolysis of amorphous and crystalline cellulose by heterologously produced cellulases of Melanocarpus albomyces. Journal of Biotechnology. 2008; 136: 140-147. DOI: 10.1016/j.jbiotec.2008.05.010

[25] Gonçalves GAL, Takasugi Y, Jia L, Mori Y, Noda S, Tanaka T, Ichinose H, Kamiya N. Synergistic effect and application of xylanases as accessory enzymes to enhance the hydrolysis of pretreated bagasse. Enzyme and Microbila Technology. 2015; 72: 16-24. DOI: 10.1016/j.enzmictec.2015.01.007

[26] Miettinen-Oinonen A, Londesborough J, Joutsjoki V, Lantto R, Vehmaanpera J. 2004. Three cellulases from Melanocarpus albomyces with applications in the textile industry. Enzyme and Microbial Technology. 2004; 34: 332-341. DOI: 10.1016/j.enzmictec.2003.11.011

[27] Sofia M, Brown R, Tsuzuki T, Rainey T. A comparison of cellulose nanocrystals and cellulose nanofibres extracted from bagasse using acid and ball milling methods. Advances in Natural Sciences: Nanosciences and Nanotechnology. 2016; 7: 1-9. DOI: 10.1088/2043-6262/7/3/035004

[28] Mohammadkazemi F. 2015. Surface properties of bacterial nanocellulose using spectroscopic methods and X-ray diffraction. American Journal of Applied and Industrial Chemistry. 2015; 1(2): 10-13. DOI: 10.11648/j.ajaic.20170101.13

[29] Aditiawati P, Dungani R, Amelia C. Enzymatic production of cellulose nanofibers from oil palm empty fruit bunch (EFB) with crude cellulase of *Trichoderma* sp. Material Research Express. 2018; 5: 1-7. DOI: 10.1088/2053-1591/aab449

Chapter 9

Electrodeposition of Nanoporous Gold Thin Films

Palak Sondhi and Keith J. Stine

Abstract

Nanoporous gold (NPG) films have attracted increasing interest over the last ten years due to their unique properties of high surface area, high selectivity, and electrochemical activity along with enhanced electrical conductivity, and chemical stability. A variety of fabrication techniques to synthesize NPG thin films have been explored so far including dealloying, templating, sputtering, self-assembling, and electrodeposition. In this review, the progress in the synthetic techniques over the last ten years to prepare porous gold films has been discussed with emphasis given on the technique of electrodeposition. Such films have wide-ranging applications in the fields of drug delivery, energy storage, heterogeneous catalysis, and optical sensing.

Keywords: nanoporous gold, electrodeposition, surface area, potential, thin films

1. Introduction

Over the last two decades, nanotechnology and nanoscience have generated great scientific interest focusing mainly on the development of nanomaterials with specific and tunable properties and their applications in various areas [1]. Nanotechnology offers the ability to design, synthesize, and control length scales ranging from <1 to >100 nm. In the literature, reports of discoveries based on novel properties arising from these small size features have been increasing and nano-sized noble metal particles have occupied a central place [2]. Also, nanotechnology has grown in significance in the study of fibrous materials, namely nanofibers and silicate nanocomposites wherein the synthesis and characterization along with the unique properties have been studied [3]. An emerging area of great interest is that of nanowire research which will interface with living cells for precise delivery of small molecules, proteins, and deoxyribonucleic acid (DNA) [4]. From the viewpoint of the relationship between nanostructures and properties, remarkable advances have been made in the commercial use of thin films that find wide-ranging applications in almost all the industrial fields such as optics, electronics, mechanics, and even biotechnology [5]. There is a surge of interest seen in the scientific community when it comes to NPG due to its intriguing material properties arising from its high specific surface area, high electrical conductivity, reduced stiffness, and the prospect of easy surface modification. NPG has controllable pore morphology and ligament size that opens up a wide range of studies of its mechanical and surface properties [6]. Compared to regular gold thin films which are dense inside, NPG films have interconnected ligaments with nanometers-sized

IntechOpen

gaps throughout the bulk of the film. The pore size can be modulated depending on the type of synthesis protocol followed ranging from typically 20–50 nm in size but to as small as 5 nm [7]. Additionally, the porous structure of the NPG electrode tremendously increases the number of adsorption sites for various molecules of biological interest making it an attractive candidate in the field of biosensors [8]. Gold electrodes with nanoporous structures possess a higher roughness factor (the ratio between the real surface area and the geometrical area of the electrode) and better electron transport in comparison with their counterparts with smooth surfaces [9]. Metal nanoporous films have been prepared by various methods of high productivity and controllability of which chemical and electrochemical dealloying laid the foundation for other methods [10]. Moreover, dealloying is a potent approach for the fabrication of both monoporous (i.e., nanoporous or microporous) and hierarchical (i.e., possessing both microporosity and nanoporosity) porous metal structures with novel properties [11]. Multimodal pore size distribution on the nanometer and micrometer scale is highly desirable. The presence of larger size pores enables fast transport of the reactants, while the nanopores are responsible for providing high surface area thereby increasing the rate of electrochemical reactions. High surface area gold could be prepared by the electrodeposition technique, illustrated in **Figure 1**. Porous metals prepared via dealloying often contain some amount of residual less noble metal and therefore other fabrication techniques were explored [12].

The electrochemical deposition of NPG on a solid substrate has been extensively researched in recent years. This facile technique enhances the electrochemical activity of the nanoporous film by offering fine control over the growth and nucleation mechanism which in turn determines the morphology of the deposited film [13]. The three-dimensional (3-D) nanoporous films, membranes or powders of large surface area have received great attention and it has been seen that the templating strategy is the most popular method for their preparation using polycarbonate membranes, colloidal crystals, lyotropic liquid crystalline phases of surfactants, and echinoid skeletal structures as the templates and will be discussed in this chapter [14, 15]. Electroplated gold continues to play an integral role in modern electronics technology, and it is hard to find an equivalent substitute due to the unique combination of properties of the metal. It is speculated that as information technologies continue to expand, the quantity of gold used will continue to

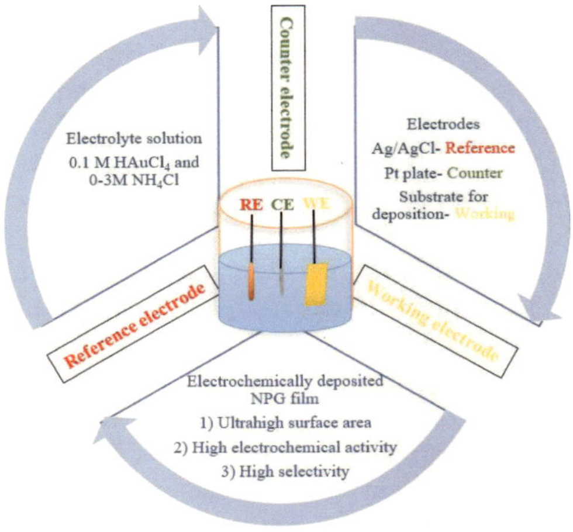

Figure 1.
A representation of the electrochemical deposition set up.

increase [16]. Experimental parameters have been seen to influence the morphology of gold and therefore, this chapter will give insights into the various methods used for fabricating NPG thin films with special emphasis on electrodeposition strategies. Along with the synthetic approaches, applications and the characterization of the NPG film will be discussed.

2. Fabrication techniques

There are various methods for fabricating porous gold films, and these are categorically described below.

2.1 Dealloying methods

De-alloying is an effective corrosion method for the fabrication of NPG films wherein the presence of less noble metals in the gold alloy has been exploited in a way that they are chemically or electrochemically dissolved to produce monolithic metal bodies with nanoscale pore structure. Au-Ag alloys are considered ideal due to their similar atomic volumes and continuous solid solubility allowing for coherent transformation from the master alloy to the nanoporous structure, see **Table 1** [17, 18]. Chemical dealloying has been studied employing Metropolis Monte Carlo simulations wherein the simulation of the dealloying process in the first stage describes the equilibrated systems followed by the second stage of dealloying with the exclusion of interaction parameters [19]. A simple method to dealloy the precursor alloy is to immerse it in nitric acid leading to selective etching of silver forming a 3-D pattern resulting in the formation of an open, bicontinuous highly porous network of gold with tunable ligament and channel width by varying the alloy composition, electrochemical potential, or by thermal annealing after dealloying [20]. **Figure 2** depicts the outcome of NPG structures upon a change in the experimental parameters.

Fabrication techniques	Advantages	Application
Dealloying	This approach enables the fabrication of NPG thin films, either free-standing or supported on a conductive substrate	Enhanced electrocatalytic activity towards methanol oxidation, potential in the field of catalysis, optics, and sensor technology
Self-assembly	Thin films formed exhibit enhanced thermal stability and capability of electron transfer. This method can be applied to various conductive surfaces without harmful pretreatment	Energy storage, photovoltaics, sensing, and electrochemical usage
Sputter deposition	Simple green method, reproducible and generates low-price final product	Electrochemical biosensing, used as supercapacitors, in microelectronics and photovoltaics
Electrodeposition	One-step fabrication of thin films directly on a substrate, control of particle morphology, size, and density is relatively easy. Uniform deposition is seen along with good stability	Electroanalytical and catalytic field

Table 1.
Summary of the fabrication techniques used to synthesize NPG thin films.

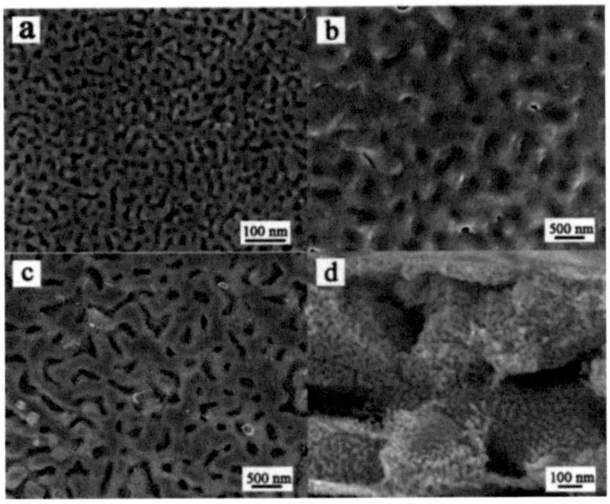

Figure 2.
SEM images of NPG structures. (a) NPG membrane made by dealloying 1 μm thick gold leaf in nitric acid for 1 h. (b) Coarsened NPG structure generated via annealing of (a) at 400°C for 8 h. (c) Hierarchical porous membrane structure produced by performing second dealloying (of sample b) in nitric acid for 5 min and annealing at 400°C for 4 h. (d) Cross-section micrograph of sample (c). Reproduced with permission from reference [20], Copyright 2003, American Chemical Society.

Variables such as acid concentration, etching time, and solution temperature are known to influence the size of pores and ligaments. NPG films with ultrafine pores have been produced by a pulse electrochemical dealloying carried out at a potential of 0.6 V with 50 ms on-time and 10 ms off-time in 8 M HNO_3 at 23°C [21]. It has been seen that with the use of strong acid or alkali corrosion, the rearrangement is very rapid leaving little scope for porosity adjustment in such nanoporous products. Therefore, instead of the traditional corrosive acid etching, a two-step dealloying method has been reported which utilized $FeCl_3$ to synthesize NPG wherein the pore size was easily tunable by using a surfactant like polyvinyl pyrrolidone (PVP) or replacing corrosion reaction solvent with ethylene glycol (EG) [22]. Dealloying can be further classified into three categories namely, (i) chemical dealloying, (ii) electrochemical or potentiostatic dealloying, and (iii) liquid metal dealloying. The driving force towards the selective dissolution of the active component is varied in the above-mentioned categories. It is a corrosive solution such as an acid in the case of chemical dealloying, the constant potential for electrochemical dealloying and the nature of liquid metal medium and temperature are the crucial deciding factors in the emerging field of liquid metal dealloying [11]. Extensive investigations have demonstrated that nanoporous metals with 3-D bicontinuous structures fabricated by the dealloying method, has many active sites for the excitation of localized surface plasmons and can, therefore, serve as a potential substrate for practical surface-enhanced Raman scattering (SERS) application [23]. Unlike single-sized porous materials, a hierarchical (bimodal) porous structure can impart novel properties to the material wherein large pores can favor increased mass transport and small pores can impart high specific surface area [24].

2.2 Self-assembly

Self-assembly is one of the most versatile, simple, and inexpensive methods aiding the formation of porous polymer films with finely controlled topography [25]. The self-assembly strategy is a powerful approach to create functional nanomaterials due to the high selection capability of the precursor materials and inclusion of functional groups, and nanoarchitectonics. The process is usually carried out under ambient and mild conditions, making the process suitable for biological materials, see **Table 1** [26]. The self-assembled nanostructured film is a rapidly emerging field of great fundamental and practical interest due to the prospective applications in the field of microelectronics, non-linear optics, catalysis, and sensor science. Controlled covalent attachment of nanoparticles to functionalized surfaces is a versatile approach for producing thin-film structures. It has been seen that self-assembly of gold nanoparticles (AuNPs) on thiol modified surfaces exhibited non-metallic optical and electronic properties [27]. The self-assembly process of the AuNPs via aggregation and coalescence leads to the porous structure directly without relying on external assistance. Zhang and coworkers successfully fabricated porous gold films from a colloidal gold solution by evaporation induced self-assembly method (EISA) [28]. A facile synthetic approach for the self-assembly of AuNPs on sulfide functionalized polydopamine surface in high-density has been reported. AuNPs were seen to self-assemble strongly on the modified surface due to the strong interaction between gold and sulfur atoms [29]. Nanostructural control is very critical in material chemistry to bring out unique physical and chemical properties. So far, many mesoporous materials with varying compositions have been reported via self-assembly of amphiphilic organic molecules and have attracted keen interest due to their wide range of potential applications in energy storage, separation, catalysis, ion exchange, sensing, and drug delivery [30]. Another useful technique to immobilize gold thin films on micro- and nanopatterns is via the use of directed self-assembly on templates that are prepared by phase- separated mixed Langmuir Blodgett (LB) films. Atomic force microscopy (AFM), Auger electron spectroscopy, and scanning Auger electron mapping of the gold thin films revealed that the immobilized layer was following the patterns of the original mixed LB films [31]. Systematic analysis of the thermodynamics and kinetics of self-assembly in thin films of supramolecular nanocomposite has been studied in detail to extract information about the interfacial area defects, chain mobility, and activation energy needed for the diffusion of materials. Co-assemblies of nanoparticles and organic moieties are promising, and the resultant material will combine the properties of both the families of building blocks. Hierarchically structured nanocomposite thin films over macroscopic distances have been created by a fast ordering process with minimal usage of solvent via a deep understanding of the kinetic pathways [32].

2.3 Sputter deposition

Sputter deposition is an industry-relevant, high-rate, large-scale, and well-controllable deposition technique used to prepare gold films and other device fabrication processes. It is a large-scale deposition method, allowing high-rate vacuum coating with nanoscale precision, see **Figure 3**. Sputter deposition is an ideal method for preparing nanometer-thick films along with precise control over the deposition rate, highly adhesive films, large area uniformity, uniform temperature, and the ability to coat a variety of substrates including the non-heat resistant [33].

Sputtering is a pollution-free ("green") technique that offers the advantage of reproducibility and low price of the final product. Nanostructured gold film adhesion and electrical contact properties are strongly influenced by interface structure, see **Table 1** [34]. Thin gold films prepared by this strategy are found to be continuous and relatively homogeneous, with distinct grain surfaces without showing the formation of islands. Films of thickness ranging between 5 and 52 nm have been prepared for the study of third-order non-linear properties where the third-order susceptibility was found to be of the order of 10^{-9} esu. This non-linear optical interaction enhancement has helped to construct nanodevices to be used in metrology, sensing, imaging, and telecommunications [35]. It has been shown that pure

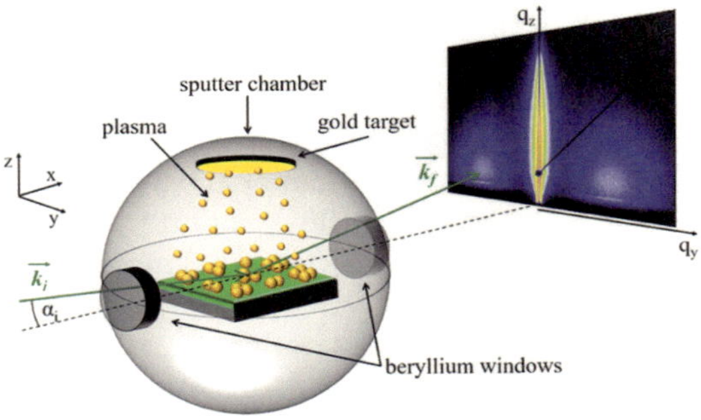

Figure 3.
Schematic illustration of the in situ *grazing-incidence small-angle X-ray scattering (GISAXS) set up showing sputtering of gold atoms on spiro-OMeTAD film. A two-dimensional detector is detecting the signal produced by scattered X-rays. Reproduced with permission from reference [33], Copyright 2020, American Chemical Society.*

sputtered gold films can match the hardness of gold electrodeposits. The hardness of sputtered films and the grain size of the deposit is controlled by maintaining the temperature of the substrate during deposition [36]. It has been reported that ultrathin semi-transparent gold films have been deposited using radio frequency (RF) magnetron sputtering at room temperature over a small area (23 mm^2) of the porous silicon layer. Various film thicknesses were obtained by changing the sputtering time from 5 to 20 s at constant chamber pressure and argon gas flow rate [37]. Thin films have been deposited by varying the experimental conditions concerning the substrate tilt angle and background pressure. Growth regimes of thin gold films deposited via magnetron sputtering at oblique angles and low temperatures have been studied from both theoretical and experimental points of view [38]. The growth and morphology of a room temperature sputter-coated thin gold film on a soft polymeric substrate from nucleation to thin film formation has been investigated using AFM. It was observed that an initial 3-D island-type growth starts with the deposition and with increasing time the morphology evolved from hemispherical islands to partially coalesced worm-like island structures, to percolation, and finally to a rough and continuous film [39]. In a study, it was seen that after the stage of nucleation, the growth of gold clusters proceeds mainly in the lateral direction. As the discontinuous islands change into a continuous thin film, a rapid decline in the resistance of the gold layer has been observed [40]. A unique method for synthesizing porous gold films by co-deposition of Au-Cu alloy has been done via co-sputtering Au and Cu using a multi-target sputtering system at room temperature. Selective removal of Cu was done via corrosive dealloying leading to the formation of the porous gold film via the physical–chemical combination method [41]. Sputter deposited thin porous gold films find applications in the field of electrochemical biosensing for enhancing redox signals by modulating the nanopore size and film thickness. Higher detection resolution is exhibited to that obtained by conventional bulk gold electrodes [42]. The atom sputtering deposition technique has allowed scientists to study the optical and electrical properties, density, and crystalline structure of gold nanostructures sputtered on glass [40].

2.4 Electrodeposition

Electrodeposition is a simple, controllable, and cost-effective method that can control the growth process by varying deposition current density. Kinetic control over the growth process can give insights into the mechanisms in the synthesis of deposits with various sizes and shapes. The mass of gold on the electrode is controlled by the electric charge passed during the electrodeposition process, based on Faraday's laws of electrolysis [43]. Deng and co-workers gave a mechanistic viewpoint to the facile method of electrodeposition in the electrochemical cell, where a process of electrodissolution-disproportion-deposition is involved. Upon the potential step, the gold substrate undergoes active electrodissolution in HCl providing the diffusion control, forming $AuCl_2^-$ and $AuCl_4^-$. With the progress of the reaction, HCl depletes and $AuCl_2^-$ starts accumulating near the gold surface immediately giving off Au atoms and $AuCl_4^-$ wherein the newly formed Au atoms aggregate and deposit on the substrate to form NPG film [44, 45]. NPG morphology evolution has a direct relationship with the topography of the underlying substrate. A study has very well demonstrated that micropattern widths that are within the magnitude of film thickness produce tunable pores with less cracking [46]. Porous nanomaterials particularly NPG has drawn tremendous research interests mainly in the development of electrochemical sensors due to the high surface area and the trait of being electrocatalytically active. Electrosynthesis can impart precise control over the bicontinuous network of interconnected nanometric gold grains and multiple sized pores [47].

Conductive substrates coated with AuNPs can be exploited as biochemical sensors and electrodes owing to their excellent electrocatalytic activity. A simple, powerful, and cost-effective way for attachment of AuNPs onto a substrate has been electrodeposition with pulsating current resulting in smoother, brighter, finer, and less porous Au grains concerning the direct current [48]. A simple template-less, surfactant-less and effective electrochemical method for the preparation of AuNP arrays with an average diameter of ~14 nm onto indium tin oxide (ITO) glass has been reported. The system exhibits excellent catalytic properties due to the presence of large active sites on the surface [49]. The deposition of thin metallic films onto non-conductive surfaces by the wet processing technique of electrodeposition is another route for fabricating various unique electronic devices and systems. Amine terminated self-assembled monolayer (SAM) modified insulator surface has been employed for the lateral growth of electrodeposited gold. Controlled morphology and thickness of laterally grown metal films has paved a way to create nanogap electrodes [50].

Electrochemical deposition methods have been classified into two groups namely, templated electrodeposition and self-templated electrodeposition. Template technology offers advantages for designing new types of electrode materials aided by "hard" or "soft" templates that give rise to functional materials with diverse structures and morphologies, such as, one-dimensional nanostructures, two-dimensional films, and 3-D porous frameworks. In templated electrodeposition, porous gold is deposited on a template by reducing gold near the electrode surface via fixed potential application followed by the removal of the template. The self-templated approach initializes with the generation of H_2 bubbles in the solution containing a supporting electrolyte by applying a potential of at least −2 V vs. saturated calomel electrode (SCE), allowing a gold reduction in the interstitial space [51, 52]. The monolithic NPG film has been synthesized using a bottom-up synthesis from a bicontinuous microemulsion (BME) acting as a dynamic soft template, depicted in **Figure 4**. The aqueous phase of BME acted as a medium for gold electrodeposition and the intertwined structure came from the aqueous and oil phases of BME compartmentalized by surfactant and cosurfactant. The resultant film nanostructure has been controlled by adjusting the initial BME composition and by varying kinetic parameters such as deposition potential and time [53].

The role of electrodeposition potential on the growth morphologies has been seen in a study where gold deposition from sulfite electrolyte exhibits a range of geometries from vertically oriented nanowires to lenticular grains and dendrites at more negative potentials [54].

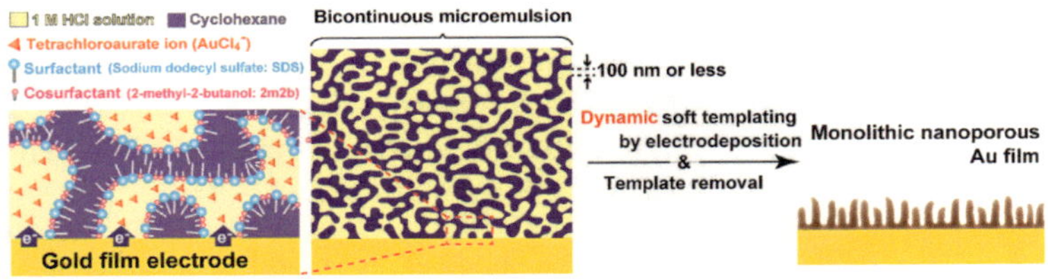

Figure 4.
Schematic representation of the dynamic soft templating of a nanoporous thin gold film from BME by electrodeposition. Reproduced with permission from reference [69], Copyright 2013, American Chemical Society.

In recent years, studies have been performed to produce 3D porous materials that have the advantage of increasing the mass transport for electrolytes but also allow rapid electrochemical reactions due to considerable active surface area. A new way of producing such materials is via concurrent generation of hydrogen bubbles with simultaneous metal deposition at high cathodic current densities. The reduction of H^+ giving rise to hydrogen bubbles acts as a dynamic template for metal electrodeposition [15]. The electrodeposition parameters such as time and potential are the deciding factors for the pore size and film density. The dynamic hydrogen bubble template (DHBT) method results in a gold nanoporous structure with outstanding properties, like high specific surface area, large pore volume, uniform nanostructure, good conductivity, and enhanced electrochemical activity [55]. A simple method has been proposed for the one-step electrodeposition of NPG-islands films on the surface of the glassy carbon (GC) electrode via the hydrogen bubble template approach. The 3-D structure generated via the DHBT method is more attractive as it is clean and porous in an efficient way in the absence of inorganic and organic templates and exhibited improved electrocatalytic activity for oxygen reduction and hydrogen evolution reactions (HER) [56]. Recently, self-supported 3D metal foams of copper, tin, and silver have been reported. 3D porous noble metals such as gold have higher equilibrium potential and lower overpotential for HER and therefore, a two-step rout was taken involving the deposition of the less noble element followed by its galvanic displacement. Chung and coworkers have therefore shown a fast one-step preparation of high surface area NPG with a multimodal pore-size distribution utilizing a DHBT [12].

Another study produced porous gold incorporating nanocorrals on gold screen-printed electrodes (SPEs) utilizing hydrogen bubbles as a dynamic template. The structure produced using this one-step electrodeposition using high overpotential had a high roughness factor [57]. Moreover, distinct pore morphology can be obtained by taking advantage of electrical conductivity and morphological plasticity of NPG. Pore morphology can be tuned using the novel and versatile technique of electro-annealing on NPG thin films at low temperatures [58]. The microstructure of electrodeposited gold films and their deposition characteristics are affected by base metal ions. It has been seen that gold deposits obtained from Co and Ni-containing electrolytes are generally hard, while Pb and Tl-containing electrolytes tend to give soft gold deposits [59]. The crystallographic structure of gold films is also dependent on the current density at which electrodeposition has been carried out. It has been seen that Au electrodeposited at a current density less than 0.25 mA cm^{-2} from dicyanoaurate baths with or without Cu^{2+} or Tl^+, gave rise to the formation of hexagonal structure [60].

The electrochemical potential and the concentration of $HAuCl_4$ have been modified to create a variety of morphologies in the final structure. Nanopyramidal, nano rod-like, and spherical gold nanostructures were fabricated on polycrystalline gold substrates via one-step, non-templated electrochemical overpotential deposition (OPD) [61]. Moreover, the chemical nature of the organic ions present in the organic electrolytes has a huge role to play on the morphology of the deposits produced. The organic species adsorb onto the electrodeposits due to high surface energy and finally influence the shape of the growing grains and the roughness of gold electrodeposits [62]. Recently, a study described the electrodeposition of gold nanostructures at the interface of a Pickering emulsion. The controlled electrodeposition of AuNPs on the surface of an emulsion droplet gave rise to intricate structures with fine control over the locus or duration of nanoparticle growth. Decamethylferrocene present in the emulsion droplet acted as a heterogeneous electron transfer agent for the reduction of aqueous phase Au (III) resulting in its deposition as nanoparticles [63]. In one of the studies, it was found that SERS of the electrodeposited gold was correlated to the roughness and the size of surface nanostructures where these two parameters were largely controlled by the applied potential during deposition [64]. Over the last decade, researchers have focused on improving methods for fabricating reproducible substrates for surface-enhanced resonance Raman scattering (SERRS). Amongst these, deposited gold films have been the most heavily researched, and therefore, Bartlett and his group fabricated ordered-spherical-cavity gold films using colloidal templated electrodeposition method for the first time. The net enhancements are found to be ~10^9 for SERRS over normal Raman [65]. Nanostructured gold films for use in localized surface plasmon resonance (LSPR) spectroscopy have been prepared from flat gold film substrate made by stripping off epoxy coated glass slides off the gold-sputtered silicon wafers and then subsequent two-step chronoamperometry to electrodeposit gold from potassium dicyanoaurate solution [66].

Solid thin film deposition on the soft ionic liquid (IL) substrate was used for L-arginine detection. A gold monolayer was deposited on the surface of the IL substrate using the conventional electro-co-deposition (CECD) technique using an electrochemical workstation with three electrodes system in 0.5 M NaCl and 3 mM $KAuCl_4$ electrolyte solutions [67]. A new study described the bottom-up approach of lithographically patterned nanowire electrodeposition (LPNE) for synthesizing noble metal nanowires on glass or oxidized silicon surfaces. The process of LPNE starts with the preparation of a nickel nanoband electrode which acts as the surface for electrodeposition [68].

LPNE has been combined with colloidal lithography to create a novel low-cost method for the lithographically patterned electrodeposition of metallic nanoring close-packed arrays over large areas. By altering the width and radius of the nanoring during the fabrication, near-infrared (NIR) plasmonic resonances could be tuned from 3500 to 8000 cm^{-1} with potential applications in the fabrication of plasmonic antennae, plasmonic semiconductors, and negative-index metamaterials [69]. Direct electrodeposition of porous gold nanowire arrays has been developed utilizing a one-step electrodeposition methodology utilizing nanochannel alumina templates. Current density during deposition is the deciding factor for the microstructure of gold nanowires and the resulting structure has shown excellent electrochemical biosensing ability towards the detection of glucose [70].

It has been known that the surface roughness and the structure of gold deposits can be influenced by periodically reversed or pulsed current and in some electronic applications pulse plated gold deposits are considered superior to DC plated deposits. Electrodeposition of gold using pulse plating produces a dense fine-grained structure with half the resistivity of normal DC plated films [71]. The pulse potentiostatic method (PPSM) offers numerous advantages in terms of controllability, particle size, stronger adhesion, and uniform film morphology. PPSM has been used to deposit gold nanoflowers (AuNFs) onto a polymer film. Through this simple and rapid method, a new organic–inorganic hybrid film has been fabricated with superior electroactivity, electrochemical, and interfacial characteristics suggesting the suitability of such electrodes for sensors, electrocatalysis, and diode applications [72]. Epitaxial, ultrathin, semitransparent, and catalytic gold films were electrodeposited on n-type silicon (n-Si) to protect the substrate from photo-passivation, see **Figure 5**. In addition to being a good redox catalyst, the ultrathin gold layers serve to induce band bending in n-Si thereby making it a diode of an ideal quality factor due to minimal electron–hole recombination [73].

Fabrication methods have been used in combination in the past to generate a promising eco-friendly biosensor platform with advantages coming from both the techniques. One such study has been done to fabricate a glucose biosensor wherein AuNPs and glucose oxidase multilayer films were generated via electrodeposition and self-assembly respectively [74]. As an eco-friendly alternative, deep eutectic solvents (DESs) are a new class of green and sustainable solvents extensively used for electrodepositions due to their intrinsic properties of good solubility, non-flammability, low toxicities, and suitable electrochemical windows. Recently, an electrochemical method has been developed to fabricate NPG electrodes by alloying and dealloying the precursor alloy in $ZnCl_2$-urea deep eutectic solvent [75]. The wide acceptance of DES in metal/alloy electrodeposition processes has been seen in recent times due to their ability to act as stabilizers and reducing agents in metal nanoparticle synthesis. Moreover, the nucleation rates are enhanced leading to a

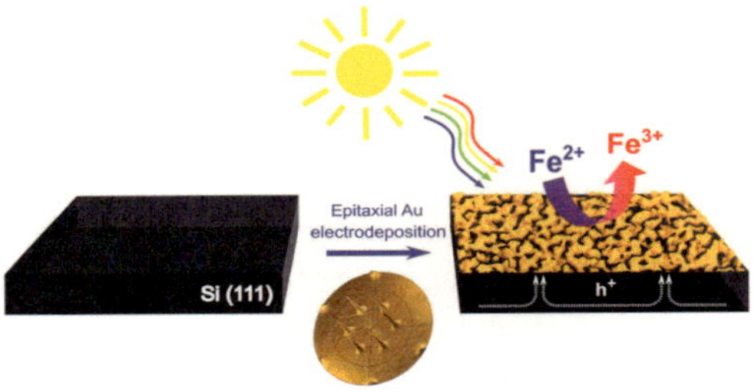

Figure 5.
Illustration showing the formation of an ultrathin epitaxial gold layer electrodeposited on n-Si (111) to form a Schottky junction in a regenerative photoelectrochemical cell. Reproduced with permission from reference [73], Copyright 2018, American Chemical Society.

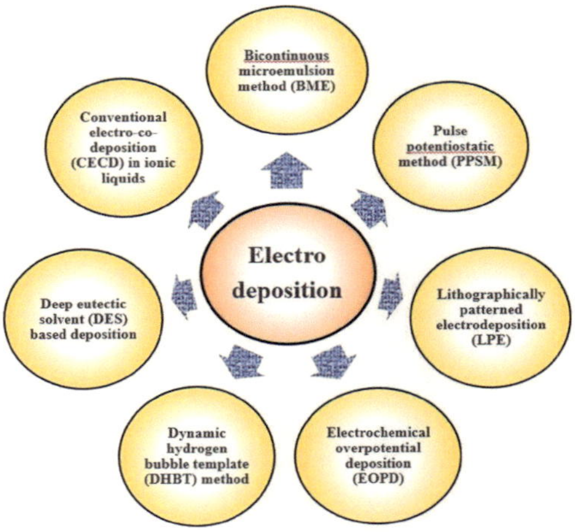

Figure 6.
A general overview of the techniques used for electrodeposition to form thin nanoporous films.

decrease in the particle size in the presence of DES as it has relatively low surface tension [76]. Remarkable optical behavior has been observed when the bulk material is transformed into nanostructured surfaces. A simple two-step electrochemical process has been developed wherein electrodeposition and anodization is used in conjunction to generate black gold surfaces that can absorb more than 93% of the incident light over the entire visible spectrum due to the canopy of dendritic nanostructures within a nanoscale roughness with potential applications in photovoltaic solar cells [77]. **Figure 6** summarizes the general methods used for electrodeposition to form thin nanoporous films.

3. Characterization techniques for NPG thin films

The physical and morphological properties of nanoporous thin films are strongly linked with material's porosity (defined as the ratio of void volume to the total volume of the film). Hence, many efforts have been devoted to developing a reliable self-consistent quantitative characterization of their porosity [78]. The physical characteristics of thin films have been characterized by a combination of X-ray photoelectron spectroscopy (XPS), grazing-incidence small-angle X-ray scattering (GISAXS) along with adsorption isotherm surface area measurements. Porosity and internal feature sizes range from a few to tens of nanometers [79]. Cyclic voltammetry (CV) and electrochemical impedance spectroscopy (EIS) experiments have the potential to find the electrochemically active surface area of the NPG film. Studies have shown that using CV and EIS it was shown that NPG films have 4–8.5 times more accessible surface area than thermally evaporated gold (EG) films [80]. Potential step (PS) chronoamperometry has also been combined with surface plasmon resonance (SPR) for probing electrochemical deposition, conformational changes linked with redox-initiated film reorganization, and the quantification of electrodeposited thin film thickness [81]. A well-recognized approach to determine the specific surface area of nanoporous materials is the Brunauer, Emmett, and Teller (BET) method which is based on physical adsorption of gas molecules to determine the specific surface area [82]. Many groups have extensively used scanning electron microscopy (SEM) for rapidly exploring pore size and ligament size due to their ease of use and applicability to varying types of samples. Elastic modulus is an important mechanical property to calculate residual stress in free-standing beams. This in turn determines film stiffness and therefore, sensor performance [83]. SEM and AFM techniques have been used by many researchers to characterize the surface morphology, size, and shape of the pores as well as the surface roughness of the porous gold films [84]. Maaroof and coworkers have measured the optical properties of porous gold film using the techniques of spectrophotometry and ellipsometry. The spectral response was delivered using a homogeneous Lorentz-Drude (L-D) model and showed that the optical properties of NPG films are dependent on void occupancies [44]. Nanoporous metals have significant geometric complexity in the form of random bicontinuous structures possessing bubbles within ligaments, regions of very high negative, positive, and saddlepoint curvature, and multiple facets. Erlebacher introduced methods to geometrically quantify the structure of nanoporous metals using large-scale kinetic Monte Carlo simulations using mesh-smoothing algorithms [85]. Atom probe tomography (APT) analysis of NPG material, produced by dealloying can give huge information regarding the compositional variation within the structure of nanoscale ligaments. 3-D analysis of materials at the sub-nanometer scale is possible by careful preparation of samples through a reproducible process for complete pore filling through electrodeposition of copper into finely sized pores. Compositional profiling and mapping of ligaments are now possible by APT analysis [86]. Electrodeposition of gold nanostructures having sharper features yield higher refractive index sensitivity and therefore, can be used as transducers in LSPR spectroscopy for probing many types of biomolecular interactions [66].

4. Applications

Nanoporous materials such as carbon nanotubes, nanoporous anodic alumina, nanotubular titania, porous silicon, and NPG have significant potential in the field of biomedicine involving high drug loading capacity and its controlled release. It was seen that a sub-micron-thick sputter-coated NPG thin films have a loading capacity of 1.12 μg/cm^2 and molecular release half-lives between 3.6 hours to 12.8 hours [87]. NPG is a promising material for drug delivery applications and for studying the influence of surface modification on the drug release kinetics due to its high effective surface area, well-known surface gold-thiol chemistry, and tunable pore morphology and is depicted in **Figure** 7 [88].

Rough and activated noble metals exhibit some unexpected properties in comparison with their smooth counterparts. The electrocatalytic activities of the porous films depend on the roughness factor and the existence of special binding sites on the surface. NPG films possess higher roughness and better electron transport leading to its distinguished performance in the field of catalysis [89]. Studies have shown that NPG is active towards catalyzing low-temperature CO oxidation which is attributed to the peculiar structure of NPG and the prevalence of step and kink sites on the surface of the material [90]. Nanoporous metals have recently attracted considerable attention fueled by their potential use in the filed od catalysis, sensor, and actuator applications by increasing the activity drastically via thick oxide film deposition to stabilize the nanoscale morphology [91]. The potential of nanostructured metallic surfaces has also been seen in optical applications and has been demonstrated for biosensing applications, SERS, guiding and manipulating light, and trapping of micro-sized particles [92]. The electrodeposition approach has been used for depositing a thin layer of AuNPs from 10 mM $HAuCl_4$ for 20 s at −0.2 V (vs. Ag/AgCl) yielding oblate particles of 200 nm average diameter that immobilized an aptamer specific for LPS detection achieving a linear range of 0.1–10.24 ng mL^{-1} [93]. Thin nanoporous membranes of gold are best suited to the examination of surface plasmons as the analyte can get into the pores of NPG and can modify their dielectric atmosphere are detectable via absorption peaks in SPR

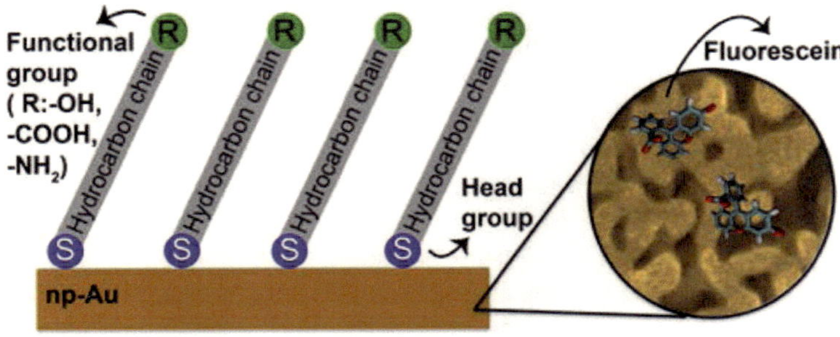

Figure 7.
Depiction of surface engineering of NPG film via immobilizing alkanethiols with varying functional groups and chain lengths to enhance the drug delivery performance monitored via fluorescein release signal. Reproduced with permission from reference [88], Copyright 2016, American Chemical Society.

whereas the species that adsorb onto the geometric surface of the pores of the film are equally detected by SPR measurements [94].

5. Conclusion

The electrodeposition of noble metals on support provides a unique possibility to obtain films of varying thickness and roughness. Many advanced electrodeposition techniques are known so far, and the field is actively developing with specific controllable nanostructural features leading to an increased number of grain boundaries for high catalytic performance [95]. It is interesting to see how the porous gold nanostructures can be electrodeposited on a solid support and the impact of electrodeposition parameters namely, potential and time of deposition, on the morphology and thickness of the film so formed. Detailed investigation on NPG surface pore size and the correlation with electrocatalytic activity has aimed to understand the growth mechanism of NPG [13]. The experimental results from the electrodeposition techniques have highlighted that the formation of a well-organized NPG film requires the appropriate electrochemistry and physics/mechanics interactions between the substrate and the deposits [96]. NPG morphology evolution is highly influenced by the topography of the substrate emphasizing the structure–property relationship and opening doors for such high-throughput combinatorial studies [46]. An exciting new field of research within the domain is the electrodeposition of hybrid thin films which has opened the gate to an unlimited number of new materials [97]. Electrodeposition, therefore, is a promising fabrication technique that has the potential of producing thin NPG films for effective drug delivery systems as precise control of the NPG pore and ligament dimensions can be achieved which in turn control the drug loading and release performance inside the body [98]. Electrodeposition is a distinct form of grain boundary engineering by which a material's property can be enhanced to synthesize advanced materials both in bulk form and as thin films. It is a technologically viable production method for upgrading the mechanical, electrical, magnetic, and corrosion properties exhibited by metals [99].

Conflict of interest

The authors declare no conflict of interest.

Author details

Palak Sondhi and Keith J. Stine*
Department of Chemistry and Biochemistry, University of Missouri–Saint Louis, Saint Louis, MO, USA

*Address all correspondence to: kstine@umsl.edu

References

[1] Njuguna J, Ansari F, Sachse S, Zhu H, Rodriguez VM. 1 - nanomaterials, nanofillers, and nanocomposites: Types and properties. In: Njuguna J, Pielichowski K, Zhu H, editors. Health and Environmental Safety of Nanomaterials: Woodhead Publishing; 2014. p. 3-27.

[2] Kung HH, Kung MC. Nanotechnology: Applications and potentials for heterogeneous catalysis. Catalysis Today. 2004;**97**(4):219-224

[3] Patanaik A, Anandjiwala RD, Rengasamy RS, Ghosh A, Pal H. Nanotechnology in fibrous materials–a new perspective. Textile Progress. 2007;**39**(2):67-120

[4] Yang P, Yan R, Fardy M. Semiconductor nanowire: What's next? Nano Letters. 2010;**10**(5):1529-1536

[5] Taga Y. Recent progress of nanotechnologies of thin films for industrial applications. Materials Science and Engineering: C. 2001;**15**(1):231-235

[6] Seker E, Reed M, Begley M. Nanoporous gold: Fabrication, characterization, and applications. Materials. 2009;**2**(4):2188-2215

[7] Kim SH. Nanoporous gold films as catalyst. In: Mishra NK, editor. Catalytic Application of Nano-Gold Catalysts: IntechOpen; 2016. pp. 1-20

[8] Jia F, Yu C, Ai Z. Zhang L. fabrication of nanoporous gold film electrodes with ultrahigh surface area and electrochemical activity. ChemMater. 2007;**19**:3648-3653

[9] Chuanfang Yu, Falong Jia, Zhihui Ai, Zhang L. Direct Oxidation of Methanol on Self-Supported Nanoporous Gold Film Electrodes with High Catalytic Activity and Stabilit. ChemMater. 2007;19:6065-7

[10] Jia F, Yu C, Deng K. Zhang L. nanoporous metal (Cu, Ag, Au) films with high surface area: General fabrication and preliminary electrochemical performance. Journal of Physical Chemistry C. 2007;**111**:8424-8431

[11] Song T, Yan M, Qian M. The enabling role of dealloying in the creation of specific hierarchical porous metal structures—A review. Corrosion Science. 2018;**134**:78-98

[12] Cherevko S, Chung C-H. Direct electrodeposition of nanoporous gold with controlled multimodal pore size distribution. Electrochemistry Communications. 2011;**13**(1):16-19

[13] Kumar A, Gonçalves JM, Sukeri A, Araki K, Bertotti M. Correlating surface growth of nanoporous gold with electrodeposition parameters to optimize amperometric sensing of nitrite. Sensors and Actuators B: Chemical. 2018;**263**:237-247

[14] He F, Qiao Z, Qin X, Chao L, Tan Y, Xie Q, et al. Dynamic gas bubble template electrodeposition mechanisms and amperometric glucose sensing performance of three kinds of three-dimensional honeycomb-like porous nano-golds. Sensors and Actuators B: Chemical. 2019;**296**:126679

[15] Li Y, Song Y-Y, Yang C, Xia X-H. Hydrogen bubble dynamic template synthesis of porous gold for nonenzymatic electrochemical detection of glucose. Electrochemistry Communications. 2007;**9**(5):981-988

[16] Christie IR, Cameron BP. Gold electrodeposition within the electronics industry. Gold Bulletin. 1994;**27**(1):12-20

[17] Xu Y, Ke X, Yu C, Liu S, Zhao J, Cui G, et al. A strategy for fabricating nanoporous gold films through chemical dealloying of electrochemically deposited Au-Sn alloys. Nanotechnology. 2014;**25**(44):445602

[18] Parida S, Kramer D, Volkert CA, Rösner H, Erlebacher J, Weissmüller J. Volume change during the formation of nanoporous gold by dealloying. Physical Review Letters. 2006;**97**(3):1-4

[19] Zinchenko O, De Raedt HA, Detsi E, Onck PR, De Hosson JTM. Nanoporous gold formation by dealloying: A Metropolis Monte Carlo study. Computer Physics Communications. 2013;**184**(6):1562-1569

[20] Ding Y, Erlebacher J. Nanoporous metals with controlled multimodal pore size distribution. Journal of the American Chemical Society. 2003;**125**(26):7772-7773

[21] Chen AY, Wang JW, Wang Y, Jia YQ, Gu JF, Xie XF, et al. Effects of pore size and residual Ag on electrocatalytic properties of nanoporous gold films prepared by pulse electrochemical dealloying. Electrochimica Acta. 2015;**153**:552-558

[22] Shao K, Fang C, Yao Y, Zhao C, Yang Z, Liu J, et al. An easily modified method using FeCl 3 to synthesize nanoporous gold with a high surface area. RSC advances. 2017;**7**(30):18327-18332

[23] Xue Y, Scaglione F, Celegato F, Denis P, Fecht H-J, Rizzi P, et al. Shape controlled gold nanostructures on de-alloyed nanoporous gold with excellent SERS performance. Chemical Physics Letters. 2018;**709**:46-51

[24] Song T, Yan M, Shi Z, Atrens A, Qian M. Creation of bimodal porous copper materials by an annealing-electrochemical dealloying approach. Electrochimica Acta. 2015;**164**:288-296

[25] Bormashenko E. Breath-figure self-assembly, a versatile method of manufacturing membranes and porous structures: Physical, chemical and technological aspects. Membranes (Basel). 2017;**7**(3):1-20

[26] Ariga K, Nishikawa M, Mori T, Takeya J, Shrestha LK, Hill JP. Self-assembly as a key player for materials nanoarchitectonics. Science and Technology of Advanced Materials. 2019;**20**(1):51-95

[27] Brust M, Bethell D, Kiely CJ, Schiffrin DJ. Self-assembled gold nanoparticle thin films with nonmetallic optical and electronic properties. Langmuir. 1998;**14**(19):5425-5429

[28] Zhang R, Olin H. Porous gold films-a short review on recent Progress. Materials (Basel). 2014;**7**(5):3834-3854

[29] Ananthi A, Phani KL. Self-assembly of gold nanoparticles on sulphide functionalized polydopamine in application to electrocatalytic oxidation of nitric oxide. Journal of Electroanalytical Chemistry. 2016;**764**:7-14

[30] Li C, Dag Ö, Dao TD, Nagao T, Sakamoto Y, Kimura T, et al. Electrochemical synthesis of mesoporous gold films toward mesospace-stimulated optical properties. Nature Communications. 2015;**6**(1):1-8

[31] Watanabe S, Shibata H, Sakamoto F, Azumi R, Sakai H, Abe M, et al. Directed self-assembly of gold nanoparticles and gold thin films on micro- and nanopatterned templates fabricated from mixed phase-separated Langmuir-Blodgett films. Journal of Materials Chemistry. 2009;**19**(37):6796-6803

[32] Kao J, Thorkelsson K, Bai P, Zhang Z, Sun C, Xu T. Rapid fabrication of hierarchically structured

supramolecular nanocomposite thin films in one minute. Nature Communications. 2014;**5**(1):1-8

[33] Song L, Niedermeier MA, Körstgens V, Löhrer FC, Chen Y, Roth SV, et al. In situ study of sputtering nanometer-thick gold films onto 100-nm-thick Spiro-OMeTAD films: Implications for perovskite solar cells. ACS Applied Nano Materials. 2020;**3**(6):5987-5994

[34] Libansky M, Zima J, Barek J, Reznickova A, Svorcik V, Dejmkova H. Basic electrochemical properties of sputtered gold film electrodes. Electrochimica Acta. 2017;**251**:452-460

[35] Xenogiannopoulou E, Aloukos P, Couris S, Kaminska E, Piotrowska A, Dynowska E. Third-order nonlinear optical properties of thin sputtered gold films. Optics Communications. 2007;**275**(1):217-222

[36] Augis JA, Lo CC, Pinnel MR. The hardness and ductility of sputtered gold films. Thin Solid Films. 1979;**58**(2):357-363

[37] Hong C, Kim H, Park S, Lee C. Optical properties of porous silicon coated with ultrathin gold film by RF-magnetron sputtering. Journal of the European Ceramic Society. 2010;**30**(2):459-463

[38] Alvarez R, García-Martín JM, Macías-Montero M, Gonzalez-Garcia L, González JC, Rico V, et al. Growth regimes of porous gold thin films deposited by magnetron sputtering at oblique incidence: From compact to columnar microstructures. Nanotechnology. 2013;**24**(4):1-10

[39] Ruffino F, Torrisi V, Marletta G, Grimaldi MG. Growth morphology of nanoscale sputter-deposited Au films on amorphous soft polymeric substrates. Applied Physics A. 2011;**103**(4):1-11

[40] Siegel J, Lyutakov O, Rybka V, Kolská Z, Švorčík V. Properties of gold nanostructures sputtered on glass. Nanoscale Research Letters. 2011;**6**(1):1-9

[41] Wi J-S, Tominaka S, Uosaki K, Nagao T. Porous gold nanodisks with multiple internal hot spots. Physical Chemistry Chemical Physics. 2012;**14**(25):9131-9136

[42] Rho S, Jahng D, Lim JH, Choi J, Chang JH, Lee SC, et al. Electrochemical DNA biosensors based on thin gold films sputtered on capacitive nanoporous niobium oxide. Biosensors and Bioelectronics. 2008;**23**(6):852-856

[43] Feng J-J, Lv Z-Y, Qin S-F, Li A-Q, Fei Y, Wang A-J. N-methylimidazole-assisted electrodeposition of Au porous textile-like sheet arrays and its application to electrocatalysis. Electrochimica Acta. 2013;**102**:312-318

[44] Zhang R, Olin H. Porous gold films—A short review on recent progress. Materials. 2014;**7**(5):3834-3854

[45] Deng Y, Huang W, Chen X, Li Z. Facile fabrication of nanoporous gold film electrodes. Electrochemistry Communications. 2008;**10**(5):810-813

[46] Chapman CAR, Daggumati P, Gott SC, Rao MP, Seker E. Substrate topography guides pore morphology evolution in nanoporous gold thin films. Scripta Materialia. 2016;**110**:33-36

[47] Kumar A, Furtado VL, Gonçalves JM, Bannitz-Fernandes R, Netto LES, Araki K, et al. Amperometric microsensor based on nanoporous gold for ascorbic acid detection in highly acidic biological extracts. Analytica Chimica Acta. 2020;**1095**:61-70

[48] Vahdatkhah P, Sadrnezhaad SK. Influence of substrate, additives, and pulse parameters on electrodeposition

of gold nanoparticles from potassium Dicyanoaurate. Metallurgical and Materials Transactions B. 2015;**46**(6):2584-2592

[49] Zhang K, Wei J, Zhu H, Ma F, Wang S. Electrodeposition of gold nanoparticle arrays on ITO glass as electrode with high electrocatalytic activity. Materials Research Bulletin. 2013;**48**(3):1338-1341

[50] Kobayashi C, Saito M, Homma T. Laterally enhanced growth of electrodeposited Au to form ultrathin films on nonconductive surfaces. Electrochimica Acta. 2012;**74**:235-243

[51] Walcarius A. Template-directed porous electrodes in electroanalysis. Analytical and Bioanalytical Chemistry. 2010;**396**(1):261-272

[52] Bollella P. Porous gold: A new frontier for enzyme-based electrodes. Nanomaterials. 2020;**10**(4):722

[53] Shiba S, Hirabayashi S, Niwa O, Kato D, Kunitake M, Matsuguchi M. Monolithic Au nanoscale films with tunable nanoporosity prepared via dynamic soft templating for electrocatalytic oxidation of methanol. ACS Applied Nano Materials. 2020;**3**(8):7750-7760

[54] Josell D, Levin I, Moffat TP. Morphological transitions during Au electrodeposition: From porous films to compact films and nanowires. Journal of The Electrochemical Society. 2015;**162**(12):1-13

[55] Regiart M, Gimenez AM, Lopes AT, Carreño MNP, Bertotti M. Ultrasensitive microfluidic electrochemical immunosensor based on electrodeposited nanoporous gold for SOX-2 determination. Analytica Chimica Acta. 2020;**1127**:122-130

[56] Hernández-Saravia LP, Sukeri A, Bertotti M. Fabrication of nanoporous gold-islands via hydrogen bubble template: An efficient electrocatalyst for oxygen reduction and hydrogen evolution reactions. International Journal of Hydrogen Energy. 2019;**44**(29):15001-15008

[57] Sanzó G, Taurino I, Antiochia R, Gorton L, Favero G, Mazzei F, et al. Bubble electrodeposition of gold porous nanocorals for the enzymatic and non-enzymatic detection of glucose. Bioelectrochemistry. 2016;**112**:125-131

[58] Dorofeeva TS, Seker E. Electrically tunable pore morphology in nanoporous gold thin films. Nano Research. 2015;**8**(7):2188-2198

[59] Bindra P, Light D, Freudenthal P, Smith D. The effect of base metal ions on the electrochemical and structural characteristics of electrodeposited gold films. Journal of The Electrochemical Society. 1989;**136**(12):3616-3626

[60] Bozzini B, Giovannelli G, Natali S, Fanigliulo A, Cavallotti PL. Crystallographic structure of gold films electrodeposited at low current densities. Journal of Materials Science. 2002;**37**(18):3903-3913

[61] Tian Y, Liu H, Zhao G, Tatsuma T. Shape-controlled electrodeposition of gold nanostructures. The Journal of Physical Chemistry B. 2006;**110**(46):23478-23481

[62] Monzon LMA, Byrne F, Coey JMD. Gold electrodeposition in organic media. Journal of Electroanalytical Chemistry. 2011;**657**(1):54-60

[63] Booth SG, Alghamdi RG, Belić D, Brust M. Electrodeposition of gold nanostructures at the Interface of a Pickering emulsion. ChemElectroChem. 2018;**5**(15):2055-2058

[64] Bechelany M, Brodard P, Elias J, Brioude A, Michler J, Philippe L. Simple synthetic route for SERS-active gold nanoparticles substrate with controlled

shape and organization. Langmuir. 2010;**26**(17):14364-14371

[65] Mahajan S, Baumberg JJ, Russell AE, Bartlett PN. Reproducible SERRS from structured gold surfaces. Physical Chemistry Chemical Physics. 2007;**9**(45):6016-6020

[66] Bhattarai JK, Sharma A, Fujikawa K, Demchenko AV, Stine KJ. Electrochemical synthesis of nanostructured gold film for the study of carbohydrate–lectin interactions using localized surface plasmon resonance spectroscopy. Carbohydrate Research. 2015;**405**:55-65

[67] Ma Q, Pang L, Li M, Zhang Y, Ren X, Liu SF. Controlled electrodeposition of Au monolayer film on ionic liquid. Applied Surface Science. 2016;**371**:258-261

[68] Menke EJ, Thompson MA, Xiang C, Yang LC, Penner RM. Lithographically patterned nanowire electrodeposition. Nature Materials. 2006;**5**(11):914-919

[69] Halpern AR, Corn RM. Lithographically patterned electrodeposition of gold, silver, and nickel Nanoring arrays with widely tunable near-infrared plasmonic resonances. ACS Nano. 2013;7(2):1755-1762

[70] Zhang X, Li D, Bourgeois L, Wang H, Webley PA. Direct electrodeposition of porous gold nanowire arrays for biosensing applications. ChemPhysChem. 2009;**10**(2):436-441

[71] Raub CJ, Knödler A. The electrodeposition of gold by pulse plating. Gold Bulletin. 1977;**10**(2):38-44

[72] Sai-Anand G, Gopalan AI, Kang S, Lee K. Fabrication of gold Nanoflower anchored conducting polymer hybrid film electrode by pulse Potentiostatic deposition. IEEE Electron Device Letters. 2013;**34**(8):1065-1067

[73] Chen Q, Switzer JA. Photoelectrochemistry of ultrathin, semitransparent, and catalytic gold films electrodeposited epitaxially onto n-silicon (111). ACS Applied Materials & Interfaces. 2018;**10**(25):21365-21371

[74] Wu B, Hou S, Xue Y, Chen Z. Electrodeposition–assisted assembled multilayer films of gold nanoparticles and glucose oxidase onto polypyrrole-reduced graphene oxide matrix and their electrocatalytic activity toward glucose. Nanomaterials. 2018;**8**(12):1-12

[75] Rong K, Huang L, Zhang H, Zhai J, Fang Y, Dong S. Electrochemical fabrication of nanoporous gold electrodes in a deep eutectic solvent for electrochemical detections. Chemical Communications. 2018;**54**(64):8853-8856

[76] Renjith A, Roy A, Lakshminarayanan V. In situ fabrication of electrochemically grown mesoporous metallic thin films by anodic dissolution in deep eutectic solvents. Journal of Colloid and Interface Science. 2014;**426**:270-279

[77] Zheng H, Picard C, Ravaine S. Nanostructured gold films exhibiting almost complete absorption of light at visible wavelengths. Frontiers of Chemical Science and Engineering. 2018;**12**(2):247-251

[78] Shpigel N, Sigalov S, Malchik F, Levi MD, Girshevitz O, Khalfin RL, et al. Quantification of porosity in extensively nanoporous thin films in contact with gases and liquids. Nature Communications. 2019;**10**(1):1-9

[79] Dixon MC, Daniel TA, Hieda M, Smilgies DM, Chan MHW, Allara DL. Preparation, structure, and optical properties of nanoporous gold thin films. Langmuir. 2007;**23**(5):2414-2422

[80] Ahl S, Cameron PJ, Liu J, Knoll W, Erlebacher J, Yu F. A comparative plasmonic study of nanoporous and evaporated gold films. Plasmonics. 2008;**3**(1):13-20

[81] Liu Q, Wang X, Benedict A, Janibekyan L, Su SW, Wang Y, et al. Surface plasmon resonance coupled with potential-step Chronoamperometry: Theory and applications for quantitative measurements of electrodeposited thin films. Electroanalysis. 2019;**31**(11):2155-2161

[82] Detsi E, De Jong E, Zinchenko A, Vuković Z, Vuković I, Punzhin S, et al. On the specific surface area of nanoporous materials. Acta Materialia. 2011;**59**(20):7488-7497

[83] Seker E, Reed ML, Begley MR. Nanoporous gold: Fabrication, characterization, and applications. Materials. 2009;**2**(4):2188-2215

[84] Shulga OV, Jefferson K, Khan AR, D'Souza VT, Liu J, Demchenko AV, et al. Preparation and characterization of porous gold and its application as a platform for immobilization of acetylcholine esterase. Chemistry of Materials. 2007;**19**(16):3902-3911

[85] Erlebacher J, McCue I. Geometric characterization of nanoporous metals. Acta Materialia. 2012;**60**(17):6164-6174

[86] El-Zoka AA, Langelier B, Botton GA, Newman RC. Enhanced analysis of nanoporous gold by atom probe tomography. Materials Characterization. 2017;**128**:269-277

[87] Kurtulus O, Daggumati P, Seker E. Molecular release from patterned nanoporous gold thin films. Nanoscale. 2014;**6**(12):7062-7071

[88] Polat O, Seker E. Effect of surface–molecule interactions on molecular loading capacity of nanoporous gold thin films. The Journal of Physical Chemistry C. 2016;**120**(34):19189-19194

[89] Yu C, Jia F, Ai Z, Zhang L. Direct oxidation of methanol on self-supported nanoporous gold film electrodes with high catalytic activity and stability. Chemistry of Materials. 2007;**19**(25):6065-6067

[90] Detisch MJ, Balk TJ, Bhattacharyya D. Synthesis of catalytic nanoporous metallic thin films on polymer membranes. Industrial and Engineering Chemistry Research. 2018;**57**(12):4420-4429

[91] Biener MM, Biener J, Wichmann A, Wittstock A, Baumann TF, Bäumer M, et al. ALD functionalized nanoporous gold: Thermal stability, mechanical properties, and catalytic activity. Nano Letters. 2011;**11**(8):3085-3090

[92] Oo SZ, Silva G, Carpignano F, Noual A, Pechstedt K, Mateos L, et al. A nanoporous gold membrane for sensing applications. Sensing and Bio-Sensing Research. 2016;**7**:133-140

[93] Sondhi P, Maruf MHU, Stine KJ. Nanomaterials for biosensing lipopolysaccharide. Biosensors (Basel). 2019;**10**(1):2

[94] Yu F, Ahl S, Caminade A-M, Majoral J-P, Knoll W, Erlebacher J. Simultaneous excitation of propagating and localized surface plasmon resonance in nanoporous gold membranes. Analytical Chemistry. 2006;**78**(20):7346-7350

[95] Napolskii KS, Barczuk PJ, Vassiliev SY, Veresov AG, Tsirlina GA, Kulesza PJ. Templating of electrodeposited platinum group metals as a tool to control catalytic activity. Electrochimica Acta. 2007;**52**(28):7910-7919

[96] He F, Qiao Z, Qin X, Chao L, Tan Y, Xie Q, et al. Dynamic gas bubble template electrodeposition mechanisms

and amperometric glucose sensing performance of three kinds of three-dimensional honeycomb-like porous nano-golds. Sensors and Actuators B: Chemical. 2019;**296**:1-10

[97] Yoshida T, Zhang J, Komatsu D, Sawatani S, Minoura H, Pauporté T, et al. Electrodeposition of inorganic/organic hybrid thin films. Advanced Functional Materials. 2009;**19**(1):17-43

[98] Nana ABA, Marimuthu T, Kondiah PP, Choonara YE, Du Toit LC, Pillay V. Multifunctional magnetic nanowires: Design, fabrication, and future prospects as cancer therapeutics. Cancers. 2019;**11**(12):1-23

[99] Erb V. Electrodeposited nanocrystals: Synthesis, structure, properties and future applications. Canadian Metallurgical Quarterly. 1995;**34**(3):275-280

Chapter 10

Micro Nano Manufacturing Methods for Chemical, Gas and Bio Sensors, Water Purification and Energy Technologies

Amos Adeleke Akande, Aderemi Timothy Adeleye, Abraham Abdul Adenle and Bonex Wakufwa Mwakikunga

Abstract

This chapter reports on the various methods of fabricating and manufacturing micro and nano sensor, membrane and energy devices. Firstly, the characteristic often sought after by scientists and engineers for effective and efficient performance of these technologies were thoroughly discussed in details together with the characterization techniques for evaluating them. Several state-of-the-art fabricating techniques for sensor devices, water and medical based-membranes, solar cells and batteries were also discussed.

Keywords: micro-nano device, fabrication, sensors, manufacturing, membrane, battery, solar cell

1. Introduction

1.1 Nanoscience and nanotechnology

Nanoscience can be described as the study of the phenomena and manipulation of materials at atomic, molecular and macromolecular scales, where properties differ specifically from those at a larger scale (macro scale). The macroscopic objects we see around us in our day-to-day activities are the products made from bulk materials. These objects possess physical properties that are in some way different from nano and the intermediate scale called micron-sized material (such as grains of sand or dust produced during volcano eruption). However, bulk and nanomaterial may share the same constituent but the dimension or length scale usually distinguishes between the two groups [1, 2]. Nanometer scale is conventionally defined as 1 to 100 nm which simply means one billionth of a metre (10^{-9} m). The lowest limit of nanometer size range is normally set to 1 nm which is very close to the length of a single atom since the atomic radius is just by a little femtometre less than 1 nm. However, nanoscience is not just the science of small-scale material but also the science in which materials with small dimension (in other words shape) show new physical phenomena. For instance, the principles of classical physics such as energy, force, momentum, space, time, and so on, that govern the behavior of macroscopic and microscopic systems (bulk material) are no longer applicable to nanoscale materials [3–4]. This Nanoscience is not new per se, it is a name that was given to a number of fields of research that share common principles, and hence it is referred to as an interdisciplinary science. Nanotechnology integrates a wide range of sciences which includes; Physics, Chemistry, Biology, Microbiology, Engineering, Surface Science, and Biotechnology, and apply them to practical devices [5]. There are two major approaches normally employ in fabrication techniques namely; **top-down approach** (Larger to smaller: a materials perspective) and **bottom-up approach** (Simple to complex: a molecular perspective). Top-down approach involves creating Nano-scale materials by physically or chemically breaking down larger materials. These include statistical mechanical effects, as well as quantum mechanical effects. Solid-state techniques can also be used to create devices known as nanoelectromechanical systems or NEMS, which are related to micromechanical systems or MEMS [6] while bottom-up approach simply involves simple to complex: i.e. a molecular perspective technique. These techniques are used today to manufacture a wide variety of useful chemicals such as pharmaceuticals or commercial polymers. Molecular nanotechnology, sometimes called molecular manufacturing, describes engineered nanosystems (nanoscale machines) operating on the molecular scale. Molecular nanotechnology is especially associated with the molecular assembler, a machine that can produce a desired structure or device atom-by-atom using the principles of mechanosynthesis [7].

2. Significance of micro and nano fabrication in novel devices and technologies

Micro and nano fabrication is an essential process in the manufacturing of novel devices and technologies. Many sciences, technology and engineering oriented products are developed using the concept of micro and nano fabrication. From radio transistors, integrated circuits, personal computers, to micromechanical systems (MEMS), transducers, sensors, batteries and super capacitors, solar cells, water treatment membranes and filters and other novel devices, micro and nano techniques have played significant and important role in realizing reliable technology. However, huge credit relating to the success of these technologies must be ascribed to the materials development and analyses techniques such as the analytical, macroscopic, microscopy and spectroscopy ones.

For instance, before one can realize product of gas sensor, chemical sensor and biosensor device, especially in the case of metal oxide semiconductors, carbon materials and polymers, critical studies and analyses of the materials properties is required to qualify the performance of the sensing element. The first set of investigation which must be performed on the materials intended to build these devices are crystal structure and microstructures, morphological and surface roughness studies, defects studies, thermal stability and adsorption property [8–10]. Hence, the material must be thoroughly characterized with X-ray diffraction (XRD) and high-resolution transmission electron microscopy (HR-TEM) methods to study its crystal structure. X-ray diffraction spectroscopy which is commonly used technique for characterization of crystalline materials provides information about elemental analyses such as structures, phases and preferred crystal orientations. Physical measurements like average particle size of material, homogeneous and inhomogeneous strain and crystal defect could also be estimated from the data collected using XRD technique [8–11]. The HRTEM approach has been severally employed in sensors material research to unveil material's crystallographic structures at an atomic scale [8–9]. The scanning electron microscope (SEM), scanning tunneling microscope (STM) and atomic force microscope (AFM) are important for all surface structure studies such as morphology, particles distribution, nanoscale topography and surface roughness [9–13]. These are also essential properties needed to be analyzed for chemical, gas and bio sensors devices fabrication. Other properties necessary to be investigated for these types of application include quantitative analysis of the material's elemental composition and chemical state. This study is often achieved using X-ray photoelectron spectroscopy (XPS) [9–11]. The adsorption ability and properties of the sensors materials are usually evaluated using the popular Brunauer–Emmett–Teller (BET) technique which relies on the physical adsorption (physisorption) of gas molecules on the surface of solid-state materials. With this technique, information about the specific surface area, microporous, nanoporous and mesoporous of a sensor's material can be acquired [9, 14–16].

In the same way, materials for fabricating water and medical membranes and energy devices such as solar cell, lithium and sodium ion batteries also required critical studies with the above materials characterization techniques before the manufacturing process could be initialized. The properties which Materials Scientist and Engineers are usually sough for when building in solar cell architecture are crystal and microstructures of all the semiconductors and polymers involved. These techniques are necessary to unveiled the effect of the grain's boundaries on the charge transfer of the cell, especially when device is of a p-n junction or multi-junction type [17]. This often help materials engineers in proper understanding of interfacial properties of the cell [18]. Studies of morphology, surface roughness and topology is also of a great importance when evaluating the solar cell materials for prototyping and manufacturing. This is needed to ensure a homogeneous film surface in a bid to enhance the transport of the charges for an improve energy conversion efficiency (ECE) [19]. Thermal stability studies with Thermogravimetric analysis (TGA) are another important method adopted by materials scientist to study the degradation of solar cell device [8, 20]. Lithium and sodium ion batteries are not an exception when it comes to their materials development and analyses. TGA techniques are often used to study the thermal stability, XRD and HRTEM for crystal, micro structures, particles size analysis and how monodisperse the particle are before fabricating the device. The SEM, STM and AFM techniques are being employed for particles morphology, surface roughness and topography [21].

XPS is another important technique for qualifying materials for lithium and sodium ion battery application. XPS is suitable to give important information about the interaction of membrane-based materials with electrolyte materials and further assist to develop a definite insight of interfacial structure and as well performance of the battery. From one of the previous published studies XPS was employed and useful information of the membrane interaction with vanadium electrolytes was revealed which led to understanding of interfacial structure and battery performance [22]. Nanosized fibers have great advantages owing to their high specific surface area to volume ratio, electrospun nanofibers have find their useful applications in the field of clean energy (solar cells, fuel cells and batteries), electronics, health (biomedical scaffolds, artificial organs), and environment (filter membranes) [23].

3. Advance manufacturing methods for chemical, gas and bio sensor applications

Prototyping and manufacturing sensor devices (Gas, Chemical or Bio sensor) required that the sensors materials be deposited or coated on an electrode for easy contact and connection to the device electronic circuitry or source measuring unit of the gas sensing and test station. Interdigitated electrode (IDE) type have been widely used for sensors laboratory research, prototyping and manufacturing of sensors and related products. This is a cost-effective method which often made from an aluminum oxide (Al_2O_3) substrate whose front-side surface is coated with platinum (Pt) metal in comb-like structures for sensors electrical signal measurement and the rear-side coated with nickel (Ni) metal as Microheater [9, 24]. **Figure 1** showed a schematic layout of KSGA565 KENOSTATIC gas detection station where μ-nano IDE was used as the sensor's electrode. The layout consists of an enclosed chamber called sensing chamber containing IDE with deposited sensor material. The front-side of the IDE is connected to the Keithley pico-meter source meter and the rear-side to the power supply.

During fabrication, sensor materials are usually deposited onto the IDE using micro-nano deposition technologies such as chemical vapor deposition (CVD), pulse laser deposition (PLD), physical vapor deposition (PVD) and magnetron sputtering technique [25–27]. These technologies are physical methods which have been reported to offer thin and homogenous film surface with excellent gas, chemical and bio molecule sensing properties. These technologies have also been used severally to deposit non-IDE pattern like glass, silicon wafer etc. for sensors and related device fabrication [28].

The printed patterned substrate and Lab-on-a-chips are another micro-nano contacting and printing technology commonly used when manufacturing Gas, Chemical or Bio sensor devices. These techniques are expensive and regarded as state of the art method which required specialized equipment like photolithography (PL), plasma enhanced chemical vapor deposition (PECVD) and electron beam lithography (EBL). The methods offer patterned deposition of nanostructures such as nanowires, nano-rods, nano-tubes etc., high precision contacting, highly aligned printing and deposition onto flexible substrates as advantages over others [25–30].

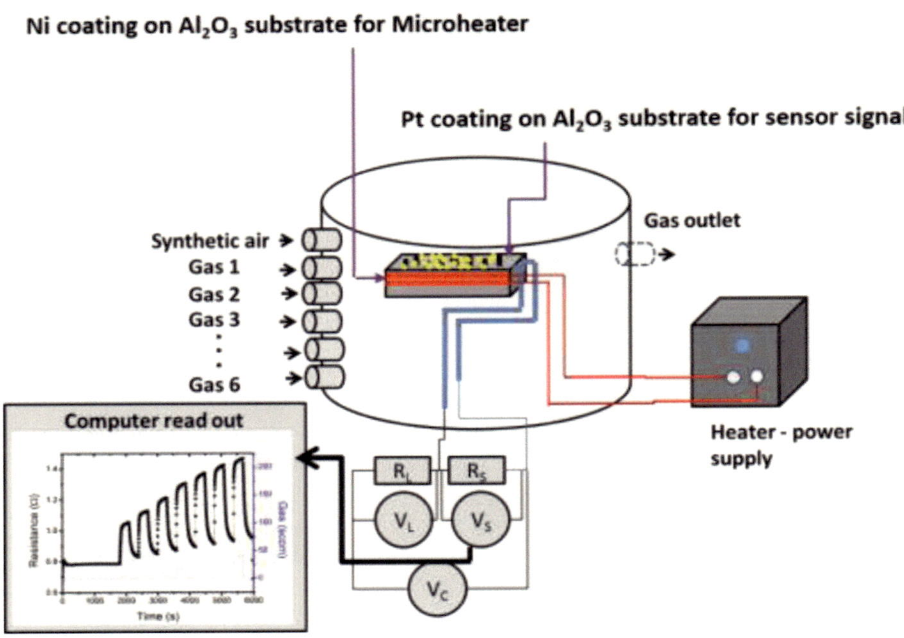

Figure 1.
Schematic diagram of KSGA565 KENOSISTEC gas sensing station illustrating how the μ-nano IDE sensor can be tested. The electronic circuit displays of the gas sensor's element showed R_L, which is the load resistor connected in series with the sensor's element ($R_L = (V-V_S)/I$). V_L is the voltage on the R_L, $V_S = V_C - IR_L$, represent the sensor's signal voltage. V_C is a constant voltage applied on the R_L and sensor's element and finally, R_S is the sensor's resistance ($R_S = V_S/I$). Adapted from Ref. [24].

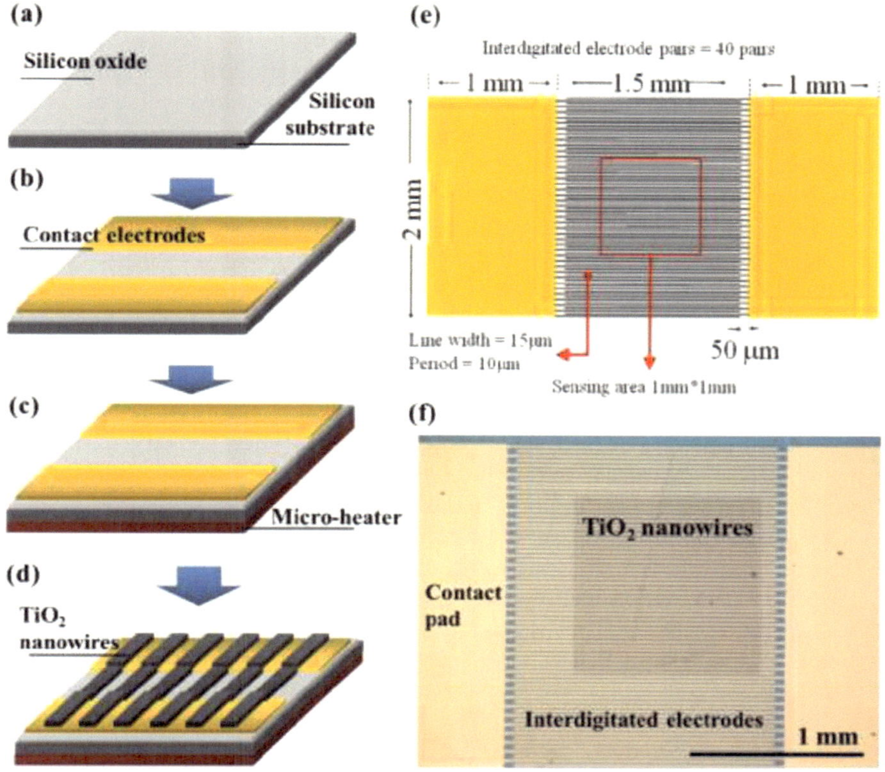

Figure 2.
Fabrication process of electron beam patterned TiO2 based gas sensor; (a) oxidation process of Si wafer, (b) Cr/Au contact fabrication, (c) microheater fabrication, (d) photoresist deposition, lift-off and TiO2 nanowire arry deposition, (e) showed the dimensions of each section of the device and (f) the optical image of the entire device. Adapted from Ref. [30].

A typical process involving the fabrication of TiO2 nanowires-based gas sensor is shown in **Figure 2**. The materials used for the fabrication are; p-type silicon wafer (**Figure 2(a)** and interdigitated Cr/Au electrodes which was initially fabricated using PL process on an oxidized Si substrate (**Figure 2(b)** [30]. The Cr and Au thin films were also blank deposited on the rear-side of the silicon wafer in an interdigitated fashion to make heating element (Microheater) (**Figure 2(c)**). Thereafter, EBL approach was used to pattern the chip surface and to produce photoresist on the film before depositing the p-type TiO2 on the top of the chip with aid of sputter machine. The photoresist was later lift-off to form the TiO2 nanowire array as shown in **Figure 2(d)**. **Figure 2(e)** showed the dimensions of each section of the device and **Figure 2(f)** the optical image of the entire device.

4. Advance manufacturing methods for water purification, lithium ion batteries and medical applications

Electrospinning is one of the techniques suitable for the fabrication of materials through innovative technology. Membrane-based technologies through electrospinning have been employing for the fabrication of both nano- and micro-based materials which finds useful applications in various fields such as in the water purification, lithium ion batteries, medical applications etc.

4.1 Applications of electrospun fibers in water purification

Electrospinning is a fabrication technique that involves application of a high electric field to generate nanofibers from a charged polymer solution or melt. it is a useful method for the fabrication of complex structures consisting on continuous fibers. The morphology of electrospun fibers can be controlled by adjusting experimental parameters, such as precursor solution concentration, type of spinneret, voltage and the spinneret-collector distance. Using this technique affords us numerous benefits such non-complicated and inexpensive equipment, easy to modify, ability to carefully monitor the morphology of materials, and as well almost all polymers with even high molecular are applicable in the synthesis [23]. The chemical properties of electrospun fibers are mainly influenced by two factors: hydrophilicity and chemical composition of the fibers. The characterization of the mechanical features is critical for the electrospun nanofibers. It can be stated that the electrospun nanofiber membranes are appropriate for the pressure driven membrane procedures where the target product is mainly the permeate phase, for example, water/wastewater treatments [31]. Water purification is mostly defined by filtration through size exclusion or adsorption. The water purification process is classified according to the average pore size of the materials and applications include microfiltration (MF) (0.1-10 μm), ultrafiltration (UF) (0.001–0.1 μm), nanofiltration (NF) (0.001–0.01 μm), reverse osmosis (RO) (0.0001–0.001 μm), and forward osmosis (FO) (0.0001–0.001 μm) [32]. In a study conducted by Mahadevappa Y et.al, where electrospinning was used to fabricate nanofibrous membranes for MF applications using polyvinyl alcohol. Owing to its cost-effectiveness, stability (thermally and chemically) and non-degradability, poly (vinyl alcohol) was selected as a precursor in the fabrication process [33]. However, the poly (vinyl alcohol) nanofiber membranes, produced from electrospinning process, must be treated with cross-linking agents for preparing a 3-D waterproof system before being utilized as water filters [31]. Liu's team has introduced a nanofiber MF membrane that required doping with copper oxide (CuO) nanosheets (**Figure 3**). The fabricated membrane has a separation efficiency of >99.89% for polystyrene (PS) microspheres with a diameter > 300 nm in water [34]. The introduction of such functional materials can not only achieve the corresponding modification purpose, but also enhance static electricity to improve the strength of individual nanofibers. Stable high porosity, good interconnectivity, and ultra-thin membrane thickness are key major factors responsible for its strong permeate flux and excellent bacteria rejection efficiency [35].

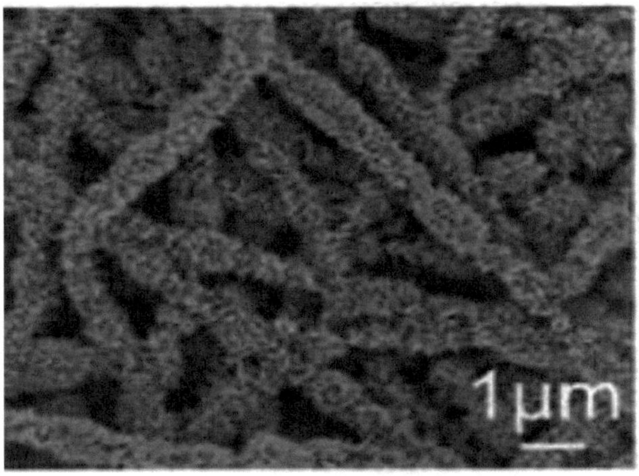

Figure 3.
Morphology of PVDF/CuO nanosheet nanofiber MF membrane. Adapted from Ref. [34].

4.2 Applications of electrospun fibers in lithium ion batteries

Electrospun carbon nanofibers exhibit favorable properties, such as nanometer-sized diameters, high specific surface areas, and web morphologies, making them highly suitable for an anode material. Electrospinning has been identified as the most promising route for designing novel anode materials and structures, owing to its simple process setup. The electrospinning technique is suitable for the implementation of existing anode material research based on the process being able to mass-produce anodes [36]. In a study conducted by Peng et al. and co-workers, the porous carbon nanofibers were synthesized using a PAN/polymethyl methacrylate (PMMA) precursor solution with the aid of electrospinning technique. PMMA is immiscible with PAN, during the course of preparation macro phase separation was observed and was then thermally treated at high reaction temperature-800°C which caused elimination of PMMA while creating pores on the surface of the fiber. In order to investigate the fiber morphology and the electrochemical performance of carbon nanofibers, the author varied the concentration effect of PMMA in the precursor solution. The variation of PMMA showed that its addition significantly improves the surface area and pore volume of the prepared fibers.

The morphologies of the electrospun fibers after carbonization are shown in **Figure 4**. In **Figure 4(a)**, the carbon nanofibers prepared using neat PAN exhibited long and bead-free morphology. By contrast, the PAN/PMMA-derived carbon nanofibers were uneven and interconnected, particularly for 5:5 PAN/PMMA-derived carbon nanofibers (**Figure 4(c, e)**). The interconnected structure was attributed to the presence of PMMA. PMMA is a thermally liable polymer, which melts during pyrolysis. **Figure 4** also provides the inner structure of the nanofibers. As observed in **Figure 4(b)**, neat PAN-derived carbon nanofibers were internally nonporous. The introduction of PMMA in precursor solution facilitated the development of pores and channels inside the carbon nanofibers (**Figure 4(d, f)**). The availability of the fiber morphology consequently resulted to highly efficient discharge capacity compared to counterpart neat PAN-prepared carbon nanofibers. Therefore, the 5:5 PAN/PMMA-derived carbon nanofibers exhibited a discharge capacity of 446 mAh/g at a current density of 150 mA/g. They exhibited a discharge capacity of 354 mAh/g after 100 cycles at a current density of 200 mA/g equivalent to 67% retention, demonstrating the favorable cycle stability. The significance of their study was based on the manipulation of morphology of electrospun carbon nanofibers for the use as anode materials for lithium ion batteries application to secure good performance. Therefore, it can be said that the superior electrochemical performance of the PAN/PMMA-derived carbon nanofibers was mainly attributed to the prevalent mesopore volume and the high-specific surface area which earned them desired contact between the fibers and electrolyte and consequently improved the diffusion of electrolyte ions into the material [37].

4.3 Electrospun fibers in biomedical applications

Electrospun nanofibers are materials of multi-applications, hence have been widely studied in the field of biomedical and tissue engineering owing to their good characteristic properties and suitability to be incorporated into various morphologies to stir the desired influence in them, such as nonwoven form, aligned nanofibers, core–shell structure, and hybrid nanocomposites. The interesting characteristic properties of electrospun nanofibers- loose structure, high porosity, and superb flexibility possess perfect features to mimic the extracellular matrix (ECM) for cells to grow and, therefore, they have been employed in tissue engineering applications [38]. In a study, a composite nanofiber scaffold made of poly (vinyl alcohol)–poly (vinyl acetate) (PVA–PVAc) was manufactured and subsequently

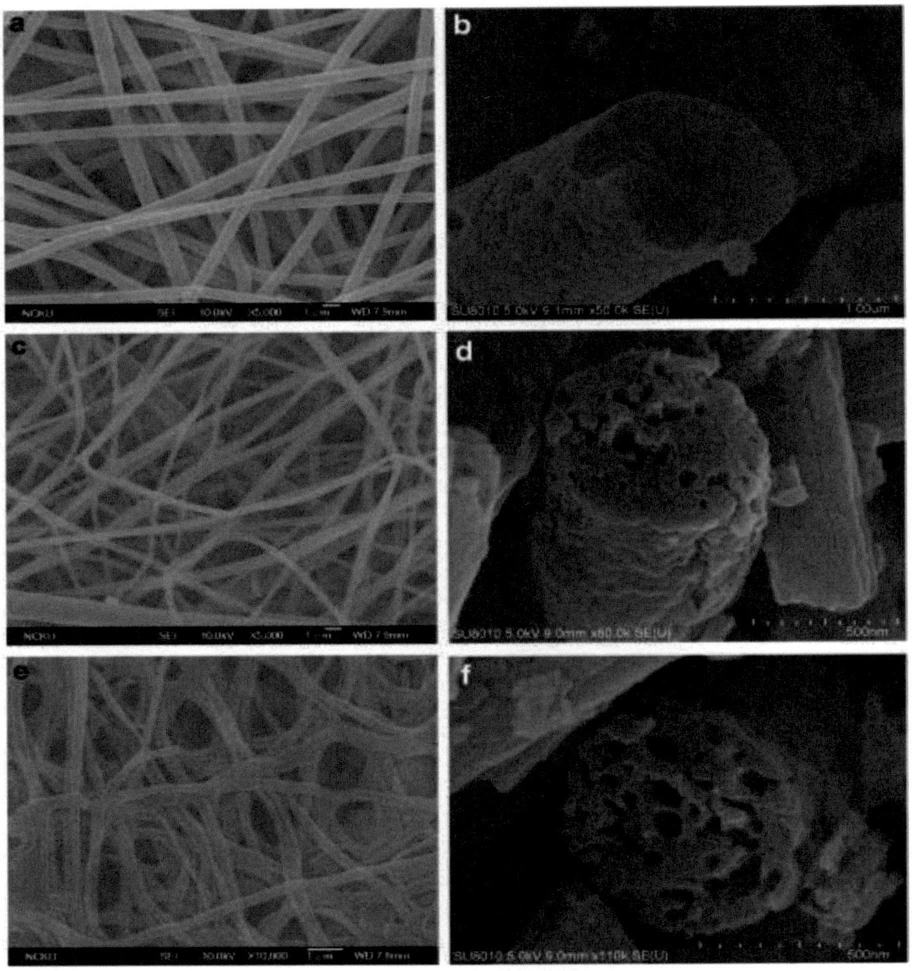

Figure 4.
SEM micrographs of electrospun fibers carbonized at 800°C. a and b PAN/PMMA = 10:0; c and d PAN/PMMA = 7:3; and e and f PAN/PMMA = 5:5. Adapted from Ref. [37].

loaded with simvastatin superficial layer to obtain an efficient osteogenesis process by the continuous release of the drug [39]. The use of PVA was attributed to its environmentally benign, elasticity, flexibility, proper mechanical properties, nontoxicity, swelling ability, and biodegradability. PVA is not stable in aqueous state, this instability however creates limitation in its use in drug delivery processes. In order to overcome instability issue, PVA was then crosslinked with biocompatible and biodegradable PVAc that possess hydrolysable groups. Afterward a simvastatin drug was loaded into the blended solution of PVA–PVAc in order to promote the efficiency of bone regeneration. The obtained results revealed good bioactivity, inducing the precipitation of bone-like apatite minerals on its surface and successfully simulating physiological conditions for cell growth [39]. Electrospun nanofibrous dressings have high surface-to-volume ratio, allow gas permeation, help to regulate wound moisture, enhance tissue regeneration, improve removal of exudates, and have high porosity, which qualifies them to be used in wound healing treatment. Previous studies have shown low inflammatory reaction and fast re-epithelization with the use of nanofiber-based wound dressing [38].

Bredigite polymer electrospun nanofibers have been widely investigated to access their suitability in wound-dressing processes. It has been reported as a

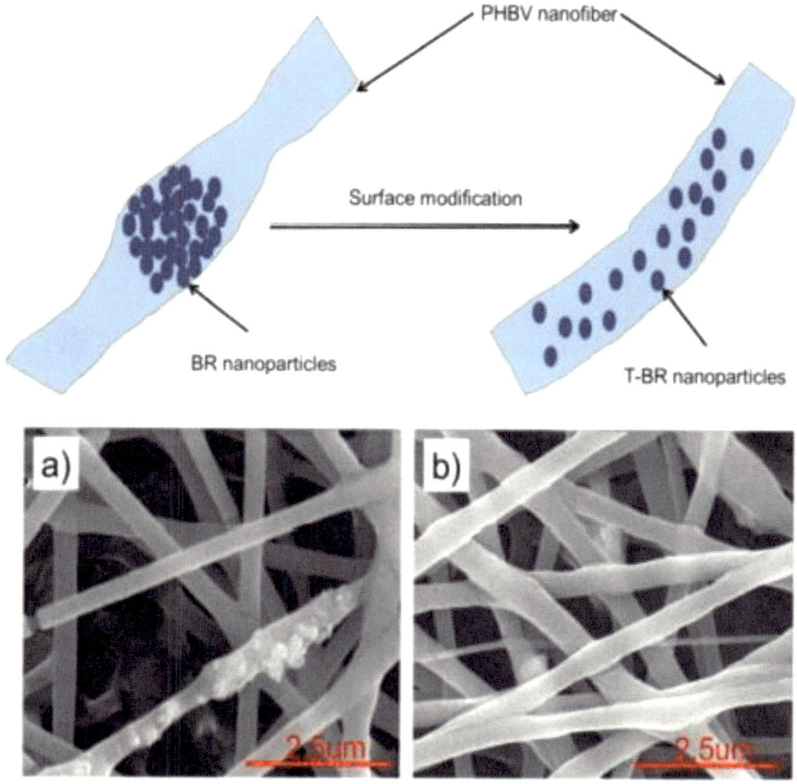

Figure 5.
Schematic illustration and SEM images of PHBV nanofibers containing 15% of (a) bredigite (BR) and (b) T-BR nanoparticles. Adapted from [40].

scaffold however, results showed that while the bioactivity of the composite nanofibers was improved, and the low dispersibility and high agglomeration of nanoparticles decrease the efficiency of prepared electrospun nanofibers [40]. In another attempt, bredigite (BR) nanoparticles were modified by an organosilane coupling agent in order to increase its dispersibility [40]. The SEM results reveal that the modified BR nanoparticles are widely dispersed in the body of the nanofibers without any agglomeration (**Figure 5**). Moreover, the mechanical and biodegradation rate of the scaffolds dramatically improved after BR modification.

5. Advance manufacturing methods for energy applications

The fabrication of energy device material such as thin film photoelectrode for splitting water into H_2 and O_2 during photoelectrochemical process and the development of photovoltaic cells, for solar energy conversion is tasking and difficult, requiring a special operational technique. For efficient solar energy capturing and conversion in photovoltaic cells, effective separation electrons and holes in photoelectrode required [41, 42]. This depend on the deposited semiconducting material ultrathin layer, evenly coated and tightly connected to conductive layer. Atomic layer deposition (ALD) as a vapor phase technique is capable of producing thin films of different materials. ALD is applicable in the fabrication of uniform and ratio structures with thickness control to Angstrom level, and tuneable film composition [43]. Due to all this advantages, ALD has emerged as a powerful tool for many

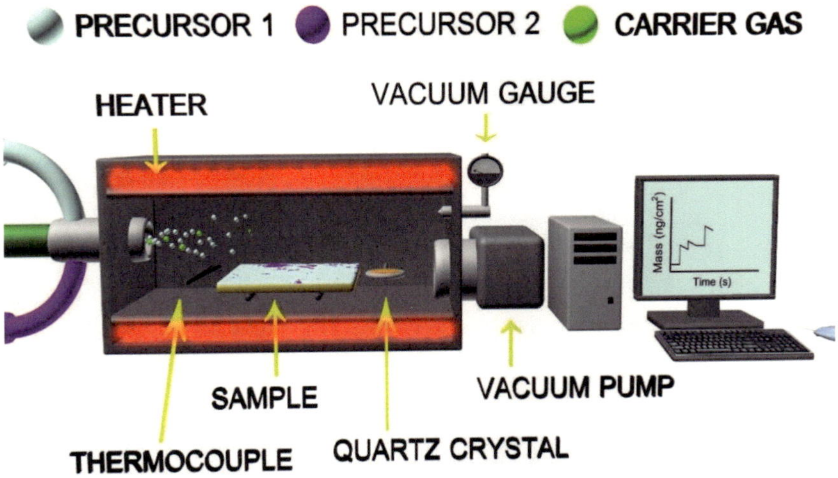

Figure 6.
Atomic layer deposition (ALD) reactor. Adapted from Ref. [43].

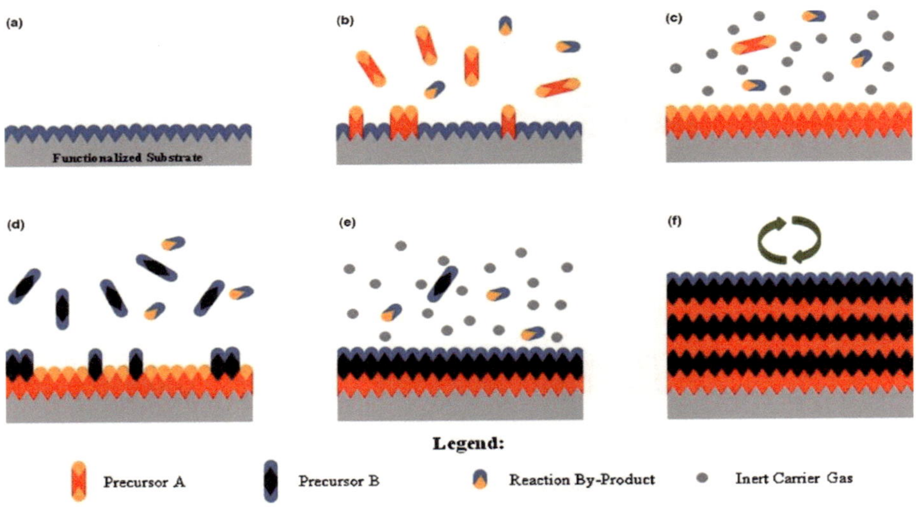

Figure 7.
Schematic illustration of ALD process schematic of ALD process. (a) Substrate surface has natural functionalization or is treated to functionalize the surface. (b) Precursor a is pulsed and reacts with surface. (c) Excess precursor and reaction by-products are purged with inert carrier gas. (d) Precursor B is pulsed and reacts with surface. (e) Excess precursor and reaction by-products are purged with inert carrier gas. (f) Steps 2–5 are repeated until the desired material thickness is achieved. Adapted from Ref. [45].

energy research material fabrications. ALD method has been a useful tool for the deposition of ultrathin-layered semiconductors on conductive substrate.

ALD process generally consists of sequential alternating pulses of gaseous chemical precursors that react with the substrate, these individual gas-surface reactions called 'half-reactions' and appropriately make up only part of the materials synthesis. During each half-reaction, the precursor is pulsed into a compartment under vacuum (< 1 Torr) over a selected extent of time to allow the precursor to fully react with the substrate surface through a self-limiting process that leaves no more than one monolayer at the surface [44, 45]. Then, the chamber is purged with an inert carrier gas (typically N_2 or Ar) to remove any unreacted precursor or reaction by-products.

This is then followed by the counter-reactant precursor pulse and purge, creating up to one layer of the desired material. This process is then cycled until the appropriate film thickness is achieved (**Figures 6** and 7).

6. Conclusion

The interdigitated electrode is reported as cost effective method for prototyping gas, chemical and bio sensor and the method is widely used for laboratory research purpose. State of the art techniques such high tech semiconductor deposition instruments, photolithography and electron beam lithography are used for commercial sensors built with printed electronics and Lab-on-a chip. Electrospinning method is highly important in the fabrication of micro and nano porous fibers for the manufacturing of membranes and battery devices. This method has also been identified for designing anode materials suitable for lithium ion battery fabrication. Atomic layer deposition is useful for producing ultrathin layer-layered semiconductors with inherent properties necessary for efficient energy capturing. This deposition technique is very useful in the manufacturing of photovoltaic cells and related devices for effective separation electrons and holes in photo-electrode.

Author details

Amos Adeleke Akande[1]*, Aderemi Timothy Adeleye[2,3], Abraham Abdul Adenle[4] and Bonex Wakufwa Mwakikunga[5]

1 Next Generation Enterprises and Institutions Cluster, EDT4IR Research Centre, Council for Scientific and Industrial Research, Pretoria, South Africa

2 CAS Key Laboratory of Science and Technology on Applied Catalysis, Dalian Institute of Chemical Physics, Chinese Academy of Sciences, Dalian, China

3 University of Chinese Academy of Sciences, Shijingshan District, Beijing, China

4 State Key Laboratory of Catalysis, Dalian Institute of Chemical Physics, Chinese Academy of Sciences, Dalian National Laboratory for Clean Energy, Dalian, China

5 National Centre for Nano-Structured Materials, Council for Scientific and Industrial Research, Pretoria, South Africa

*Address all correspondence to: aaakande@csir.co.za

References

[1] D.J. Lockwood, Nanotechnology for Electronic Materials and Devices, 4th edition, *Springer* **2007**.

[2] K. Fleming, Nanotechnology Feynman's Fancy, Chem. World. **2009**: pp58-62

[3] F. Simonis and S. Schilthuizen, Nanotechnology innovative opportunities for tomorrow's defence (*INO Science and Industry*) **2006**.

[4] Das, S.; Gates, A.J.; Abdu, H.A.; Rose, G.S.; Picconatto, C.A.; Ellenbogen, J.C. (2007)."Designs for Ultra-Tiny, Special-Purpose Nanoelectronic Circuits". *IEEE Trans. on Circuits and Systems I*. **2007**;54: 11

[5] Vanadium Dioxide (VO2) Nanostructure Production and Applications in Sensors. A dissertation submitted in partial fulfilment of the requirements for the degree of Master of Science in Physics in the Faculty of Science and Agriculture School of Physical and Mineral Sciences University of Limpopo. **2014 (unpublished thesis)** By Amos Adeleke Akande

[6] Aviram, A.; Ratner, M. A. "Molecular Rectifier". *Chemical Physics Letters*. **1974**; 29: 277-283.

[7] Aviram, A. "Molecules for memory, logic, and amplification". *Journal of the American Chemical Society*. **1988**;110: 5687-5692.

[8] A.A. Akande, E.C Linganiso, B.P Dhonge, K.E. Rammutla, A. Machatine, L. Prinsloo, H. Kunert, B.W. Mwakikunga, "Phase evolution of vanadium oxides obtained through temperature programmed calcinations of ammonium vanadate in hydrogen atmosphere and their humidity sensing properties" *Journal of Materials Chemistry and Physics*. **2015**; 151:206-214.

[9] Matin Roshanzamir Modaberia, Reza Rooydella, Sanjaya Brahmaa, Amos A. Akande, Bonex W. Mwakikungab, Chuan-Pu Liu, Enhanced response and selectivity of H_2S sensing through controlled Ni doping into ZnO nanorods by using single metal organic precursor, *Sen & Act B: Chemical*. **2018**; 273:1278-1290.

[10] Chimowa. G., Tshabalala, ZP., Akande, A.A., Bepete, G., Mwakikunga, B., Ray,S.S., Improving methane gas sensing properties of multi-walled carbon nanotubes by vanadium oxide filling, *Sens & Act B: Chemical* **2017**; 247:11-18.

[11] A. A Akande, A. G. J. Machatine, B. Masina, G. Chimowa, B. Matsoso, K. Roro, M-M Duvenhage, H. Swart, S.S. Ray, B.W. Mwakikunga, Blue- and red- shifts of V_2O_5 phonons in NH_3 environment by in situ Raman Spectroscopy, *J. Phys. D: Applied Physics*. **2017**;51 (1): 105106.

[12] S. Surnev, M. G. Ramsey, F. P. Netzer, Vanadium oxide surface studies, *Prog. Surf. Sc*. **2003**;73(4-8):117-165

[13] J. Tao, T. Luttrell, J. Bylsma, M. Batzill, Adsorption of acetic acid on rutile TiO2 vs (011)-2 x 1 Surfaces, J. Phy. Chem. C. **2011**;115(8):3434-3442.

[14] A.A. Akande, B.W. Mwakikunga, K.E. Rammutla, B.P Dhonge, A.G.J, Machatine "Room temperature Methane (CH_4) sensing by Vanadium oxide (VO_x) nanoparticles" *Adv. Sc. Lett.***2016**; 22, 4

[15] A.A. Akande, B.W. Mwakikunga, K.E. Rammutla, B.P Dhonge, A.G.J, Machatine, Gate Voltage Controlled Humidity Sensing Using MOSFET of VO_2 Particles, *Int. J. Chem. Mol. Nucl. Mat. Met. Eng.* **2017**;11(1): 78-81

[16] Amos Adeleke Akande, Bonex Wakufwa Mwakikunga, Koena Erasmus Rammutla, Augusto Machatine "Larger Selectivity of the V2O5 nano-particles Sensitivity to NO2 than NH3" *Sensors & Transducers*. **2015**;192(9): 61-65

[17] I. Vladimirov, M. Kühn, T.Geßner, F. May, R.T.Weitz, Energy barriers at grain boundaries dominate charge carrier transport in an electron-conductive organic semiconductor, *Scientific Reports* **2018**; 8:14868, 1-10.

[18] Lothar Weinhardt, Dirk Hauschild, Clemens Heske, Surface and Interface Properties in Thin-Film Solar Cells: Using Soft X-rays and Electrons to Unravel the Electronic and Chemical Structure, *Adv. Mater*. **2019**, 1806660 1-9.

[19] Yaxian Pei, Xiaoping Zou, Xiaolei Qi, Gongqing Teng, Qi Li, Dongdong Guo, and Shuangxiong Zeng, Effect of Perovskite Film Preparation on Performance of Solar Cells, *J Che* **2016**;1-10,

[20] S. A. Olaleru, D. Wamwangi, J. K. Kirui, K. T. Roro, B. Mwakikunga, N. Palaniyandy3, N. El-Mahallawy, M. Rashad, The impact of synthesis techniques on the properties of hybrid perovskite materials for photovoltaic application, *Mater. Express*, **2020**;10:7.

[21] M. Lupiwana, R. Taziwa, E. Meyer, D. Katwire, Structural, Morphological, Topographical Characterization of Titanium Dioxide Nanotubes Metal Substrates for Solar Cell Application, *J. Mat. Sc. Tech. Res*. **2017**; 3:17-31

[22] Shutthanandan, V., Nandasiri, M., Zheng, J., Engelhard, M. H., Xu, W., Thevuthasan, S., & Murugesan, V. Applications of XPS in the characterization of Battery materials. *Journal of Electron Spectroscopy and Related Phenomena*. **2018** ;231: 2-10.

[23] Chen, Z., Chen, Z., Zhang, A., Hu, J., Wang, X., & Yang, Z. Electrospun nanofibers for cancer diagnosis and therapy. *Biomaterials Science*, **2016**;4(6):922-932.

[24] A.A. Akande, T. Mosuang, E.M. Benecha, C. N. M. Ouma, K. T. Roro, A.G.J, Machatine, T. Tesfamicheal, B.W. Mwakikunga, "Ammonia gas sensing characteristics of V_2O_5 nanostructure: A combined experimental and *ab initio* density functional theory approach" *J. Alloy and Compound* **2020**;821 : 153565

[25] Stella Vallejos, Francesco Di Maggio, Tahira Shujah, Chris Blackman, Chemical Vapour Deposition of Gas Sensitive Metal Oxides, *Chemosensors*. **2016**;4(4): 1-18

[26] M. Horprathum, P. Eiamchai, J. Kaewkhao, C. Chananonnawathorn, V. Patthanasettakul, S. Limwichean, N. Nuntawong, P. Chindaudom, *AIP Conference Proceedings*. **2014**;7: 1617.

[27] M. D. Nguyen, R. Tiggelaar, T. Aukes, G. Rijnders, G. Roelof, Wafer-scale growth of highly textured piezoelectric thin films by pulsed laser deposition for micro-scale sensors and actuators, *Journal of Physics: Conf. Series*. **2017**; 922: 012022

[28] X. Ming Hu, Photolithography technology in electronic fabrication, International Power, Electronics and Materials. *Engineering Conference* (IPEMEC **2015**)

[29] L. Xu, Z. Dai, G. Duan, L. Guo, Y. Wang, H. Zhou, Y. Liu, W. Cai, Y. Wang, T. Li, Micro/Nano Gas Sensors: A New Strategy Towards In-Situ Wafer-Level Fabrication of High-Performance Gas Sensing Chips, *Scientific Reports*. **2015**; 5(10507):1-12

[30] W-H. Tian, Y-H. Ho, C-H. Chen, C-Y, Kuo, Sensing Performance of Precisely Ordered TiO2 Nanowire Gas Sensors Fabricated by Electron-Beam Lithography, *Sensors* **2013;**13:865-874

[31] Saleem, H., Trabzon, L., Kilic, A., & Zaidi, S. J. Recent advances in nanofibrous membranes: Production and applications in water treatment and desalination. *Desalination*. **2020**; 478:114178.

[32] Mariela Toriello , Morteza Afsari, Ho Kyong Shon and Leonard D. Tijing. Progress on the Fabrication and Application of Electrospun Nanofiber Composites. *Membranes* **2020**; 10(9), 204.

[33] Mahadevappa Y. Kariduraganavar, Srikant S. Kulkarni, Arjumand A. Kittur, Pervaporation separation of water-acetic acid mixtures through poly (vinyl alcohol)-silicone based hybrid membranes, *J. Membr. Sci*. 2005;246 (1):83-93.

[34] Lin, Y.-Z., Zhong, L.-B., Dou, S., Shao, Z.-D., Liu, Q., Zheng, Y.-M. Facile synthesis of electrospun carbon nanofiber/graphene oxide composite aerogels for high efficiency oils absorption. *Environ. Int*. 2019; 128:37-45.

[35] Chen, H., Huang, M., Liu, Y., Meng, L., & Ma, M. Functionalized electrospun nanofiber membranes for water treatment: A review. *Science of The Total Environment*. **2020**;139944.

[36] Zhu, M., Liu, H., Cao, Q., Zheng, H., Xu, D., Guo, H., Zhou, J. Electrospun Lignin-based Carbon Nanofibers as Supercapacitor Electrodes. *ACS Sustainable Chemistry & Engineering.* **2020**; 8:12831-12841

[37] Peng, Y.-T., & Lo, C.-T. Electrospun porous carbon nanofibers as lithium ion battery anodes. *Journal of Solid-State Electrochemistry*, **2015**;19(11):3401-3410.

[38] Tijing, L.D.; Woo, Y.C.; Yao, M.; Ren, J.; Shon, H.K. 1.16 Electrospinning for Membrane Fabrication: Strategies and Applications. In Comprehensive Membrane Science and Engineering; *Elsevier: Oxford, UK*, **2017**; pp. 418-444.

[39] Rezk, A.I.; Unnithan, A.R.; Park, C.H.; Kim, C.S. Rational design of bone extracellular matrix mimicking tri-layered composite nanofibers for bone tissue regeneration. *Chem. Eng. J.* **2018**; 350:812-823.

[40] Kouhi, M.; Fathi, M.; Jayarama Reddy, V.; Ramakrishna, S. Bredigite Reinforced Electrospun Nanofibers for Bone Tissue Engineering. *Mater. Today Proc*. **2019**; 7:449-454.

[41] Wang, W.-C.; Tsai, M.-C.; Yang, J.; Hsu, C.; Chen, M.-J., Efficiency enhancement of nanotextured black silicon solar cells using Al_2O_3/TiO_2 dual-layer passivation stack prepared by atomic layer deposition. *ACS applied materials & interfaces* **2015**;*7* (19): 10228-10237

[42] Chen, H.-Y.; Lu, H.-L.; Sun, L.; Ren, Q.-H.; Zhang, H.; Ji, X.-M.; Liu, W.-J.; Ding, S.-J.; Yang, X.-F.; Zhang, D. W., Realizing a facile and environmental-friendly fabrication of high-performance multi-crystalline silicon solar cells by employing ZnO nanostructures and an Al_2O_3 passivation layer. *Scientific reports* **2016,** *6* (1), 1-11. doi: 10.1038/srep38486

[43] Dasgupta, N. P.; Meng, X.; Elam, J. W.; Martinson, A. B., Atomic layer deposition of metal sulfide materials. *Accounts of chemical research* **2015**;*48*(2):341-348.

[44] Marchack, N.; Chang, J. P., Chemical processing of materials on silicon: more functionality, smaller features, and larger wafers. *Annual review of chemical and biomolecular engineering* **2012**;3: 235-262.

[45] Johnson, R. W.; Hultqvist, A.; Bent, S. F., A brief review of atomic layer deposition: from fundamentals to applications. *Materials today* **2014;**17(5):236-246.

Chapter 11

Microwave Assisted Synthesis of Organic Compounds and Nanomaterials

Anjali Jha

Abstract

In the Conventional laboratory or industry heating technique involve Bunsen burner, heating mental/hot plates and electric heating ovens. To produce a variety of useful compounds for betterment of mankind, the Microwave Chemistry was introduced in year 1955 and finds a place in one of the Green chemistry method. In **Microwave chemistry** is the science of applying microwave radiation to chemical reactions. Microwaves act as high frequency electric fields and will generally heat any material containing mobile electric charges, such as polar molecules in a solvent or conducting ions in a solid. Polar solvents are heated as their component molecules are forced to rotate with the field and lose energy in collisions i.e. the dipole moments of molecules are important in order to proceed with the chemical reactions in this method. It can be termed as microwave-assisted organic synthesis (*MAOS*), Microwave-Enhanced Chemistry (*MEC*) or Microwave-organic Reaction Enhancement synthesis (*MORE*). Microwave-Assisted Syntheses is a promising area of modern Green Chemistry could be adopted to save the earth.

Keywords: Green synthesis, Microwave synthesis, Organic reactions, Nanomaterials

1. Introduction

Chemistry play major part in our daily life in the form of medicines, food colours, soaps, detergents, sunscreen lotions, toothpaste, pharmaceuticals, etc. what we use in our daily life literally everything. Scientist from various disciplines had used their knowledge and skill to produce these materials. To produce these chemicals in industries and laboratory several techniques are used like stirring, heating, refluxing, protection of specific functional group and deprotection of them. Several harmful materials were also generated during the above processes causing environment pollution. To combat these problems, Green Chemistry has emerged as a challenge for scientist from industries and academia to develop the synthetic process for sustainable development of society in last few decades. In this context, several green synthetic techniques were developed to generate a vast library of organic compounds.

In the Conventional laboratory or industry heating technique involve Bunsen burner, heating mental/hot plates/furnace/oil bath and electric heating ovens. These process comprise the heating of walls of vessels first by convection or conduction of energy then the reaction content, leading to much longer time to achieve the target temperature. These techniques not only give useful materials but also give so many by products in the form of solid, liquid and gases, resulting to

IntechOpen

lot of chemical pollution to mankind as well as for other living creatures on Earth. Furthermore, to produce these useful materials, several resources are also going to be consumed. To combat these problems the various arena of Scientist are trying to develop new Green techniques.

Conventional heating usually involves the use of a furnace or oil bath, which heats the walls of the reactor by convection or conduction method. The core of the sample takes much longer to achieve the target temperature, e.g. when heating a large sample of ceramic bricks (**Figure 1**). The microwave portion of electromagnetic radiation is emerged a significant area in the acceleration of chemical reactions in 1980s. A number of reviews were reported by various scientists [1–10]. There are three main reasons for its special place of microwave (MW) assisted synthesis. The first point is high heating efficiency caused by MW energy, which reaches and absorbed directly by substance. The second is selectivity i.e. having dielectric polar molecule will absorb this energy. And thirdly enhancement of chemical syntheses by the MW effect or non-thermal effect. In other sense, the microwave irradiation energy acts as internal heat source, which is able to heat the target compounds without heating the entire furnace or oil bath, consequently, saves time and energy. It produces more uniform heating. This unconventional energy source (MW) has been used to heat food stuff since more than 5 decades and now the concept is also utilized to accelerate a wide range of chemical reactions. The organic synthesis has very special and significant role in the improvement of day to day life of human being from morning toothpaste to all useful medications/drugs, plastic/polymer materials etc. To produce these useful substance in reasonable time and environment friendly manner is the prime responsibility of the scientist of several discipline. In this regard Microwave heating technique has blossomed in a variety of applications in Organic Synthetic methods. Although the initial synthetic work was performed in domestic Microwave ovens but recently most organic reactions are carried out in dedicated high speed specialized Microwave equipment with appropriate pressure and temperature control.

Microwaves (MW) are the electromagnetic waves, having shortest wavelength of radio spectrum. The word microwave means "a very short wave". Microwave irradiation is a low energy phenomena as compared to other radiations. The energy of MW is about 0.0016 eV at 2450 MHz, which is very low to break any chemical bond. However, a microwave energy can effect a number of chemical transformation. When a substance is exposed to microwave energy, then three process can occur. Firstly, if it is a conductor metal, then MW will be reflected and it cannot be heated. Secondly, if insulator material is exposed to MW it will be not heated since it is MW transparent. Thirdly if material is dielectric in nature then MW energy will be absorbed and gets heated. This case is very interesting for a Chemist.

A microwave works by directly coupling with polar molecule in a reacting species. Microwaves radiation are a form of non-ionizing electromagnetic radiation

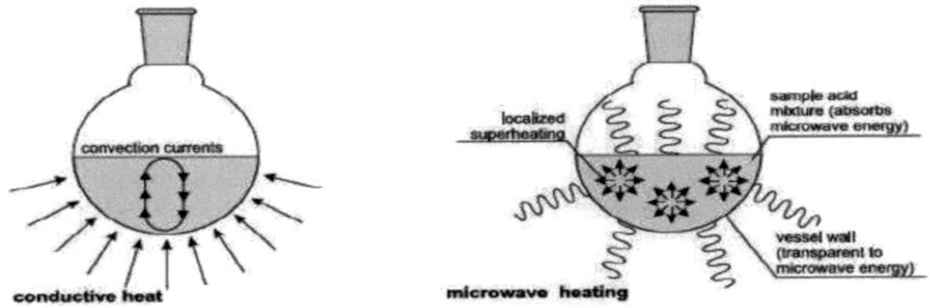

Figure 1.
Process of heating.

with 300 MHz to 300 GHz frequency with the corresponding wavelength 1 mm to 1 m, which places MW in between Infrared radiations and radio waves. The molecules get heated in microwave irradiation process by a dielectric heating phenomenon. The electric dipoles, i.e. a partial positive charge at one end and a partial negative charge at the other, rotate as they try to align themselves with the alternating electric field of the microwaves. Additionally, molecular rotation occurs in a material/food (containing polar molecules) causing electric dipole moment which can partially/completely aligned themselves with an applied electromagnetic field. Since electromagnetic field alternates, the rotating molecule resulting they change their directions by pushing/pulling/colliding. Rotating molecules hit other molecules and put them into motion, thus dispersion of energy happens. This energy, dispersed as molecular rotations, vibrations and/or translations in solids and liquids, raises the temperature of the food, in a process similar to heat transfer by contact with a hotter body. It is a common misconception that microwave ovens heat food by operating at a special resonance of water molecules in the food. As noted microwave ovens can operate at many frequencies.

Additionally, nanomaterials or nanoparticles (NPs) are the materials having a single unit small sized between 1 and 100 nm. During the past few decades nanoparticles research have become a subject of intensive interest because of its potential applications in various fields i.e. biosensing, drug delivery, bioimaging, catalysis, lubrication, electronics, textile manufacturing, water treatment systems. Nanoparticles (NPs) are synthesized from several types of materials i.e. inorganic, organic, hybrid and biological by adopting various methods. The main methods are ball milling, vapor deposition, electro-spraying, reduction of metal salts, sol–gel, coprecipitation and thermal decomposition for inorganic NPs. While organic NPs are synthesized by microemulsion, nanoprecipitation, dialysis and rapid expansion of supercritical solutions. Hybrid NPs are synthesized from both organic and inorganic materials. However, the present trend for the synthesis of nanomaterials endures a challenge to redesign the synthetic strategies offering the use of less hazardous chemicals and reduction in the reaction time and required energy. In this context, microwave (MW)-assisted methods can be considered as a promising green approach for synthesis of nanomaterials and nanocomposites. Besides, MW-assisted strategies offer a homogenous heating with reliable nucleation and growth environment leading to the formation of NPs with uniform size. Several reviews were reported on microwave assisted synthesis of NPs [11, 12].

2. Microwave equipment

Though the range of MW frequencies is 0.3 to 300 GHz, out of this most of the frequencies are used in Radar and Telecommunication equipment. However very limited frequency range can be used in microwave heating equipment i.e. 2.45 GHz corresponding to 12.2 cm wave length. So almost all domestic and commercial microwave heating equipment either for domestic or scientific purpose has a fixed frequency 2.45 GHz. In general MW accelerated reactions are carried out either domestic microwave oven or especially designed microwave equipment. These equipment works in between 500 to 1500 W power. In these ovens Microwaves are generated by magnetron, and the temperature is maintained by turning this on and off. The generated microwaves are travelled into vessel (cavity) and reflected back by the walls of cavity. If the generated microwaves are not absorbed, it may reflect back down and damage the magnetron. Thus it is essential to have microwave active dummy load should be used during the process, which will absorb excess microwave and prevent such damage.

3. Solvent used in MOAS

In Microwave solid and liquid phase reactions can be performed at various conditions. The boiling point of solvents will be raised by 25^0 C with their actual boiling points during the process of microwave heating [13]. Although all type of solvents can be used in MOAS, however DMF, ethanol and water are the best solvents for this process, since they absorb MW radiation and heat the content efficiently. All polar solvents comprising OH groups are suitable for MW reactions, since they can able to absorb MW energy. Water, alcohols (ethanol & isopropanol), amides, acetonitrile and acids (acetic and formic) are some of the commonly used solvents in this process. The polarity order of these solvents are like

$$\text{Water} > \text{Acids} > \text{Amides} > \text{Alcohols} \quad (1)$$

The less polar or non-polar solvents are transparent to microwave irradiation i.e. could not absorb MW energy and consequently will not get heated in the pure form. However in general the reaction mixture content contains enough polar or ionic substance, which can absorb microwave irradiation and generate heat to complete the reactions. Moreover small amount of polar or ionic additives could be added to enhance the absorption capacity of reaction content having less dielectric properties of less polar solvents. The energy transfer between polar molecule and non polar solvents under microwave irradiation is so rapid and efficient leading to proceed a variety of chemical transformation and thus a very good alternative for conventional techniques in tune with green chemistry [14].

4. Types of microwave energy assisted synthesis

In general two types of Microwave assisted synthesis were reported in literature: -

I. Microwave assisted synthesis using solvents

II. Microwave assisted synthesis without solvents or solvent free conditions

The reports on Microwave assisted synthesis are well documented in the literature since last 5 decades, which includes mainly following reactions **Figure 2.**

There are several synthesis which can be performed in presence of solvent under microwave conditions:.

a. **Hydrolysis:**

The hydrolysis of benzyl chloride can be successfully done within 3 minutes in MW while normally it takes 35 min.

$$C_6H_5CH_2Cl + H_2O \xrightarrow{\text{MW, 3min}} C_6H_5CH_2OH \quad (2)$$

Similarly hydrolysis of benzamide in presence of sulphuric acid can be completed in 7 mins instead of one hr. under MW.

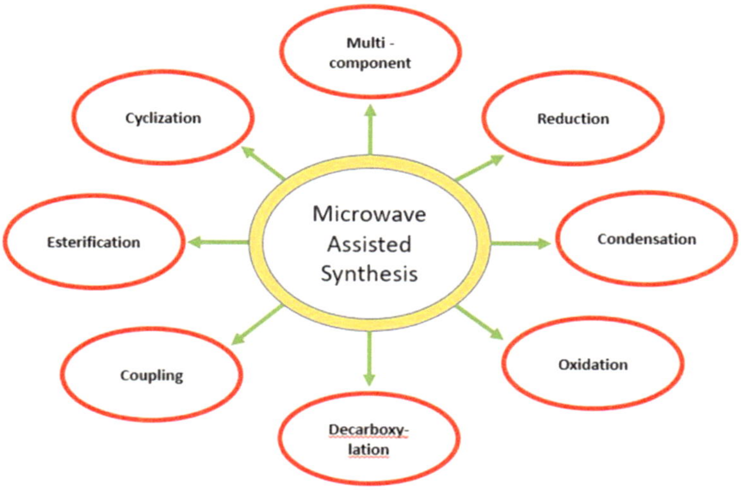

Figure 2.
Various types of Microwave assisted synthesis.

$$\underset{\text{Benzamide}}{C_6H_5CONH_2} \xrightarrow[\text{MW 7 min}]{20\%H_2SO_4} \underset{\text{Benzoic acid 99\%}}{C_6H_5COOH} \tag{3}$$

The acid-sensitive 3-hydroxyacetals and 3-methoxyacetals being hydrolyzed efficient method in minutes in good yields. In this paper reports on efficiently hydrolyzed acetals with silica gel-supported pyridinium tosylate moistened with water in solvent-free conditions under microwave irradiation were reported [15]. Since the hydrolysis of carboxylic acid esters is one of the most studied chemical reactions Safari et al. had reported Microwave-assisted expeditious hydrolysis of isobenzofuranone derivatives using silica supported acid under organic solvent-free conditions [16].

b. **Oxidation:**

The oxidation of Toluene with potassium paramagnet usually takes 10–12 hrs in normal condition, however under MW condition within 5 mins it is converted to benzoic acid [17].

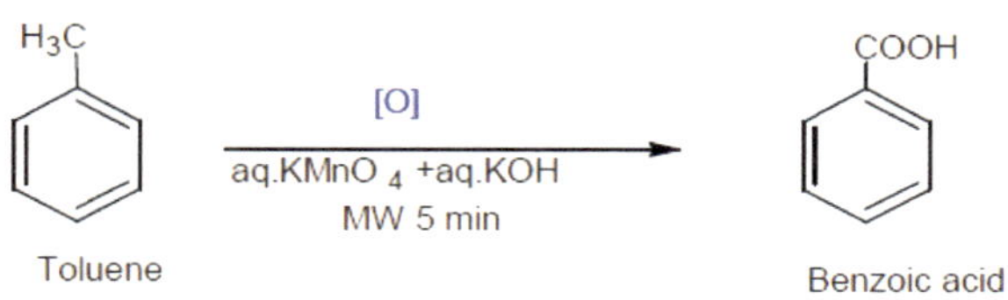

(4)

Since the oxidation of alcohols plays a significant role in organic synthesis and as carbonyl compounds are widely used as intermediates both in manufacturing and laboratory. The progress of new oxidative methods continues to be a focus of research area in spite of the availability of several protocols to achieve such objectives. Bogdał and Łukasiewicz reported [18] an interesting protocol of oxidation of primary and secondary alcohols into equivalent carboxylic acids and ketones within 10–20 min using 30% aqueous hydrogen peroxide and commercially available catalysts under microwave irradiation. The application of hydrogen peroxide as an oxidant is appreciated because water is the sole expected side product.

Further, alcohols are adsorbed on clayfen easily oxidized to carbonyl compounds under solvent-free conditions in microwaves. By this rapid, selective and environmentally benign method, the use of excess solvents and toxic oxidants usually employed can be avoided. Varma and Dahiya reported the oxidation of alcohols to the corresponding carbonyl compounds in presence of clay-supported iron(III) nitrate, under solvent-free dry microwaves conditions [19].

c. **Esterification:**

The esterification of benzoic acid can be achieved in 6 mins only [20] in the presence of catalytic amount of sulphuric acid under MW conditions.

$$\underset{\text{Benzoic acid}}{C_6H_5COOH} + nC_3H_7OH \xrightarrow[\text{MW 6 min}]{\text{Conc } H_2SO_4} \underset{\text{Propylbenzoate}}{C_6H_5COOC_3H_7} \qquad (5)$$

Since esters are one of the important products for all industrial applications. In this context convenient and feasible technique for synthesis of esters was reported [21, 22] in dry media and using heteropoly acid as catalyst respectively under microwave irradiation in the absence of organic solvents.

Recently Reilly et al. [23] demonstrated the microwave assisted esterification reaction to undergraduate student, which is a good initiative to develpo the interest among students about green chemistry.

d. **Decarboxylation:**

The conventional decarboxylation of carboxylic acid usually carried out in presence of quinolone by refluxing the acid however in microwave it can be done in very less time [24].

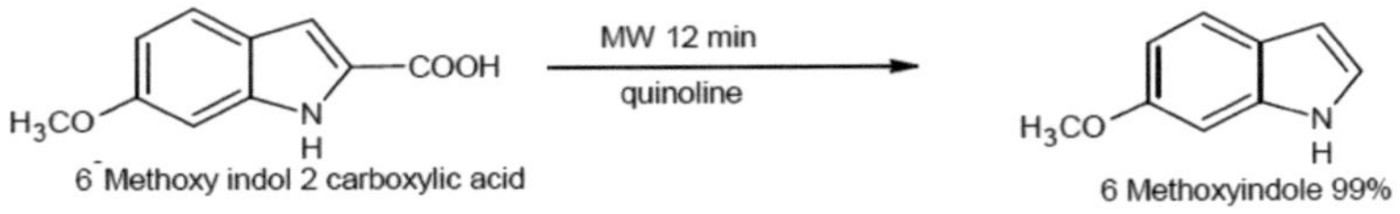

(6)

The reports on microwave assisted decarboxylation of malonic acid derivatives are available in the literature [25]. They carried out these reactions under solvent-free and catalyst free conditions. This new method produces the corresponding carboxylic acid in a pure manner and with a high yield within short time: 3–10 min.

e. **Multicomponent reactions**

The multi component reaction (MCR) approach has emerged as a promising alternative to incorporate the various diversity in a single step reaction to produce diverse library of important compounds. A rapid one-pot synthesis of imidazo [1,2-a] annulated pyridines, pyrazines and pyrimidines was described by Varma and Kumar, in the presence of recyclable montmorillonite K 10 clay under solvent-free conditions using microwave irradiation [26, 27]. They condensed three components viz.: aldehydes(aliphatic/ aromatic/vinylic); isocyanides (aliphatic/aromatic/cyclic) and amines (2-aminopyridines, 2-aminopyrazines and 2-aminopyrimidines in presence of catalytic amount of clay in solvent free microwave irradiation.

Mahmood et al. 2017 conveniently synthesized thiazolidinones from aldehydes, thiosemicarbazide and maleic anhydride in the presence of KSF@Ni as heterogeneous catalyst in one-pot three-component reaction under microwave irradiation in good yield in short reaction time [28].

The metal free the microwave assisted synthetic multicomponent approach for N-unsustituted-1,2,3-triazoles was recently reported by Garg et al. [29]. They used aromatic aldehydes, nitroalkane and sodium azide for synthesis, which demonstrates great scope to be utilized by future generation for production of novel biological active molecules.

Since acridines and acridinium ions are one of the important compounds, as they are used in pharmaceuticals, materials, dyes and photo-catalysis. Mandal et al., 2020, reported [30] an unconventional $FeCl_3$-alcohol catalysed one-pot synthesis of these materials by aldehydes, 1,3-cyclohexanedione and amines under microwave conditions. They beautifully merged high atom-economy and multicomponent reaction with novel iron catalyzed dehydrogenation, using aerobic oxygen as the terminal oxidant, in alcoholic solvent to produce water as the by-product.

f. **Condensation reaction**

Several known transformation such as hydrolysis of ester or amide, Diels-Alder cycloaddition, Fisher Indole synthesis were investigated in a microwave reactor, in near critical water (NCW) under microwave assisted synthesis and reported [31].

Ashry and Kassem, 2006 has given a comprehensive account [32] of synthesis of Six, Seven membered, Spiro and Fused Heterocycles under microwave conditions.

Naqvi et al. 2009 were developed and synthesized some Schiff bases under microwave irradiation by taking 3-chloro-4-fluoro aniline and various substituted benzaldehydes [33]. They used other green chemistry approach also for their synthesis.

The thermal effect of microwaves were shown in 2006 by condensing ortho-diphenylamine in acetic acid and observed that by conventional means it took 9 weeks while in MW only tem minutes [34]. It was a great findings and opportunities for synthetic chemists.

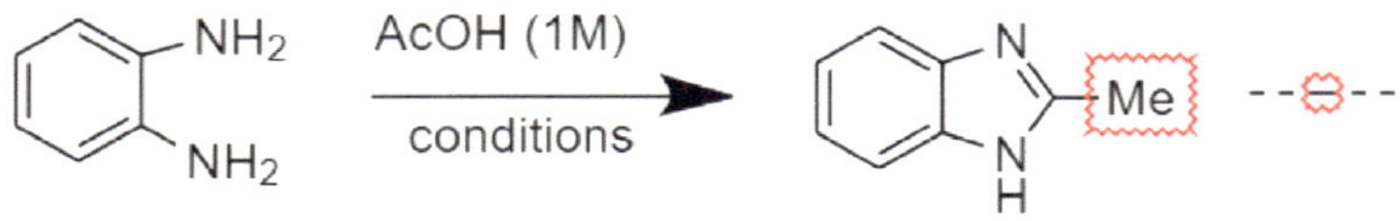

(7)

Conditions	Temperature/Pressure	Time
Conv	25°C	9 weeks
Conv	60°C	3 days
Conv	100°C	5 hrs
MW	130°C /2 bar	1 h
MW	160°C/ 4 bar	10 min
MW	200°C /9 bar	3 min
MW	270°C / 29 bar	1s

Sondhi et al., 2010; have developed microwave assisted synthesis of complex tri- an d tetracyclic heterocyclic molecules in a simple, high yielding, one step process [35]. They synthesized these important moieties by condensing of dicarboxylic acids with diamines under solvent-free condition.

The design and synthesis of novel structures having α, β-unsaturated ketone moieties, responsible for the bioactivities of the Mannich bases were carried out by Mete et al. [36]. The target compounds were synthesized by the conventional heating method and also by the microwave irradiation method. They optimized the conditions which can be explored by future researcher.

The reports on the condensation reaction between active methylene group and various aldehydes are well documentated in literature since several decades for their potential diverse biological activities especially for anticancer activities [37]. Furthermore, the various substituted aromatic aldehydes and Thiazolidine-2,4-dione were reacted under microwave irradiation using water as green solvent by simple Knoevenagel condensation. They tried various solvent conditions aqueous ethanol, H2O, DMF, DMSO, and solvent less under MW irradiation [38] and found the water as the best suitable solvent.

The 1,3,5-trisubstituted benzene was synthesized by self-condensation of enamines, enaminones and enaminoester in the presence of pyridinium chloride ([PyH]Cl) in short time in domestic MW oven and ionic liquids in year 2010 with appreaciable yield [39]. Brun et al. 2015 prepared boric acid catalyzed novel acetophenone derivatives with methylene compounds under solvent free microwave assisted condensation reactions [40]. They prepared a large number of compounds having sensitive acid and base function groups in substrates. Knoevenagel Condensation.

A comprehensive review on reactions of cyano malononitrile-derivatives under Microwave Radiation was also reported recently [41]. The Knoevenagel condensation between aromatic aldehydes or ketones and malonic acid in the presence of tetrabutyl ammonium bromide and K2CO3 was carried out and reported by Gupta and Wakhloo 2007 under microwave irradiation in water. The yields of products were excellent with high purity [42].

g. **Cycloaddition reactions**

Since Computational calculations emerged as a significant tool to study reaction kinetics and strategies in last couple of decades. de Cozar et al. [43] studied the thermal and non-thermal effects of microwave irradiation and determined the thermodynamic and kinetic parameters of the reaction. They performed the computational study of two previously reported cycloaddition reactions and analysed the presence of a thermal effect which is responsible for the microwave irradiation that produces changes in the regioselectivity or in the reaction mechanism.

The 1,2,3- triazoles are one of the most active moieties presents in several types of drug candidate. So, the convenient synthesis is a hot topic among various researchers. The catalyst and solvent free synthesis of substituted triazoles via cycloaddition of trimethylsilylazide and acetylenes were carried out recently by Roshandel et al. [44] under microwave conditions In the same year Expeditious microwave assisted synthesis and bio-evaluation of novel bis (trifluoromethyl) phenyl-triazole-pyridine hybrid analogues by click chemistry approach under microwave irradiation was reported from our group [45].

h. **Coupling reactions**

The efficient synthesis of 2,3 disubstituted-6-aminoquinoxaline derivatives were prepared efficiently by applying Microwave assisted Sonogashira Coupling by Lee et al. in 2013. They prepared a library of compounds by very region-selective MW conditions [46]. The mini-review on the Suzuki-Miyaura and on the Heck cross-couplings of nucleosides under microwave irradiations reported as an alternative technology in tune with green chemistry [47]. A simple microwave assisted synthesis of isoindolinones bearing a quaternary C-atom, introducing a Pd-mediated tandem-coupling reactions are reported recently [48]. As Cross-coupling reactions furnishing carbon–carbon (C–C) and carbon–heteroatom (C–X)bond is still one of the most challenging area in organic syntheses. Younis Baqi has reported [49] a comprehensive review on Recent Advances in Microwave-Assisted Copper-Catalyzed Cross-Coupling Reactions very recently. The microwave energy was also used in the synthesis of Ribonucleosides using natural phoshphate as catalyst in 2013 by Ouzebla et al. [50]. The hydrogenation of Levulinic acid to γ-Valerolactone in presence of Ru/TiO2 catalysts was reported very recently by Howe et al. 2019 [51]. The one-pot microwave-assisted polyol method was found to be highly efficient. The preparation temperature, microwave irradiation time and choice of Ru precursor have a significant effect on catalyst activity.

The extraction of bioactive components from fruits of *Ficus racemosa* was recently carried out by Sharma et al. 2020 under the microwave – assisted extraction (MAE). They developed the optimum condition for MAE at 3.5 pH,360.55 W microwave power in 30.01 s, which can be used as promising green technology to extract other useful materials from other natural sources [52]. The microwave synthesis of hybrid inorganic–organic porous materials was reported by Jhung et al. 2006. They observed phae selective rapid crystallization of hybrid material by MW method, which could be better alternative method to be explored further [53].

The comparative study of conventional and Microwave induced organic reaction enhancement (MORE) methodologies was carried out to synthesize some selected heterocyclic molecules and found more efficient over conventional classical methods in 2010 [54]. Kitchen et al. 2014 reported [55] a comprehensive review on the significant advances in the area of solid-state MW synthesis in context with future scope in the 21st century. They also presented the merits of MW heating along with understanding of MW heating and interaction mechanisms. The major focus of the review is on the use of MW heating to make novel improved materials with MW heating. They analytically discussed the developments in MW techniques and instrumentation and their potential to foster interest among the materials chemistry research and industrial communities and to explore the access to MW method.

However, Guadino et al. 2013, reported a comprehensive review on potential application of MW heating to accelerate heterogeneous reactions under gaseous reagent i.e. CO and CO_2, which gave a new insight to various scientist to explore the MW heating method. Since the poisonous nature of CO and high cost of CO_2 storage restricted various resercher to use as heterogeneous medium, so the concept of dielectric heating can be further explored viaMW heating protocol [56]. Recently Darekara et al. 2020, synthesized a series of derivatives of thiadiazoles and triazoles under microwave conditions. They also explored their antibacterial activities [57].

5. Synthesis of nano particles

As Nanoscience and nanotechnology are new frontiers of this century in all facet of human life and comfort. Nanoparticles are important components in a wide type of applications, including medicine, semiconductors, catalysis, and energy. Microwave energy has the potential to selectively heat either the solvent or the precursor molecules for nanomaterial preparation. In this process of heating rapid temperature rise allows nanomaterial synthesis to take place in a homogeneously mixed precursor solution To produce nano materials or particles by using microwave technology is a need of hour in context of green chemistry also. The silver nanoparticle has an importance place in biomedical and catalytical field. Its easy synthesis is a hot topic of research. The template MW Synthesis of silver nanoparticles were carried out [58] from silver nitrate and starch solution. They performed the synthesis under direct heating, controlled heating, and microwave irradiation. The Starch acted as a reductant as well as a capping material to protect the nanoparticles surfaces and prevents the particles from aggregation.

Additionally, the MW heating technique is also useful for the preparation of superior nanomaterials from direct heating of the various molecular precursors. The microwave-assisted process involve reduction of metal salts either in organic solvents or in aqueous medium in presence or absence of surface-directing agents in one-pot synthesis [59].

As Ionic liquids, due to their high ionic conductivity and polarizability, were an excellent microwave-absorbing agent, have also been used for the synthesis of metal nanostructures. By changing the various parameters i.e. precursor, precursor concentration, solvent, microwave power, pressure and temperature different metallic nanostructures in numerous sizes and morphologies have been prepared. Many simple and complex metal oxides have been also synthesized by using the microwave-assisted route. The microwave hydrothermal method has been used to prepare a large variety of binary and ternary oxides such as:

Sl No	Types of nano particles	References
1	CuO	Zhao Y, et al. [60]
2	ZnO	Huang J et al. [61]
3	PdO	Wang et al. [62]
4	In2O3, Tl2O3	Patra CR et al. [63]
5	SnO2	Jouhannaud J et al. [64]
6	HfO2	Eliziário SA et al. [65]
7	BiVO4	Zhang HM et al. [66]
8	ZnAl2O4	Zawadzki M et al. [67]
9	BaTiO3	Nyutu EK et al. [68]
10	CaTiO3	Moreira ML et al. [69]

Motshekga et al. reported [70] a comprehensive review on the synthesis and applications of carbon-nanotube-coated metal/oxides nanoparticles under the microwave-assisted method. They explained various studies in which the microwave-assisted synthesis method of the composites formation will be completed in a shorter reaction time with uniform and well-dispersed nanoparticle.

A simple and microwave irradiation method for the biosynthesis of silver nanoparticles (SNPs) using aqueous leaf extract of *Origanum majorana* and *Citrus sinensis* as a novel bio source of cost-effective, non-hazardous reducing, and stabilizing agents was reported [71]. The microwave heating made the synthesis of SNPs fast, uniform, and reproducible. They also exhibited good antibacterial activity against *E. coli* and *B. subtilis* pathogens.

Similarly five plant extracts of *Syzygium aromaticum, Origanum vulgare, Origanum majorana, Theobroma cacao* and *Cichorium intybus* were used [72] for microwave assisted synthesis of alumina nanoparticles from Aluminum nitrate.

On analysis of XRD pattern of particles synthesized with *S. aromaticum* showed semi-crystalline structure however others showed nano dimension of particles or amorphous structure nanoparticles.

The microwave-assisted sol–gel synthesis of high-quality and uniform ZnFe2O4 nanocrystals were performed by Suchomski et al. [73]. They described these novel nano particles. Additionally, the synthesis of magnetite and maghemite nanoparticles with well-controlled size, high crystallinity and good magnetic properties, in less reaction time was developed under microwave irradiation method [74]. It was observed that microwave power and heating time are the main parameters controlling the size of the nanoparticles and the presence of maghemite, while ammonia concentration does not strongly affect it.

The magnetic metallic nickel nanoparticles were synthesized by reacting simple nickel chloride as precursor, ethylene glycol as solvent and ethanol used as reducing agent following microwave-assisted method using a monomode microwave reactor by Zuliani et al. [75]. They achieved metallic nickel in five minutes at 250°C under MW conditions, which showed high catalytic activity for the hydrogenolysis of benzyl phenyl ether.

Further, fabrication of small anatase titanium dioxide (TiO2) nanoparticles (NPs) attached to larger anisotropic gold (Au) morphologies by a very fast and simple two-step microwave-assisted synthesis was performed by May-Masnou et al. [76]. They also evaluated the photocatalytic activity of the two Au/TiO2 NPs, in the photoproduction of hydrogen from gaseous water/ethanol mixtures at ambient temperature and pressure.

The amazing utilization of microwave energy and nanotechnology to improve the printability and performance of cotton prints via screen printing technique was reported [77]. They pre-treated each cotton sample separately by microwave power in the range of 300 to 700 watts for a period of 1 to 9 min. Then the optimum sample was printed by printing paste comprising a reactive dye Remazol™ and silver nanoparticles (Ag-NPs) with diverse concentrations. Further, the printed samples were fixed using MW energy lateron subjected to steaming or thermo-fixation. On the analysis of results indicated that, the prints obtained using microwave and Ag-NPs were found to have better colour strength, fastness properties, antibacterial behaviour and surface morphology as compared to the conventional techniques.

Though the conventional wet chemical synthetic method is still the most widely used method for nanoparticle synthesis due to its simplicity, inexpensive processing, and scalable nature. But nanoparticles produced by this route tend to agglomerate after some time, in particular at high temperature and pressure, resulting to a gradual loss of activity. To tackle these challenges a facile and scalable thermal shock synthesis method based on microwave irradiation for the rapid synthesis of nanoparticles on a reduced graphene oxide (RGO) substrate was demonstrated by Xu et al. 2019 very recently [78]. By loading precursors onto RGO, with medium amount of defects can efficiently absorb microwaves leading to rapid temperature change resulting uniform nanoparticles synthesis. This technique is also known as defect engineering. The beauty of this technique is scalable i.e. has potential in large-scale production nanoparticles.

6. Conclusion

In last couple of decades, the application of Microwave enegy has emerged a promising tool for synthesis of novel molecule in the area of drug research, in synthesis of nano material, in catalyst field, in inorganic synthesis, in CO and CO2 capture reactions, paint industry, in extraction of products from natural sources etc. It blossemed as a reliable alternavite of conventional heating due to high speed or shortening of reaction time; simplicity i.e.reaction can be performed in sealed and open glass vials with wide range of options; with high productivity i.e. almost 200–400% more than conventional means. The MW assisted reactions will be a one of the best method to synthesise diverse type of molecules of interest.

Conflict of interest

None.

Author details

Anjali Jha
Department of Chemistry, GIS, GITAM (deemed to be University) Rushikonda, Visakhapatnam, India

*Address all correspondence to: ajhamani@gitam.edu

References

[1] Lidstrom P, Tierney J, Wathey B, Westman J. Microwave assisted organic synthesis-a review . Tetrahedron 2001, 57 9225-9283

[2] Kappe O C. Controlled Microwave heating in Modern Organic Synthesis. Angew. Chem.Int.Ed. 2004, 43, 6250-6284; DOI:10.1002/anie.200400655

[3] Wolkenberg SE, Shipe WD, Lindsley CW, Guare JP, Pawluczyk JM. Applications of microwave-assisted organic synthesis on the multigram scale. Current Opinion in Drug Discovery and Development, 2005, 8(6): 701-708

[4] Kappe O C, Dallinger D, The impact of microwave synthesis on drug discovery, Nature Reviews Drug Discovery, 2006, 5, 51-63

[5] Man A K, Shahidan R. Microwave-assisted Chemical Reactions. J Macromolecular Science Part A, 2007, 44(6) 651-657. doi:10.1080/10601320 701285136

[6] Rajak H, Jain D P, Dewangan P K, Patel V, Agrawal N. application of Microwaves in Organic Synthesi: speeding up the Process of drug Discovery. RGUHS JmPharm Sci. 2013, 3(1) 16-23.

[7] Lukasik N, Wysiecka EW. A Review of Amide Bond Formation in Microwave Organic Synthesis. Current Organic Synthesis. 2014, 11(4).

[8] Frecentese F, Saccone I, Caliendo G, Corvino A, Fiorino F, Magli E, Perissutti E, Severino B, Santagada V. Microwave Assisted Organic Synthesis of Heterocycles in Aqueous Media: Recent Advances in Medicinal Chemistry. Medicinal Chemistry, 2016, 12(7)

[9] Naina S, Singha R, Ravichandran S. Importance of Microwave Heating in Organic Synthesis. Advanced Journal of Chemistry-Section A, 2019, 2(2) 94-104

[10] Kumara A, Kuang Y, Liang Z, Sun X, Microwave chemistry, recent advancements, and eco-friendly microwave-assisted synthesis of nanoarchitectures and their applications: a review. Materials Today Nano, 2020, 11, 100076

[11] Anuj Kumarabd Yun Kuangab ZhengLiangc Xiaoming Sun; Materials Today Nano, Volume 11, August 2020, 100076.

[12] M. Gnana Ruba Priya, Bincy Raj and Rashmi T. A REVIEW ON MICROWAVE ASSISTED SYNTHESIS OF SOME NANOPARTICLES, European Journal of Biomedical and Pharmaceutical Sciences,2020,7(7), 691-700.

[13] Baghurst D R, Mingos DMP. Superheating effect associated with microwave dielectric heating. J Chem Soc, Chem, Commun. 1992, 674-677

[14] Colombo M, Peretto I Chemistry stratergies in early drug discovery: an overview of recent trends. Drug Discov. Today, 2008, 13, 677-684

[15] He Y, Johansson M, Sterner O. Mild Microwave-Assisted Hydrolysis of Acetals Under Solvent-Free Conditions. Synthetic Communication, 2004, 4: 22, 4153-4158, DOI:10.1081/SCC-200036616

[16] Safari J, Banitaba S H, Khalili S D. Microwave-assisted expeditious hydrolysis of isobenzofuranone derivatives using silica supported acid under organic solvent-free conditions Arabian Journal of Chemistry, 2011, 4(1),11-15

[17] Gedye R N, Smith F E, Westaway K C, The rapid synthesis of organic compounds in microwave ovens, Canadian Journal of Chemistry, 1988, 66(1),17

[18] Bogdał D, Łukasiewicz M, Microwave-assisted oxidation of alcohols using aqueous hydrogen peroxide. Synlett 2000, 1(1): 143-145

[19] Varma R S, Dahiya R, Microwave assisted oxidation of alcohols Under solvent free conditions using clayfen, Tetrahedron Letters; 1997, 38 (12) 2043-2044. doi:10.1016/S0040-4039(97)00262-1

[20] Gedye, R., Smith, F., Westaway, K., Ali, H., Baldisera, L., Laberge, L. and Rousell, J. ; The Use of Microwave Ovens for Rapid Organic Synthesis. Tetrahedron Letters, 1986, 27, 279-282. doi:10.1016/S0040-4039(00)83996-9

[21] Loupy A, Petit A, Ramdani Md, Yvanaeff C, Majdoub M, Labiad B, Villemin D. Can. J. Chem. 1993, 71, 90

[22] Ozturk G, Gumgum B, Akba O. Synthesis of esters under microwave irradiation using heteropolyacids as catalysts, Catalysis Letters, 2002, 82 (3-4), 233

[23] Reilly M K, King R P, Wagner A J, King S M, Microwave-Assisted Esterification: A Discovery-Based Microscale Laboratory Experiment, J. Chem. Educ. 2014, 91, 10, 1706-1709. doi:10.1021/ed400721p

[24] Jones G B, Chapman B J, Decarboxylation of indole-2-carboxylic acids: improved procedures, J. Org. Chem. 58, 20, 5558-5559, 1993; doi: 10.1021/jo00072a052

[25] Cabrera-Rivera F A, Hernández-Vázquez L G, Flores-Sánchez P, Durán-Galván M, Escalante J. Solvent- and Catalyst-Free Microwave-Assisted Decarboxylation of Malonic Acid Derivatives. Green and Sustainable Chemistry, 2017, 7, 270-280, http://www.scirp.org/journal/gsc

[26] Varma, R S, Dahiya R.; Kumar S; Clay catalyzed synthesis of imines and enamines under solvent-free conditions using microwave irradiation, Tetrahed. Lett. 1997, 38, 2039-2042.

[27] Varma R S, Kumar D; Microwave-accelerated three-component condensation reaction on clay: solvent-free synthesis of imidazo [1, 2-a] annulated pyridines, pyrazines and pyrimidines, Tetrahedron lett., 1999 40 (43),7665-7669

[28] Mahmood N O, Mohammadgholipour S, Pirbasti F G; Microwave –assisted one-pot three-component synthesis thiazolidinones using KSF@Ni as an efficient heterogeneous catalyst, J of Sulfur Chemistry, 38, 668-678, 2017

[29] Garg A, Sarma D, Ali A A; Microwave assisted metal free approach to access 1, 2, 3- triazoles through multicomponent synthesis, Current Research in Green chemistry, 2020, 3, 100013

[30] Mandal S, Chhetri K, Bhuyana S, Roy B G, Efficient iron catalyzed ligand-free access to acridines and acridinium ions; Green Chemistry, 2020, 22, 3178-3185; doi:10.1039/D0GC00617C

[31] Kremsner J M, Kappe C O; Microwave Assisted Organic synthesis in Near Critical water at 300°C-A proof –of –Concept study, Eup. J Org Chem 3672-3679, 2005.

[32] Ashry El S H El, Kassem A A, Account of microwave irradiation for accelerating organic reactions, ARKIVOC 2006 (ix) 1-16

[33] Naqvi A, Shahnawaaz Md, Rao A V, Seth D S, Sharma N K; Synthesis of Schiff Bases via Environmentally Benign and Energy-Efficient Greener Methodologies E-Journal of Chemistry, 2009, 6(S1), S75-S78

[34] Damm M, Glasnov T N, Kappe C O, Org. Process Res. Dev. 14, 215, 2010

[35] Sondhi S M, Rani R, Microwave-mediated one step synthesis of tri- and tetracyclic heterocyclic molecules, Green Chemistry Letters and Reviews, 2010, 3(2), 115-120

[36] Mete E, Gul H I, Bilginer S, Algul O, Topaloglu M E, Gulluce M, Kazaz C, Synthesis and Antifungal Evaluation of 1-Aryl-2-dimethylaminomethyl-2-propen-1-one Hydrochlorides, Molecules, 2011, 16, 4660-4671; doi:10.3390/molecules16064660

[37] Zhu C, Zuo Y, Wang r, Liang B, Yue X, Shang N, Huang L, Chen Y, Du J et al; Discovery of potent cytotoxic ortho-ary chalcones as new scaffold targeting tubulin and mitosis with affinity-based fluorescence. J Med Chem. 2014, 57, 6364-6382

[38] Bhat A R, Najar Md H, Dongre R S, Akhter Md S, Microwave assisted synthesis of Knoevenagel Derivatives using water as green solvent, Current Research in Green and Sustainable Chemistry. 2020, 3, 100008

[39] Khadijah M. Al-Zaydia, Laila M. Nharia, Rita M. Borika and Mohamed H. Elnagdi, Green technologies in organic synthesis: self-condensation of enamines, enaminones and enaminoesters under microwave irradiation in ionic liquid; Green Chemistry Letters and Reviews, 2010, 3(2), 93-99

[40] Brun E, Safer A, Carveaus F, Bourahla K, Lhelgouach J M, Bazureau J P, Villalgordo J M; Microwave-Assisted Condensation Reactions of Acetophenone Derivatives and Activated Methylene compounds with aldehydes Catalyzed by Boric acid under Solvent Free Conditions. Molecules. 2015, 20, 11617-11631

[41] Zanin L L, Jimenez D E Q, Fonseca L P, Porto A L M, Knoevenagel Condensation Reactions of Cyano Malononitrile-Derivatives Under Microwave Radiation. Current Organic Chemistry, 2018, 22(6). DOI:10.2174/1385272822666180123145819

[42] Gupta M, Wakhloo B P, Tetra butyl ammonium bromide mediated Knoevenagel condensation in water: synthesis of cinnamic acids ARKIVOC, (i) 94-98, 2007; DOI: https://doi.org/10.3998/ark.5550190.0008.110

[43] de Cozar A, Millan M C, Cebrian C, Prieto P, D'ıaz-Ortiz A, de la Hoza A, Coss'ıob F P, Computational calculations in microwave-assisted organic synthesis (MAOS). Application to cycloaddition reactions. Org. Biomol. Chem., 2010, 8, 1000-1009

[44] Roshandel S,Suri S C, Mascischak J C, Rasul G, Prakash G K S, Catalyst and solvent free microwave –assisted synthesis of substituted 1,2,3-triazoles. Green Chem, July 2018, DOI: 101039/c8gc01516c .

[45] Jha A, Ramarao T A, Expeditious microwave assisted synthesis and bio-evaluation of novel bis (trifluoro-methyl) phenyl-triazole-pyridine hybrid analogues by click chemistry approach, Research on Chemical Intermediates. 2018, 44, 585-599, 2018, DOI: 10.1007/s11164-017-3121-2

[46] Lee D, Seo Y Ho, Bae J S, Lee S, Lee T I, Gong Y D, Lee T; Bull. Korean Chem. Soc. 2013, 34(8), 2523-2524

[47] Hervé G, Christophe Len C, Aqueous microwave-assisted cross-coupling reactions applied to unprotected nucleosides. Front. Chem., 2015, 17; doi:10.3389/fchem.2015.00010

[48] Eduardo Mendoza-Pérez E, Vázquez A, Microwave-assisted synthesis of isoindolinones via Pd-mediated tandem coupling reactions, Heterocycl. Commun. 2018; 24(3): 129-133

[49] Younis Baqi, Recent Advances in Microwave-Assisted Copper-Catalyzed Cross-Coupling Reactions, Catalysts 2021, 11, 46. doi:10.3390/catal11010046

[50] Ouzebla D, Lazrek H B, Smietana M, Vasseur J J, Applications of Microwave in

Organic Synthesis; A one –step Synthsis of Ribonucleosides using natural phoshphate as solid catalyst, Chemistry and Material Research, 2013, ,3(4), 84.

[51] Howe A G R, Maunder R, Morgan D J, Edwards J K, Rapid Microwave-Assisted Polyol Synthesis of TiO2-Supported Ruthenium Catalysts for Levulinic Acid Hydrogenation, Catalysts 2019, 9(9), 748; doi:10.3390/catal9090748

[52] Sharma B R, Kumar V, Kumar S, Panesar; Microwave assisted extraction of phytochemicals from *Ficus racemosa*, Current Research in Green Chemistry, 2020, 3, 100020

[53] Jhung S H, Lee J H, Forster P M, Ferey G, Cheetham A K, Chang J S, Microwave Synthesis of Hybrid Inorganic-Organic Porous Materials: Phae-Selective Rapid Crystallization; Chem. Eup. J; 2006, 12, 7899-7905. https://doi.org/10.1002/chem.200600270]

[54] Rajasekhar K K, Ananth V S, Nithiyananthan T S, Hareesh G, Kumar P N, Reddy R S P, International Journal of Chem Tech Research, 2010, 2(1), 592-597

[55] Kitchen H J, Vallance S R, Kennedy J L, Tapia-Ruiz N, Carassiti L, Harrison A, Whittaker A G, Drysdale T D, Kingman S W, Gregory D H; Modern Microwave Methods in Solid-State Inorganic Materials Chemistry: From Fundamentals to Manufacturing, Chem. Rev. 2014, 114, 2, 1170-1206

[56] Guadino E C, Rinaldi L, Rotolo L, Carnaroglio D, Pirola C, Cravotto G, Heterogeneous Phase Microwave Assisted Reactions under CO_2 or CO Pressure; Molecule, 2016, 21, 253.

[57] Darekara N R, Karalea B K, Akolkara H N, Burungaleb A S, Microwave Assisted Synthesis and Antibacterial Activity of New 1, 3, 4-Thiadiazoles and 1,2,4-Triazoles Derived from 2-{2-[2-(4-Fluorophenyl)-4-methylthiazol-5-yl]-1H-benzo [d]imidazol-1- yl} aceto-hydrazide Russian Journal of General Chemistry, 2020, 9(9) 1721– 1726

[58] Sreeram K J, Nidhin M, Nair B U, Microwave assisted template synthesis of silver nanoparticles, Bull. Mater. Sci., 2008 31(7) 937-942

[59] Polshettiwar V, Nadagouda M N, Varma R S, MW-assisted chemistry: a rapid and sustainable route for synthesis of organics and nanomaterials Aust. J. Chem., 2009, 62, 16-26.

[60] Zhao Y, Zhu J, Hong, Bian J N, Chen H, Microwave-Induced Polyol-Process Synthesis of Copper and Copper Oxide Nanocrystals with Controllable Morphology Eur. J. Inorg. Chem., 2004, 2004, 4072– 4080

[61] Huang J, Xia C, Cao L, Zeng X, Facile microwave hydrothermal synthesis of zinc oxide one-dimensional nanostructure with three-dimensional morphology Mater. Sci. Eng. B, 2008, 150, 187-193

[62] Wang K, Huang T, Liu H, Zhao Y, Liu H, Sun C, Colloids Surfaces A Physicochem. Eng. Asp., 2008, 325, 21-25

[63] Patra C R, Gedanken A, Rapid synthesis of nanoparticles of hexagonal type In2O3 and spherical type Tl2O3 by microwave irradiation, New J. Chem., 2004, 28, 1060-1065

[64] Jouhannaud J, Rossignol J, Stuerga D, Rapid synthesis of tin (IV) oxide nanoparticles by microwave induced thermohydrolysis, J. Solid State Chem., 2008, 181, 1439-1444

[65] Eliziário S A, Cavalcante L S, Sczancoski J C, Pizani P S, Varela J A, Espinosa J W M, Longo E, Morphology and Photoluminescence of HfO (2) Obtained by Microwave-Hydrothermal Nanoscale Res. Lett., 2009, 4, 1371-1379.

[66] Zhang H M, Liu J. B, H. Wang, W. X. Zhang, and H. Yan, Rapid

microwave-assisted synthesis of phase controlled BiVO4 nanocrystals and research on photocatalytic properties under visible light irradiation J. Nanoparticle Res., 2008, 10, 767-774.

[67] Zawadzki M, Synthesis of nanosized and microporous zinc aluminate spinel by microwave assisted hydrothermal method (microwave–hydrothermal synthesis of ZnAl2O4) Solid state Sci., 2006, 8, 14-18

[68] Nyutu E K, Chen C-H, Dutta P K, Suib S L, Effect of Microwave Frequency on Hydrothermal Synthesis of Nanocrystalline Tetragonal Barium Titanate, J. Phys. Chem. C, 2008, 112, 9659– 9667

[69] Moreira M L, Paris E C, do Nascimento G S, Longo V M, Sambrano J R, Mastelaro V R, Bernardi M I B, Andrés J, Varela J A, Longo E, Structural and optical properties of CaTiO3 perovskite-based materials obtained by microwave-assisted hydrothermal synthesis: An experimental and theoretical insigh Acta Mater., 2009, 57, 5174-5185.

[70] Motshekga S C, Pillai S K, Ray S S, Jalama K, Krause R W M, Recent Trends in the Microwave-Assisted Synthesis of Metal Oxide Nanoparticles Supported on Carbon Nanotubes and Their Applications, Journal of Nanomaterials Volume 2012, Article ID 691503, 15 pages, doi:10.1155/2012/691503

[71] Singh D, Rawat D and Isha, Microwave-assisted synthesis of silver nanoparticles from Origanum majorana and *Citrus sinensis* leaf and their antibacterial activity: a green chemistry approach. Bioresour. Bioprocess. (2016) 3: 14 DOI 10.1186/s40643-016-0090-z

[72] Hasanpoor M, Nabavi H F and Aliofkhazraei M, Microwave-assisted Synthesis of Alumina Nanoparticles Using Some Plants Extracts J Nanostruct, 2017, 7(1): 40-46

[73] Suchomski C, Breitung B, Witte R, Knapp M, Bauer S, Baumbach T, Reitz C and Brezesinski T. Microwave synthesis of high-quality and uniform 4 nm ZnFe2O4 nanocrystals for application in energy storage and nanomagnetics. Beilstein J. Nanotechnol. 2016, 7, 1350-1360.doi:10.3762/bjnano.7.126

[74] Aivazoglou E, Metaxa E, and Hristoforou E, Microwave-assisted synthesis of iron oxide nanoparticles in biocompatible organic environment AIP Advances 8, 048201 (2018); doi:10.1063/1.4994057

[75] Zuliani A, Balu A M, Luque R. Efficient and Environmentally Friendly Microwave-Assisted Synthesis of Catalytically Active Magnetic Metallic Ni Nanoparticles. ACS Sustainable Chem. Eng. 2017, 5(12), 11584-11587

[76] May-Masnou A, Soler L, Torras M, Salles P, Llorca J and Roig A, Fast and Simple Microwave Synthesis of TiO2/Au Nanoparticles for Gas-Phase Photocatalytic Hydrogen Generation, Front. Chem. 2018, 6: 110. doi:10.3389/fchem.2018.00110

[77] Elmaaty T A, El-Nagar Kh, Zaghloul D N, Sayed-Ahmed Kh, ElKadi Sh, Abdelaziz E, Microwave and Nanotechnology Advanced Solutions to Improve Eco-friendly Cotton's Coloration and Performance Properties. Egypt. J. Chem. 2018, 61 (3) 493-502

[78] Xu S, Zhong G, Chen C, Zhou M, Kline D J, Jacob R J, Xie H, He S, Huang Z, Dai J, Brozena A H, Shahbazian-Yassar R, Zachariah M R, Anlage S M, and Hu L, Uniform, Scalable, High-Temperature Microwave Shock for Nanoparticle Synthesis through Defect Engineering, Xu et al., Matter, 2019, 1, 759-769. doi:10.1016/j.matt.2019.05.022

Chapter 12

Polarization of Electrospun PVDF Fiber Mats and Fiber Yarns

Harshal Gade, Sreevalli Bokka and George G. Chase

Abstract

Electrospun fibers are of interest in a number of applications due to their small size, simplicity of fabrication, and ease of modification of properties. Piezoelectric polymers such as Polyvinylidene Fluoride (PVDF) can be charged when formed in the electrospinning process. This chapter discusses fabrication of PVDF fiber mats and fiber yarns and the measurement of their charge using a custom-made Faraday bucket. The results show the measured charge per mass of fiber mats was greater than the values measured for the yarns of the same mass. The measured charges may be related to both mass and external surface areas of the mats and yarn samples. It was observed the area/mass ratios of the fiber yarns were more than 30% less than the fiber mats.

Keywords: PVDF, Faraday bucket, electrospinning, yarns, fibers

1. Introduction

In recent years, nanotechnology has been used to develop novel materials including nano and submicron scaled materials such as nanorods, nanofoams, nanotubes, nanofilms, and nanofibers. These materials find use in various industrial applications and are the topics of many contemporary academic research efforts. Of these materials, the polymer electrospun fibers have found broad uses for catalysis, drug delivery, semiconductors and filtration [1–3].

Many polymers have been electrospun into nonwoven fiber mats. The polymer materials can have intrinsic piezo, thermal, and mechanical properties. When the polymers are formed into fiber structures such as thin mats, the high porosities and high specific surface areas of the mats can enhance the mat structural properties compared to similar mats of microfibers. The material and structural properties of these mats are ideally suited for filter media for air filtration and face masks.

Less common in the literature are discussions of the fabrication of yarns from electrospun fibers. The fabrication of yarns requires a mechanical method to entangle and interlock the intrinsic fibers, often by twisting, to form a self-supporting assembly of the fibers of an overall cylindrical shaped structure that can be characterized by a structure diameter.

Prior to electrospinning, the submicron fibers were often synthesized by techniques such as drawing, templating, solution casting, and phase separating. Most of these techniques had shortcomings including deformation failures, inability to produce continuous fibers, inability to scale-up, low production rates, or significant by-product wastes. The electrospinning method overcomes some of these shortcomings, and because of its simplicity, is a highly popular synthesis method.

Electrospinning is well-documented, established, and cost-effective, and is applied commercially. **Figure 1** shows numbers of publications, by publication year, as determined from the Scifinder ™ data base for the past 25 years. The plot shows a steady rise in numbers of papers since about 2000 when Reneker [4] published a seminal paper on electrospinning. The data search was conducted in August 2020 hence the final year was incomplete.

Electrospinning has been used to spin fibers for a wide range of polymers. One of these polymers, polyvinylidene fluoride (PVDF), is well known for its electrical properties. PVDF exhibits five known crystalline phases- α, β, γ, δ and ε. Amongst them, the β-phase has the highest permanent dipole moment due to its trans, TTT, planar zig-zag configuration. The β-phase is considered most responsible for the piezoelectric response obtained from the PVDF materials. A goal of enhancing the beta-phase contents in PVDF materials is an ongoing research pursuit [5, 6].

Electrospun fibers have been used as electrets in several applications. Electrets have a surface charge which can be exploited in capturing charged particles. Nanofibers can be converted into electrets by various methods such as corona discharging, surface fluorination and nafion functionalization. Several research groups developed custom made bench-scale procedures to produce polarized fibers which involved simultaneous stretching, heating and electrical poling. Similarly, Lolla *et al.* [6, 7] produced polarized PVDF fiber mats and tested them for aerosol filtration. The polarized fibers were observed to have higher surface charges, better capture efficiencies and lower pressure drops compared to as-spun fibers. The study was limited by measurement of localized surface potential via a hand-held electrostatic field meter [6]. **Table 1** lists several instruments reported in literature used to measure surface potential and charge. All of these instruments make localized measurements (do not measure properties over a large area of a mat) and may be impractical to use for production scale processes due to complexity and cost of operations. Measurements of the surface potential or electrical field are related to electrical charges but methods to calculate charges from the measurements are not always apparent.

Gade *et al.* [12] fabricated a custom-made Faraday bucket and a procedure to calculate the charges of fiber mat samples. The Faraday bucket overcomes some limitations or challenges of using the methods listed in **Table 1**, namely: it is non-destructive, measures large sample sizes, is easy to scale-up, and has a tractable mathematical model to convert voltage to charge value. In this chapter, the Faraday

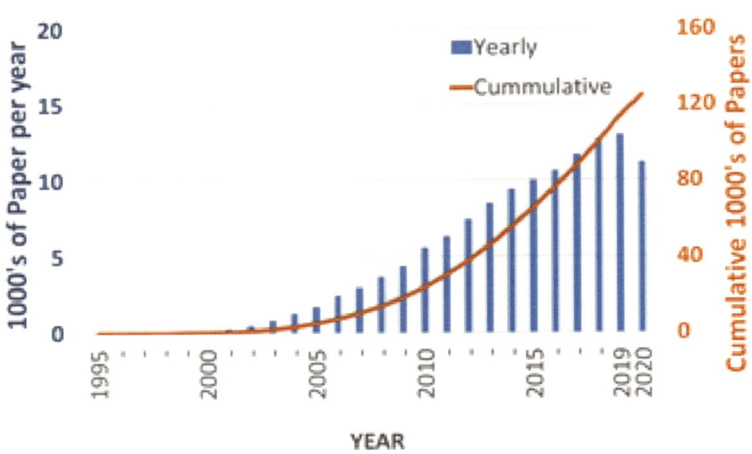

Figure 1.
Number of publications on "electrospinning" versus year of publication.

bucket is used to measure and compare charges between electrospun fiber mats and electrospun (continuous twisted fiber) yarns. Layers of fibers mats and yarns were stacked together to explore whether the Faraday bucket was more sensitive to bulk (mass) charge or more sensitive to surface charge. The charges on polarized fibers and yarns are also compared.

Many publications discuss methods to charge fibers or to modify fibers surfaces (with coatings or additives such as carbon nanotubes) to enhance performances of fiber filter media. The subject matter is broad, and the numbers of publications are too numerous for a complete list. **Table 2** lists a sample of some of the publications.

The electrospinning processes typically produce nonwoven, randomly oriented, fiber mats. These fiber mats typically have low mechanical strength (compared to microfiber mats). The electrospinning processes have a low mass production rates per nozzle that limits commercial applications from an economic standpoint. Researchers have studied various approaches to increase the mass production by increasing the number of electrospinning jets in the process [25, 26]. To overcome some of the limitations, researchers have studied electrospun yarns to improve the alignment of fibers and to increase the mechanical strength. Production of highly twisted PVDF – HFP electrospun fiber yarns using a novel ring collector was reported by Shuakat *et al.* [27]. Afifi *et al.* [28] and Teo *et al.* [29] studied methods to continuously produce electrospun yarns. In this chapter the yarns were produced by twisting and drawing the fibers in flight and the twisted yarn were wound onto

Researcher	Instrument	Materials tested	Reference
Collins *et al.*	Scanning Probe microscopy	Various dielectric surfaces	[8]
Du *et al.*	Kelvin Probe force microscopy (open and closed loop techniques)	Single and multi-layer graphene structures	[9]
Takahashi and Yoshita	Inversion algorithmic methods	DC basin-type insulator	[10]
Fatihou *et al.*	Electrostatic voltmeter	Electrospun PVDF nanofibers	[11]
Lolla *et al.*	Electrostatic field meter	Polarized electrospun nanofibers	[6]

Table 1.
List of instruments and materials tested.

Researcher(s)	Description	Reference(s)
Fredrick Brown	Fundamental physics of electrical and charge effects on filter performance	[13, 14]
Choi *et al.*	Aluminum coatings applied to micro and nanoscale fibers, modified surface charge to control filter performance	[15]
Romay *et al.* Walsh *et al.* Wang *et al.*	Quasi-permanent charges on dielectric polymer fibers	[16–19]
Liu *et al.* Khalid *et al.* Jing *et al.*	Filters made of highly polar polymer fibers showing high binding affinity to fine particulate matter in aerosols	[20–22]
Li *et al.*	Fibrous filters hybridized with carbon nanotubes (CNT) exhibiting slip flow effects at the CNT surfaces	[23, 24]

Table 2.
A sampling of literature on topics of fiber surface charge, fiber coatings, and additives.

a spool, which differs from typical electrospinning equipment that collect the fiber mats on a solid grounded surface. The resulting yarns had lengths up to tens of meters long and exhibited mechanical properties different from the electrospun mats.

2. Materials and methods

2.1 Materials

Electrospinning solutions were prepared by dissolving PVDF powder (Arkema Inc., Exton, PA, USA, Kynar® 761 grade resin with molecular weight of 500,000 g/gmol and density of 1.78 g/m^3) in co-solvents N-N-Dimethylformamide (DMF) and acetone (Sigma Aldrich, St. Louis, MO, USA). These materials were used in making the solutions without further purification.

2.2 Preparation of as-spun solutions

Observations while electrospinning the fiber mats and yarns showed different fiber diameters were obtained for the two processes likely due to variations in setup geometries and electric field strengths. By trial and error in varying solution concentrations, appropriate solution concentrations for the two processes were determined to produce fibers with diameters of 1200 nm for both processes. The comparisons of charges and properties discussed in the experiments below were obtained for fibers with these diameters.. Electrospinning solutions for producing fiber mats were prepared with 18%wt PVDF polymer by mixing the polymer with 50:50 wt.% blended DMF and Acetone solvents. The PVDF powder was added to mixture of solvents and heat-stirred for half an hour at 70 °C to attain a clear homogenous mixture. For production of fiber yarns, a 13 wt% PVDF polymer solution was prepared by mixing Acetone and N, N-Dimethylformamide (DMF) solvents at 1:1 ratio. This mixture was heated on a hot plate at 70 °C for 20 min to attain a clear homogenous mixture.

2.3 Electrospinning set-ups and mechanism

The fiber mats were synthesized by using a typical single-needle electrospinning setup as shown in **Figure 2.** The polymer solutions were loaded into 5 ml plastic syringes and fed by syringe pump (NE-1000, New Era Pump Systems, Inc., Farmingdale, NY). The metallic needles were charged by high voltage power supplies (ES30P-5 W, Gamma High Voltage Research, Ormond Beach, FL) to generate potential differences between the collector and the needle. The fiber mats were collected on rotating cylindrical drum collectors covered with 30 cm × 30 cm sheets of grounded aluminum foil. The fiber mats were electrospun for varying times to create mats of basis weights of 10, 20, 30, 40 and 50 g/m^2. In the experiments involving stacked layers of mats, all of the layers were formed of mats of 20 g/m^2 basis weights. The electrospinning conditions are listed in **Table 3**.

Figure 3 shows the experimental setup used to generate the electrospun yarn and is similar to setups reported in literature [28, 30]. The setup consisted of fiber spinning and yarn winding sections. In the fiber spinning section, a metallic conical-shaped funnel collector was connected to the motor and controller. The syringe pump and power supply were used to electrospin the polymer solution at a flowrate of 4 - 5 ml/hr. A potential difference of 10 – 20KV was applied between the

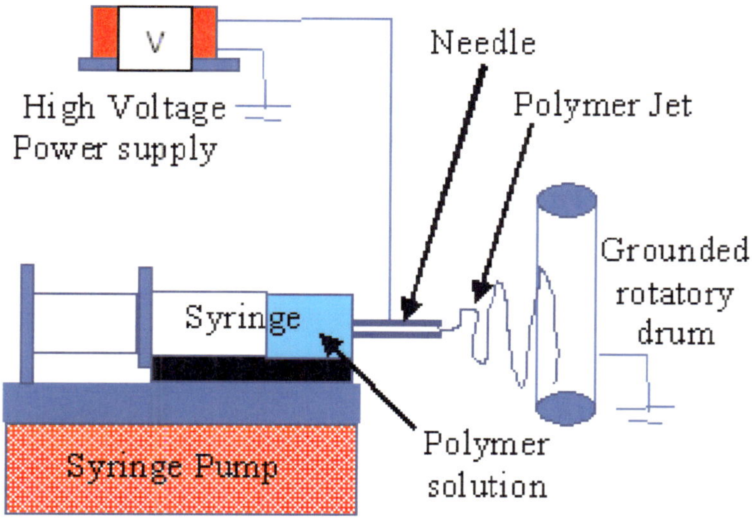

Figure 2.
Schematic of electrospinning set-up.

PVDF (wt.%)	DMF – Acetone mass ratio	Tip to Collector Distance (cm)	Applied Potential (kV)	Flow Rate (ml/hr)	Avg. Fiber Diameter (nm)	Standard Deviation (nm)	Drum Rotation Rate (rpm)
18	1:1	20	27	5	1139	654	30

Table 3.
Electrospinning conditions and fiber diameter data.

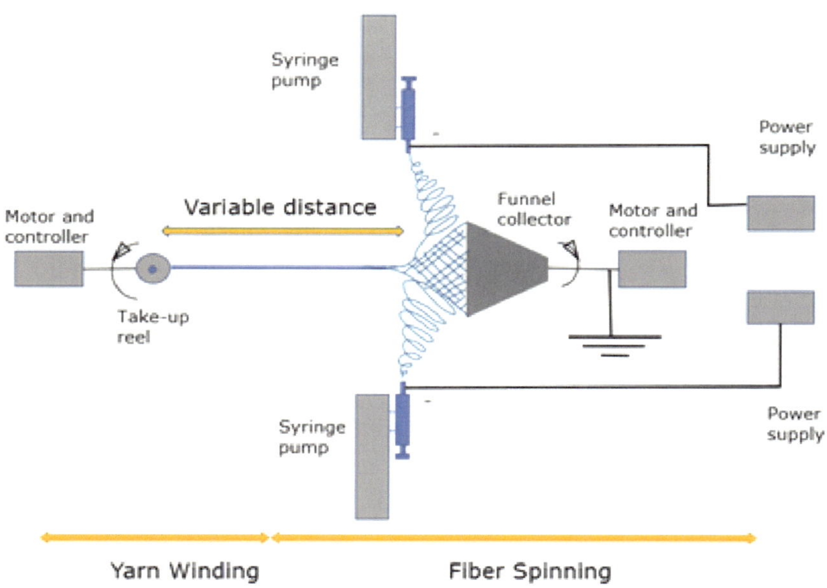

Figure 3.
Illustration of fiber yarn setup.

metal needles and the collector with an 8 cm distance between the needles and the edge of the collector.

Charged polymer jets launched from drops of polymer solution at the tips of the needles and followed the electric field gradient towards the wide neck of the conical-shaped collector. Once a substantial mat of fibers collected over the open end of the collector, the center of mat was hooked onto a wire and pulled to stretch the mat into the shape of an inverted cone.

The metal collector was rotated by the motor to twist the fiber structure into a twisted continuous yarn. The yarn gradually increased in length and was stretched and attached to the take-up reel for collection onto a spool. The rotation speeds of the metal collector and the take-up reel were adjusted by trial and error to produced yarns of uniform twist and uniform outer diameter.

In the case of electrospun mats, replicate samples were obtained at consistent basis weights by adjusting the time of fiber accumulation on the mats, so that the resulting fiber mat had uniform thickness and mass over the area of the sample. But in the case of fiber yarns a suitable length of sample was considered from each replicate run and compared for consistency by comparing the mass to length ratio of each sample. Results showed ±3% variation in mass/length for each of sample used in these experiments.

2.4 Polarization procedure

Mats and yarns were polarized by the treatments described below. The treatments were not applied to stacked layers of mats in the layered mat experiments described later. **Figure 4** (a) shows a photograph of the sample holder made of PTFE (Teflon®) for the main frame, brass bars for the clamps, and thin aluminum plates for the electrodes. The PTFE was chosen over other materials as it was easy to machine and had many desired properties such as low electrical conductivity, low dielectric constant, and relatively high melt temperature. **Figure 4** (b) shows the sample holder inside of a Fischer Scientific iso-temp oven. The aluminum plates were 19 cm × 11 cm and 1 mm in thickness. One aluminum plate was grounded and

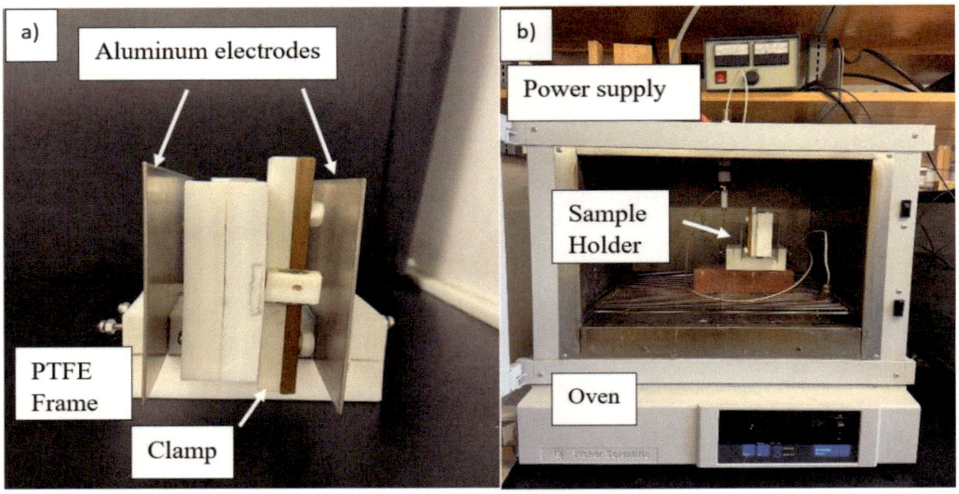

Figure 4.
(a) End view of fabricated sample holder showing the two planar electrodes used to apply the electric field for poling the sample. (b) Photo of sample holder inside of oven and high voltage power supply for charging the electrode above the oven.

the other was electrically charged to produce an electric field between the plates of 2.5 kV/cm. The distance between the electrodes was 6 cm. The fiber mat samples and yarn samples were placed in the holder to perform all polarization treatments including simultaneous heating, stretching and electrical poling.

Heat treatments were applied to change the sample temperature from room temperature to 150 °C with a temperature ramp-up rate of about 10C per min up to the soak temperature (150 C). The sample was held at the soak temperature for 5 minutes and then allowed to cool at a temperature ramp-down at rate of about 10C/min. The oven did not have ramp-rate control, so the ramp rates are estimates based on observed temperature readings.

The electric field poling was applied at field strength of 2.5 kV/cm during the heating of the oven. The poling started at the same time as the oven and stopped when the oven was turned off at the end of the soak time.

Uniaxial stretched mats and yarns were obtained by clamping the mats and yarns into the holder positioned parallel to and between the aluminum electrodes. The moveable clamp was moved to create a 10% stretch of the samples. The stretch time of the sample started when the sample was placed in the holder and stretched. The stretch time included time to place the holder into the oven, temperature ramp-up, temperature soak, temperature ramp-down, time to remove the holder, and ended when the sample was removed from the stretching mechanism in the holder, for a total of about 52 min. The as-spun and polarized samples were stored in the static shielding bags immediately after fabrication to avoid any dissipation of ions or charge.

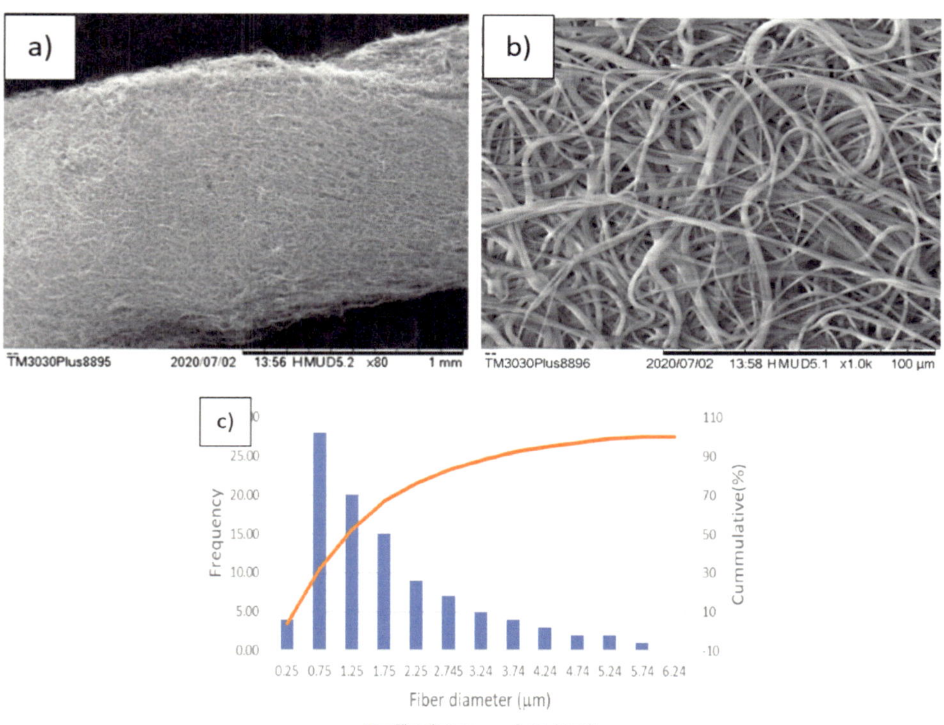

Figure 5.
SEM images of (a) fiber yarn and (b) fibers as seen on the surface of the fiber yarn with average fiber diameter of 1139 nm ± 654 nm, and average fiber yarn diameter of 900 µm ± 300 µm and c) fiber size distribution curve.

2.5 Characterization methods

The morphology characteristics of the electrospun fiber mats and yarns were observed using a scanning electron microscopy (SEM, TM3000 and TM3030 Plus, and Hitachi, Japan). SEM images were analyzed by FibraQuant 1.3 software (nano Scaffold Technologies, LLC, Chapel Hill, NC) to measure the fiber diameter distributions. **Figure 5** shows SEM images and fiber size distributions for PVDF fibers and yarns. Electric charges on the fiber mat were measured using a Faraday Bucket. A detailed description of the Faraday Bucket is given in reference [12]. The fiber mats were cut to the size needed for the measurement (4 cm by 4 cm) otherwise the measurements were non-destructive. Based on the electrostatic principles, as a sample lowered into the interior of the Faraday Bucket, the inner metallic "bucket" acquired an electric potential that was detected as a change in voltage relative to the surroundings (ground). By an appropriate circuit model of the Faraday bucket the measured potential was converted to charge.

Fiber yarns produced using setup in **Figure 3** were characterized as-spun and after polarization discussed in Section 2.4. The as-spun and polarized yarn samples were wrapped on a 'U' shaped copper wire and lowered into the Faraday bucket for measurement. The calculated charges were normalized with respect to mass of sample as discussed by Gade *et al.* [12]. The influence of U-shaped wire holding the yarn on the measured charge was found to be negligible when the wire without yarn was lowered into the Faraday bucket and produced zero measured voltage.

3. Results and discussions

3.1 Effect of stacking of fiber mats on charge measurement

Evaluation of the effects on charge measurments of stacked of mats was conducted only with as-spun mats (not with polarized mats). The purpose of this was to assess whether the Faraday bucket measurements were more sensitive to surface area or to mass of the samples.

Figure 6a shows a photograph of a single 4 × 4 cm fiber mat. **Figure 6b** shows five as-spun mats stacked on top of each other. All mats were cut to size 4 cm × 4 cm and had a 1 × 1 cm tab at one edge. **Figure 6c** shows a bar chart of calculated charges per unit mass of individual and stacked layer samples. The measurements of the five individual samples are labeled as 1 to 5. The stacked samples are labeled A to D where A was formed by stacking the mats 1 + 2 (i.e., individual mats 1 and 2 stacked), B was three mats 1 + 2 + 3, C was four mats 1 + 2 + 3 + 4, and D was five mats 1 + 2 + 3 + 4 + 5.

All the single mats in **Figure 6c** had approximately the same measured charges of about 130 nC/g. All the mats had the same basis weight (20 g/m^2), hence had the same masses.

If the Faraday bucket detected charge in bulk (i.e. per mass) then stacking the mats should not show a difference in charge/mass. If the Faraday bucket detected charge based on charge on the external mat surface area, and the charges of the mats do not transfer between the mats, then we would expect the measured charge/mass to decrease as mass increased and the surface area remained the same.

The results in **Figure 6c** shows the charge/mass linearly increased proportional to the number of mats in the stack. The charge/mass of stacked sample D (with five individual mats) was approximately double that of a single mat. Numerically this indicates that the measured charge per total mass increased over the single mat charge by about 25% for each additional mat in the stack. The increase in charge per mass indicates the

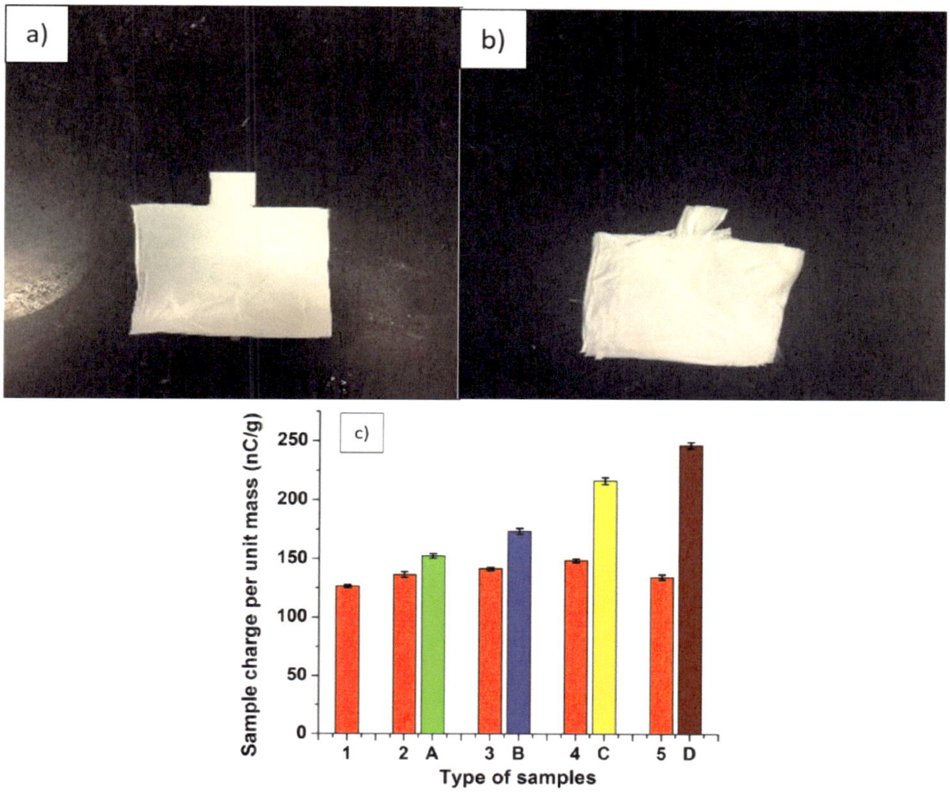

Figure 6.
(a) Photograph of example of a single fiber mat of size 4 × 4 cm with a 1 × 1 cm tab on one edge, (b) photograph of five mats stacked on top of each other. (c) Bar chart of charge/mass of various samples (1–5 = measured charge/mass of five individual samples) (a = charge of stacked mats 1 + 2, B = stacked mats 1 + 2 + 3, C = stacked mats 1 + 2 + 3 + 4, D = stacked mats 1 + 2 + 3 + 4 + 5). The error bars in (c) represent average of three charge measurements of same mats and error is one standard deviation.

bulk charge mechanism alone is unlikely. The increase in charge also strongly indicates that the measurement is not that of the charges on the external surfaces of the stacked mats assuming the charges do not migrate to the surface. Hence the mechanism is more complex. It is interesting to note that each subsequent mat added to the stack to linearly increased the measured charge by 25%. This gives the relationship

$$\frac{\left(C/M\right)}{\left(C_1/M_1\right)} = 0.25\left(\frac{M}{M_1}\right) + 0.75 \quad (1)$$

or

$$\frac{C}{C_1} = 0.25\left(\frac{M}{M_1}\right)^2 + 0.75\frac{M}{M_1} \quad (2)$$

where C is the measured charge of the stack and M is the mass of the stack. C_1 is the measured charge and M_1 is the mass of one mat. The ratio M / M_1 equals the number of mats in the stack. An interpretation of the meaning of the two terms on the right side of Eq. (2) is not apparent. Future experiments should be conducted by varying the surface areas of the mats to determine if the terms are related to area and possible migration of charges between the stacked mats.

3.2 Comparison of charge/mass of electrospun and polarized PVDF fibers and yarns

The plot in **Figure 7** compares the charges of the (a) as-spun mats and yarns, and (b) polarized mats and yarns. For comparison purposes the samples of polarized and nonpolarized mats and yarns are compared on equal mass basis. The labels A, B, C, D and E indicate the masses of fibers in the yarns and mats corresponding to 0.0058, 0.0124, 0.0196, 0.0278 and 0.0376 g respectively. The data reported in **Figure 7** are for electrospun mats of varying basis weights (not stacked layers of mats). For the given areas and masses of the mats the A, B, C, D and E mat samples correspond to 10, 20, 30, 40 and 50 g/m^2 basis weights.

In **Figure 7** the charges per mass of the mats were about 2 to 5 times the value for the yarns. The charge per mass of the as-spun and polarized mats increased as the mat mass increased with approximate slope of 17% (comparable to the 25% slope observed in the layered mats of **Figure 6**). Charges on the yarn samples did not vary as much with mass. Both the as-spun and polarized yarns tended to have a modest decrease in charge per mass as yarn mass increased. The difference in performance between the yarns and the mats is probably due to the way the yarn was folded to fit into the Faraday bucket. Increasing the mass of the yarn was obtained by increasing the length of the yarn hence the overall surface area per mass of the yarn was constant. But to fit the yarn into the Faraday bucket, the yarn was wound onto the U shaped metal wire which resulted in the first layers of the windings being covered by subsequent layers. Unlike the stacked fiber mats, the resulting charge/mass decreased with mass. This suggests that the measured charges per mass of the yarns were mostly proportional to the external area/mass ratio and may also give insight to the performance of the mats. This topic should be further explored in future work.

Figure 8 shows plots comparing the as-spun to polarized mats and yarns. The comparison of the mats in **Figure 8a** shows the polarized treatments only marginally increased the charges on the mats. This contrasts with the increases in charges reported in literature [6]. There were some differences between the treatments in this work compared to reference [6] such as the heat cycle in [6] was at a controlled ramp rate and the electrical polarization was maintained until the mat had completely cooled, while in this work the ramp rate was not controlled and the electrical polarization was for a shorter time period. It is possible, though not verified here,

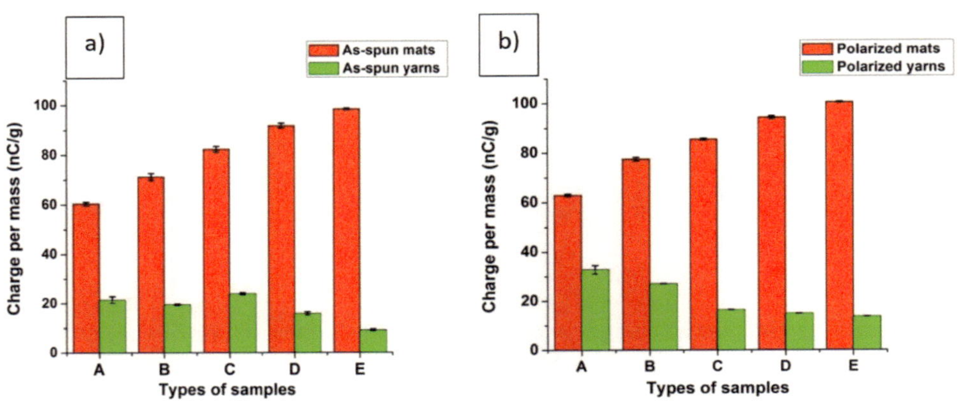

Figure 7.
Charge/mass plot for mats and yarns (a) as-spun, (b) polarized. The respective masses of the samples were a = 0.0058 g, B = 0.0124 g, C = 0.0196 g, D = 0.0278 g, and E = 0.0376 g.

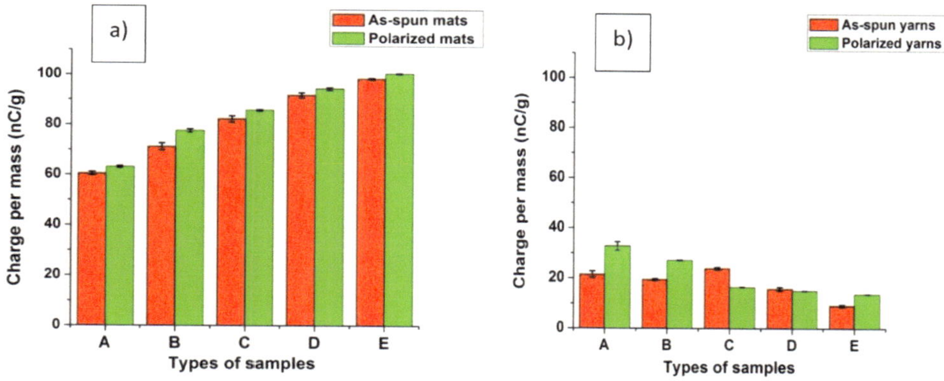

Figure 8.
Charge/mass plot for as-spun and polarized samples (a) Mats, (b) yarns the respective masses of the samples were a = 0.0058 g, B = 0.0124 g, C = 0.0196 g, D = 0.0278 g, and E = 0.0376 g.

that the beta phase content of the electrospun fibers was near its maximum in the as-spun fibers and hence the polarization treatment did not have much room to increase the beta phase content. This is left for future investigation.

The comparison of yarns in **Figure 8b** similarly show a small increase in the charge in most of the cases. Overall, the charges on the yarns did not change significantly with mass. The polarized samples A, B, and E showed greater charge compared to the as-spun samples while C and D showed less charge. These variations may be within experimental error possibly due to the hand winding of the yarns onto the U-shaped wire holder.

4. Conclusions

Polymer PVDF was electrospun to form fiber mats and continuous twisted yarns. Samples of the mats and yarns were polarized by stretching, heating and poling. The as-spun and polarized mats and fibers were measured for their charge via a Faraday bucket. The results showed the mats had significantly higher charge per mass than the yarns at the same mass. The measured charge per unit mass of the mats increased as the mass of the mat increased. The measured charge per mass of the yarns slightly decreased as mass increased. The polarization treatments used in this work did not significantly increase the charge of the mats and yarns. Charge measurements of stacked layers of mats suggest that the charge measured by the Faraday bucket is a complicated combination of surface area and bulk mass. Changing the basis weights of fiber mats (instead of stacking layers) gave similar trends suggesting the same mechanisms may apply to both stacked and directly spun mats. The nearly constant measured charges of the yarns suggest that the charge per mass may be related to the surface area per mass of the yarns.

Acknowledgements

This work was funded by Coalescence Filtration Fibers Consortium (CFNC): Parker Hannifin, Hollingsworth and Vose, and Donaldson. We acknowledge the assistance of technicians Steve Roberts and William Imes for fabrication and operation of the Faraday bucket.

Author details

Harshal Gade, Sreevalli Bokka and George G. Chase*
Department of Chemical, Biomolecular and Corrosion Engineering, The University of Akron, Akron, Ohio, United States of America

*Address all correspondence to: gchase@uakron.edu

References

[1] Damaraju SM, Wu S, Jaffe M, Arinzeh TL. Structural changes in PVDF fibers due to electrospinning and its effect on biological function. Biomedical Materials. 2013;8:045007. DOI: 10.1088/1748-6041/8/4/045007.

[2] Gaur A, Kumar C, Shukla R, Maiti P. Induced Piezoelectricity in Poly (vinylidene fluoride) Hybrid as Efficient Energy Harvester. Chemistry Select. 2017;2:8278-8287. DOI: 10.1002/slct.201701780.

[3] Li H-Y, Liu Y-L. Nafion-functionalized electrospun poly (vinylidene fluoride) (PVDF) nanofibers for high performance proton exchange membranes in fuel cells. Journal of Materials Chemistry A. 2014;2:3783-3793. DOI: 10.1039/C3TA14264G.

[4] Reneker DH, Chun I, Nanometre diameter fibres of polymer, produced by electrospinning. Nanotechnology. 1996;7:216-223.

[5] Gade H, Bokka S, Chase GG, Polarization treatments of electrospun PVDF fiber mats. Polymer. 2020;212,123152. DOI: 10.1016/j.polymer.2020.123152

[6] Lolla D, Lolla M, Abutaleb A, Renekar DH, Chase GG, Fabrication, polarization of electrospun polyvinylidene fluoride electret fibers and effect on capturing nanoscale solid aerosols. Materials. 2016;9:671-689. DOI: 10.3390/ma9080671.

[7] Lolla D, Pan L, Gade H, Chase GG, Functionalized Polyvinylidene Fluoride Electrospun Fibers and Applications, in Electrospinning Method Used to Create Functional Nanocomposite Films, T Tański editor, Intech Open Limited, London, UK, 2018; Volume 8.

[8] Collins L, Kilpatrick JI, Vlassiouk IV, Tseley A, Appl. Phys. Lett, 2014;104:133103.

[9] Du Q, Zhang Q, Wu Z, Ma J, Zhou J, Surface charge distribution on DC basin-type insulator. IEEE transactions on dielectrics and insulators. 2019;26(1): 17-25. DOI: 10.1109/TDEI.2018.007371.

[10] Takahashi T and Yoshita M, Scanning tunneling spectroscopy of-type GaAs under laser irradiation, Appl. Phys. Lett, 1997;70:2162.

[11] Fatihou A, Dascalescu L, Zouzou N, Neagoe MB, Measurement of Surface Potential of Non-uniformly Charged Insulating Materials Using a Non-contact Electrostatic Voltmeter, IEEE Transactions on Dielectrics and Electrical Insulation. 2016:23(4):2377-2384.

[12] Gade H, Parsa N, Chase GG, Renekar DH, Roberts OS, Charge measurement of electrospun polyvinylidene fluoride fibers using a custom-made Faraday bucket. Review of Scientific Instruments. 2020;91:075107. DOI: 10.1063/1.5142386

[13] Fredrick ER, Fibers, electrostatics, and filtration: a review of new technology. J Air Pollution Control Association. 1980:30(4), 426-431.

[14] Brown RC, Air Filtration, Pergamon Press, Oxford, 1993.

[15] Choi DY, An EJ, Jung S-H, Song DK, Oh YS, Lee HW, Lee HM, Al-Coated Conductive Fiber Filters for High-Efficiency Electrostatic Filtration: Effects of Electrical and Fiber Structural Properties. Scientific Reports. 2018: 8,5747.

[16] Romay FJ, Liu BY, Chae S-J. Experimental study of electrostatic capture mechanisms in commercial electret filters. Aerosol Sci. Technol. 1998:28:224-234.

[17] Walsh D, Stenhouse J, Parameters affecting the loading behavior and degradation of electrically active filter materials. Aerosol Sci. Technol. 1998:29: 419-432.

[18] Walsh D, Stenhouse J, The effect of particle size, charge, and composition on the loading characteristics of an electrically active fibrous filter material. J. Aerosol Sci. 1997:28:307-321.

[19] Wang S, Zhao X, Yin X, Yu J, Ding B, Electret Polyvinylidene Fluoride Nanofibers Hybridized by Polytetrafluoroethylene Nanoparticles for High-Efficiency Air Filtration. ACS Appl. Mater. Interfaces. 2016:8:23985-23994.

[20] Liu C, Hsu P-C, Lee H-W, Ye M, Zheng G, Liu N, Li W, Cui Y, Transparent air filter for high-efficiency PM2.5 capture. Nature Communications. 2015:6:6205.

[21] Khalid B, Bai X, Wei H, Huang Y, Wu H, Cui Y, Direct blow-spinning of nanofibers on window screen for highly efficient $PM_{2.5}$ removal. Nano Letters. 2017:17:1140-1148.

[22] Jing L, Shim K, Toe CY, Fang T, Zhao C, Amal R, Sun K-N, Kim JH, Ng YH, Electrospun Polyacrylonitrile–Ionic Liquid Nanofibers for Superior $PM_{2.5}$ Capture Capacity. ACS Applied Materials and Interfaces. 2016:8: 7030-7036.

[23] Li P, Zong Y, Zhang Y, Yang M, Zhang R, Li S, Wei F, In situ fabrication of depth-type hierarchical CNT/ quartz fiber filters for high efficiency filtration of sub-micron aerosols and high water repellency. Nanoscale 2013:5:3367-3372.

[24] Li, P., Wang, C., Zhang, Y. & Wei, F. Air Filtration in the Free Molecular Flow Regime: A Review of High-Efficiency Particulate Air Filters Based on Carbon Nanotubes. Small 2014:10:4543-4561.

[25] Varesano A, Carletto RA, Mazzuchetti G. Experimental investigations on the multi-jet electrospinning process. Journal of Materials Processing Technology. 2009;209:5178-5185. DOI : 10.1016/j. jmatprotec.2009.03.003

[26] Kim G, Cho YS, Kim WD. Stability analysis for multi-jets electrospinning process modified with a cylindrical electrode. European Polymer Journal. 2006;42:2031-2038. DOI : 10.1016/j. eurpolymj.2006.01.026

[27] Shuakat MN, Wang X, Lin T. Electrospinning of nanofiber yarns using novel ring collector, proceedings of the 2013 Fiber Society Spring conference, Fiber Society, Geelong, Australia, p. 238-239.

[28] Afifi AM, Nakano S, Yamane H, Kimura Y. Electrospinning of continuous aligning yarns with a funnel target. Macromolecular Material Engineering. 2010;295:660-665. DOI : 10.1002/mame.200900406.

[29] Teo W, Gopal R, Ramaseshan R, Fujihara K, Ramakrishna SA. dynamic liquid support system for a continuous electrospun yarn fabrication, polymer 2007;48:3400-3405. DOI : 10.1016/j. polymer.2007.04.044.

[30] Ali U, Zhou Y, Wang X, Lin T. Direct electrospinning of highly twisted. continuous nanofiber yarns. Journal of Textile Institute. 2012;103:1-9. DOI : 10.1080/00405000.2011.552254

Chapter 13

Recent Developments in All-Solid-State Micro-Supercapacitors Based on Two-Dimensional Materials

Minu Mathew, Sithara Radhakrishnan and Chandra Sekhar Rout

Abstract

Owing to their unique features such as high surface area, rich electroactive sites, ultrathin thickness, excellent flexibility and mechanical stability and multiple surface functionalities enables outstanding electrochemical response which provides high energy and power density supercapacitors based on them. Also, the Van der Waals gap between layered 2D materials encourages the fast ion transport with shorter ion diffusion path. 2D materials such as MXenes, graphene, TMDs, and 2D metal–organic frame work, TMOs/TMHs materials, have been described with regard to their electrochemical properties for MSCs. We have summarized the recent progress in MSC based on well-developed 2D materials-based electrodes and its potential outcomes with different architectures including interdigitated pattern, stacked MSC and 3D geometries for on-chip electronics. This chapter provides a brief overview of the recent developments in the field of 2D material based all-solid-state microsupercapacitors (MSCs). A brief note on the MSC device configuration and microfabrication methods for the microelectrodes have been discussed. Taking advantage of certain 2D materials such as 2D MXenes, TMDs, TMOs/TMHs that provide good surface chemistry, tunable chemical and physical properties, intercalation, surface modification (functionalization), heterostructures, phase transformations, defect engineering etc. are beneficial for enhancement in pseudocapacitance as it promotes the redox activity.

Keywords: microsupercapacitors, solid-state supercapacitors, two-dimensional materials, energy storage

1. Introduction

The popularization of portable electronic equipment has concentrated heavily on miniaturization and convergence of different technologies. While technologies such as wearable sensors and flexible displays has progressed, advances in energy storage are still lagging behind innovations in other electronic devices. Miniaturization of energy sources is also essential for environmental, medical, biological and other applications. Consequently, the reduction in size and integration of micro-power systems such as micro-batteries, micro-fuel cells, micro-supercapacitors (MSCs) and piezoelectric power harvesters are essential for the future growth of portable electronic devices [1]. MSCs have gained considerable attention among these micro-power systems due to its high power densities, fast rate capabilities, ultra-long – cycle life and simple integration into the micro-nano electronic system as energy sources [2, 3]. Three main types of device configurations for MSCs have been developed to date: in-plane architecture, fiber shape and three-dimensional (3D) type (**Figure 1**). The advantages and disadvantages of these device configurations are given in **Table 1**. The performance of MSCs is determined not only by components but also by the combination of each individual element for the development of a device.

The choice of electrode materials and electrolytes is the two critical parameters influencing the electrochemical performance of the MSCs. Two-dimensional (2D) materials with unusual properties such as ultra-thin thickness, large lateral size, excellent flexibility and tunable physicochemical properties are currently the perfect choice for MCS as an electrode material. A large number of 2D materials have been developed to date, including graphene and analog nanosheets such as transition metal dichalcogenides (TMDs), transition metal oxides/hydroxides (TMOs/TMHs), metal carbides and nitrides (MXens), boron nitride (BN), phosphorene, and so on. In addition to electrode material selection, the electrolyte selection also plays a crucial role in the performance of MSC and these electrolytes can be classified into two types (i) conventional liquid electrolytes and (ii)solid-state electrolytes [9]. Conventional liquid electrolytes have a common disadvantage in terms of their liquid nature; therefore, a strict encapsulation process is required to avoid electrolyte leakage. However, when the device is damaged, electrolyte leakage remains unavoidable. Accordingly, to overcome this disadvantage, a solid-state electrolyte was developed by blending the acids, ionic liquids and salts into a polymer matrix. Several polymer matrixes have been used in solid-state electrolytes, including poly-(vinylidene-fluoride) (PVP), polyacrylonitrile (PAN)

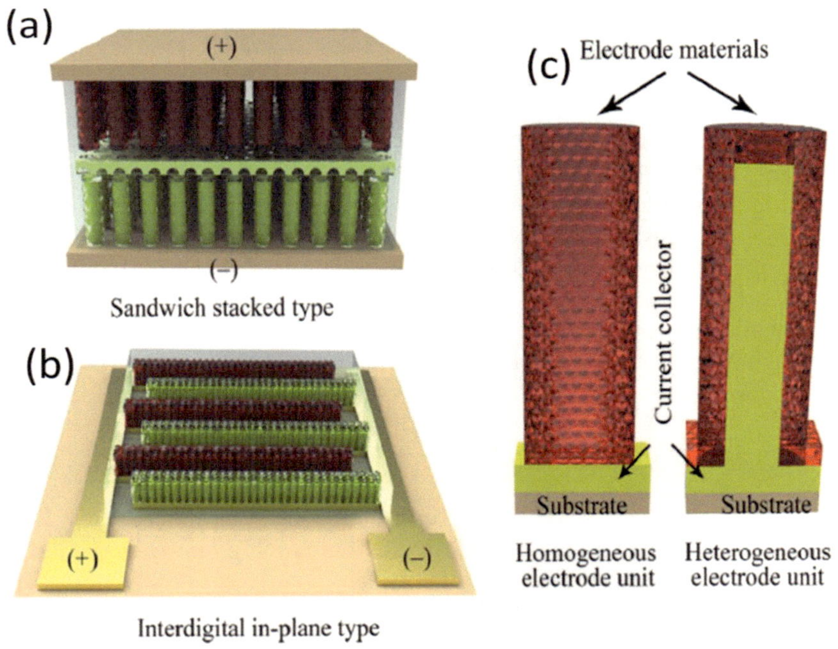

Figure 1.
Schematics of three major MSC architecture (a) stacked architecture (b) Interdigital finger electrode architecture (c) cross-section of electrodes [4].

Device configuration	Advantages	Disadvantages
In-plane architecture	• They consist of interdigital electrodes with many dense micro-fingers, where the counter electrode interspaces are small enough for the transport of ions so that the devices have little impedance, high capacity and quick frequency responses [5]. • Ease of integration with other microelectronics. • This design can make the active electrodes more accessible because they are exposed to electrolytes on the edges. Therefore high power density can be achieved with batteries and supercapacitors. • This design contributes to the separator elimination [6].	• The geometric parameters of the electrodes still have to be optimized.
Fiber-shaped	• Generally small in size and lightweight. • Thanks to their unique wire-shaped structure, they are highly flexible and can be woven or knitted with excellent wearability [7]. • Excellent versatility and can be produced in different shapes and different locations [8].	• It cannot compete in terms of energy density with micro-batteries as energy storage equipment; large scale application is a big challenge [8]
Three-dimensional type	• This design maximize the energy density of MSCs • The volume of electrolyte utilized is further reduced with this design, leading to enhanced volumetric/areal energy density [4].	• Require a series of complex micro-fabrication techniques. • Designing of 3D type MSCs with leakage-free electrolyte is still a challenge [4].

Table 1.
Advantages and disadvantages of various MSC device configurations.

and poly-(vinyl-alcohol) (PVA) [9]. These electrolytes can provide long cycle-life, low leakage current, high ionic conductivity and high mechanical flexibility. For example, the ionic conductivity of PVA/H_3PO_4 is about 10^{-3} Scm^{-1}, while the ionic conductivity PVA/H_2SO_4 can be even higher, about 7 x 10^{-3} Scm^{-1}. However, aqueous solid-state electrolytes suffer from a low voltage window at about 1 V due to the electrolysis voltage of water similar to aqueous electrolytes. A high operating voltage of 2.5 V can be achieved for micro-supercapacitors through ionic liquid solid-state electrolytes, resulting in a high energy density in sequence [10]. They also allow additional functionality, such as flexibility and stretchability, in addition to easy encapsulation. Considering these advantages, the choice of solid-state electrolytes in micro-supercapacitors is more reasonable.

2. Microfabrication technologies for microelectrodes of MCSs

This technology used for the fabrication of microelectrodes of MCSs can be grouped into two categories. The first categories include direct electrode material synthesis on the patterned current collectors using laser scribing, CVD, electrolytic deposition and pyrolysis. The second category consists of indirect manufacturing using existing electrode materials in powder or solution form.

Methods	Advantage	Disadvantage
Chemical vapor deposition (CVD)	Controlled design and structure [2]	Expensive, time-consuming process, low mass loading and vigorous reaction condition **(2)**.
Electrolytic deposition	Simple, efficient, cost-effective, environmentally friendly and large scale-production [2, 11]	Uncontrolled lateral direction growth [2].
Electrophoretic deposition	Cost-effective, simple, thickness can be controlled [2, 11].	Restricted by the species charged [2].
Inkjet printing	Cost-effective, fast process, low mass loading, precise thickness control, direct patterning, large scale production, fair resolution (around 50 μm) enhance the resolution and scalability because no manual assembly is required during device manufacturing [2, 12].	Ink preparation is a complicated process, limited by resolution, jam of nozzle [2, 10].
Screen printing	Low-cost, scalable and fast process	Low resolution
Photolithography (UV lithography)	Cost-effective with high control precision [10], simple fabrication process [9], can produce uniform and accurate large-area samples [13].	A sacrificial template is required; hence it's a complicated process, long preparation time [9].
Drop, spin and spray coating	Facile, simple, thickness-control, large-scale fabrication, time and energy saving [2].	Low-production efficiency and heterogeneous [2].
Vacuum filtration	Low- cost, simple, convenient, thickness can be controlled [2].	Shape and size is limited
Laser scribing	Cost-effective, scalable, simple, gives high throughput [9].	Non-universal [2].
Layer-by layer assembly	Multilayer films can be easily prepared, cost-effective and straightforward method, high resolution [14].	Time-consuming process
Pyrolysis	Single-step synthesis [2].	Complex and high-temperature process [2].

Table 2.
Microfabrication techniques for fabrication of MSCs.

The advantages and disadvantages of various techniques developed are explained below in **Table 2.**

No strategy in the fabrication of MSC microelectrodes is yet dominant over the others. Therefore, improving existing assembly strategies and exploring new manufacturing methods to overcome those limitations has become essential. In the meantime, to select appropriate assembly strategies to achieve high-performance MSCs, consideration should be given to overall factors such as active electrode materials, electrolytes, and interface between electrolytes and micro-electrodes [15].

3. Performance metrics of MCSs

The parameters used generally to assess supercapacitors' performance against volume and weight units are gravimetric capacitance, energy, and power. It is important to note that the supercapacitors gravimetric capacitance varies according to total density, mass and thickness of the electrode, and other components' weight. So it's hard to compare the various MSC based on gravimetric capacitance [16]. But this parameter is not suitable for planar MSC where electrode material's weight is insignificant and the device's volume and surface area are always limited. Since the overall mass load of active materials in MSCs is small, the volumetric performance and, more significantly, the areal performance are more adapted for electrochemical performance [17].

Since equipment need to be integrated with miniaturized electronic devices with limited area, a performance assessment against the footprint area of MSCs is essential. Therefore areal capacitance, power density and energy density are the more reliable parameters for MSC performance monitoring. These parameters can be calculated using the equations given below [17].

$$C_S = \frac{Q}{s\Delta V} \tag{1}$$

$$E_S = \frac{0.5\,C\Delta V^2}{3600\,s} \tag{2}$$

$$P_S = \frac{\Delta V^2}{4s\,ESR} \tag{3}$$

Where Cs is the areal capacitance in F/cm^2, 's' is the total area of microelectrode array and ΔV is the voltage range. Ps and Es is the maximum power and energy density. The essential parameter to detect the electrode's areal performance is to measure the total area accurately.

4. Two- dimensional materials for MCSs

A promising material for the production of MCSs is Planar 2D molecules of atomic thickness with a large specific surface area. The reduced dimension of these materials also satisfies the miniaturization requirements of device size, offering new possibilities for high-performance MSC development [18, 19]. The material properties of this rich family consisting of graphene, transition metal oxides (TMOs), transition metal chalcogenides (TMDs), metal carbides and nitrides (MXene), black phosphorus (BP), etc., range from superconducting, metallic, semiconducting and insulating behavior due to its diverse electronic structure, offering a wide range of material solutions to achieve high-performance. The essential reasons why 2D material based solid-state MSC are essentials is enlisted below

- Excellent electrical conductivity

The essential qualification for high-performance MCS electrode materials is the excellent electrical conductivity, which can accelerate the adsorption and desorption of the charges and increase the diffusion rate of ions. This electron transport behavior is very closely linked to the electronic crystal structure, resulting in three typical insulating, semiconducting, and metallic transport behaviors. In general, metallic 2D materials, such as TMDs, have good electrical conductivity and several other 2D semiconductors like graphenes and BPs also offer favorable electron transport characteristics [19].

- Excellent-electrochemical activities

The electrochemical activity of some 2D material would be useful for the redox reaction to further increase the pseudocapacitance. The promising electrode material of microsized pseudocapacitors, which generally display high capacitance performance, is proven to be 2D MXenes, layered double hydroxides (LDHs), metal oxides and hydroxides with excellent electrochemical properties [19].

- Large surface area

The extra-large surface area offers an energy storage platform with huge active sites to increase the electrochemical activity and charge adsorption, thus making 2D metal–organic frameworks (MOFs) and covalent organic frameworks (COFs) with inherent porosity a promising MSC functional electrode material [19].

- Mechanical flexibility

Superior mechanical flexibility at the atomic level and a diverse array of various 2D nanosheets provide desirable flexibility and multiple functionalities [15].

4.1 Graphene

Graphene is the most widely studied electrode material for MSCs due to its excellent electrical conductivity, large specific area, chemical stability, excellent intrinsic double-layer capacitance of approximately 21 $\mu F/cm^2$ and theoretical capacitance of around 550 F/g [20–22]. Several reviews of graphene-based MSCs have been published. Zhang *et al., Xiong* et al. tried to summarize the recent developments in graphene-based MCSs and the methods employed to produce high-performance MSCs [20, 21]. Similarly, Wu *et al.* classified graphene-based MSCs and provided a complete overview of on-chip graphene-based planar interdigital MSCs [21]. Gao et al. recently provided an overview of the MCS system's application, based on 1D, 2D and 3D graphene [23].

Several strategies were employed to enhance the electrochemical performance of graphene-based MSCs. The first approach was to improve the charge storage capacity of electrode materials by preparing graphene composites with other pseudocapacitive materials [11, 24, 25] or doping graphene with heteroatoms like boron [26] and fluorine [27]. The second approach consists of constructing an asymmetric structure [11, 25] and the third approach was to increase the loading quantity of active electrode materials by 3D electrode construction on the confined area of MSC [17]. To realize this 3D electrode construction, Wang *et al.* used a 3D printing technique to fabricate an all-solid-state flexible MSC using nitrogen (N)/oxygen (O)-doped graphene ink. This device shows a high power density of 0.23 mW cm^{-2} and an areal energy density of 2.59 μWh cm^{-2}, excellent cycling stability, and good mechanical flexibility. This increase in electrochemical properties observed due to three reasons (i) increased surface area due to the doping with N and O (ii) enhanced hydrophilic property of N/O doped graphene ink (iii) increased electrical conductivity as well as the pseudocapacitive effect of O and N doping [28]. Like this study, Szymon and his group developed a new type of polyaniline (PANI) anchored pseudocapacitive MnO_X passivated graphene microflake inks and manufactured a solid-state MCS using the inkjet printing technique in a 3D configuration given in **Figure 2**. This ink shows excellent stability and jetting performance that may be due to the dual-passivation process employed here. For the first time, this is to report such a fully inkjet-printed MSC with 3D electrode configurations with excellent rate performance and stability. However, these microflake inks have been developed based on the self-assembly behavior of 2D materials following Rebani et al.'s [30] 3D coarse-grained lattice gas model [29]. Toan *et al.* built an on-chip MSC where the electrode consists of silicon nanowire-graphene nanowall-PANI ternary composite. Here the silicon nanowire template with a high aspect ratio is fabricated using the metal-assisted chemical etching (MACE) technique. On this silicon nanowire, 3D hierarchical graphene nanowalls were produced using microwave plasma-enhanced chemical vapor deposition (PECVD), which significantly improves the electrochemical performance of the manufactured MSC. However, this solid-state 3D MSC delivers an areal energy density of about 10.8 $\mu Wh/cm^2$ and a power density of about 0.78 mW/cm^2 [31]. Lu *et al.* chose asymmetrical arrangement and produced an in-plane asymmetric interdigited MSC using mask-assisted vacuum filtration techniques, among the various strategies discussed to improve graphene performance. They develop this asymmetric MSC based on an all-graphene system, where both the working electrodes are graphene derivatives. Using the same chemical composition and microstructure electrodes enhances performance and helps avoid power imbalance that conventional hybrid and asymmetric systems usually encounter [32]. Therefore, the anode of this MSC is made of functional graphene oxide (FGO), while the cathode consists of functional reduced graphene oxide (FrGO). This FGO, which is electrochemically exfoliated from graphite papers, has enhanced hydrophilic properties compared to pristine graphene, difficult to disperse in water [33]. FGO has abundant functional groups responsible for the hydrophilic properties of the FGO, leading to the formation of wrinkles on its surfaces. However, this asymmetric MSC has the highest surface capacitance of approximately 7.3 mF cm^{-2} in PVA/Na_2SO_4 electrolyte. The high performance of this MSC is attributable to abundant functional group doping

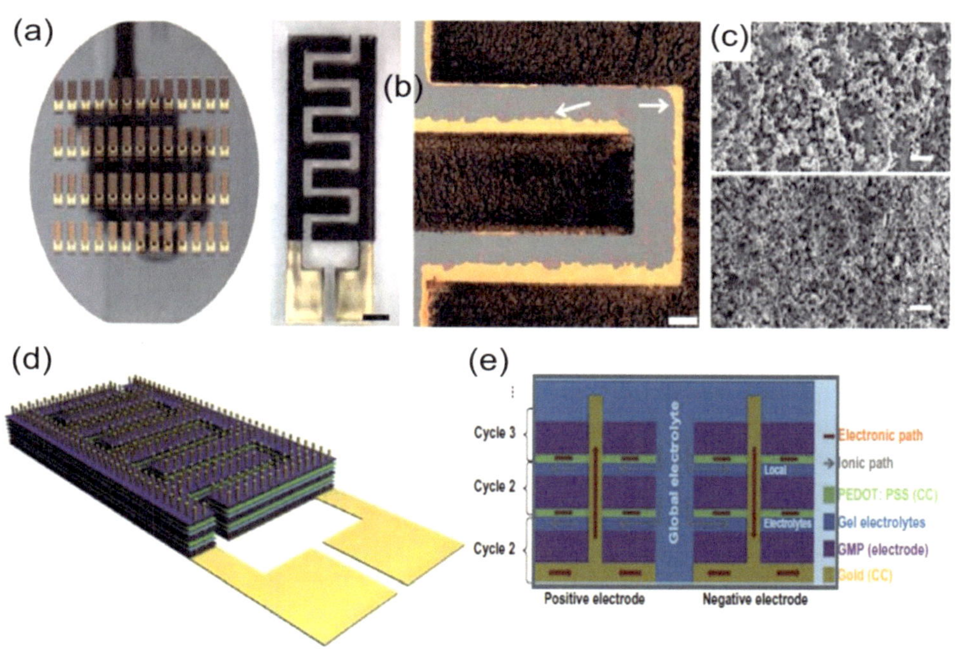

Figure 2.
Inkjet-printed GMP (doped graphene passivated flakes) MSC developed by Delekta et al. (a) Photograph of inkjet-printed MSC (b) micrographs of printed GMP MSC on glass substrate (c) SEM images of patterned GMP MSC (d and e) 3D heterogeneous GMP MSC [29].

and heteroatom substitution with graphene elements N, O, P, S, and the in-plane interdigitated architecture. This MSC has demonstrated exceptional flexibility, showing the feasibility of the wearable application of the MSCs [34].

Lochmann *et al.* reported a stamping approach combined with soft lithography for MSC production for the first time, but this approach is time-consuming and not appropriate for producing PET and paper-based MSC [35]. Zhang *et al.* subsequently made a flexible MSC based on MXene in a paper substrate using a stamping method, but even this MSC suffered from lower areal and volumetric capacitances [36]. To this end, Esfahani and Khosravi reported the manufacture of a graphene-based flexible MSC using a single-step stamping method on a PET-coated parafilm using a pre-designed pattern. This imprinted parafilm-coated PET pattern was filled with graphene oxide (GO)/MnO_2/carbon aerogel hybrid paste, which acts as a binder and additive-free active material and finally, with the help of nascent hydrogen GO reduction is made. This type of GO reduction improves stability and reduces the ohmic resistance of the prepared electrode. The active material, i.e., G/CA/MnO_2 used here, exhibits a unique morphology that enhances the electrochemical surface area and facilitates the diffusion of electrolytes ion. The edged structure of graphene also helps to improve the electrochemical activity and capacitance of the SC. Based on the Hota *et al.* [37] inferences, they also used fractal design of large and small finger widths and compared their performance with ID electrodes. The small finger width fractal design (SFWF) shows high areal and volumetric capacitance of around 14.2 mF cm^{-2} and 71.3 Fcm^{-3} among these different electrode designs. This method proves feasible in designing various low-cost architectures and improves the flexible-graphene performance based on MSC, which has been reported to date. The manufacturing process of this flexible MSC provides a new direction in the modification of the substrate and the current collector and the manufacturing method and the method of graphene reduction. This MSC fabricated by this stamping method also demonstrates excellent flexibility, reflecting its potential in future flexible electronic devices [38]. The presence and regulation of the functional group are essential for improving the performance of the graphene-based MSCs. These functional groups not only provide active pseudocapacitance sites, but they also prevent the aggregation of graphene without affecting both wettability and conductivity of graphene. To date, several strategies have been reported for regulating graphene functional groups, such as the treatment of O_2 plasma, laser power, etc. However, it is worth working on regulating the functional groups of graphene to balance the active site, electrical conductivity, and wettability, which have plenty of room for further performance improvement in MSCs. Consequently, to improve the electrochemical efficiency of reduced graphene oxide (RGO), an appropriate method of controlling the functional group is required. Based on these facts, Wu et *al.* fabricated a free planar MSC using symmetric graphene-based metal current collector, where the functional groups were regulated using both air-plasma treatment and exposure to blue violet-laser (BV-laser) as shown in **Figure 3a**. The XRD pattern of the prepared composite is shown in **Figure 3c** and it can understand from the figure that this BV-laser treatment and air-plasma treatment do not change the phase structure of RGO. The interlayer spacing calculated for this composite is higher than that of graphite structure, which indicates a π-π stacking between graphene sheets in these composites. These combined techniques balance the pseudocapacitance active sites, conductivity and wettability by tuning the functional groups on the graphene surface and the possible transformation pathway is shown in **Figure 3b**. The CV curves of this material are shown in **Figure 3d** exhibit a symmetric-quasi rectangular state that demonstrates that this material's capacitive behavior is due to the simultaneous pseudocapacitive and electrical double layer behavior (EDL). This symmetrical solid-state MSC exhibit excellent energy density

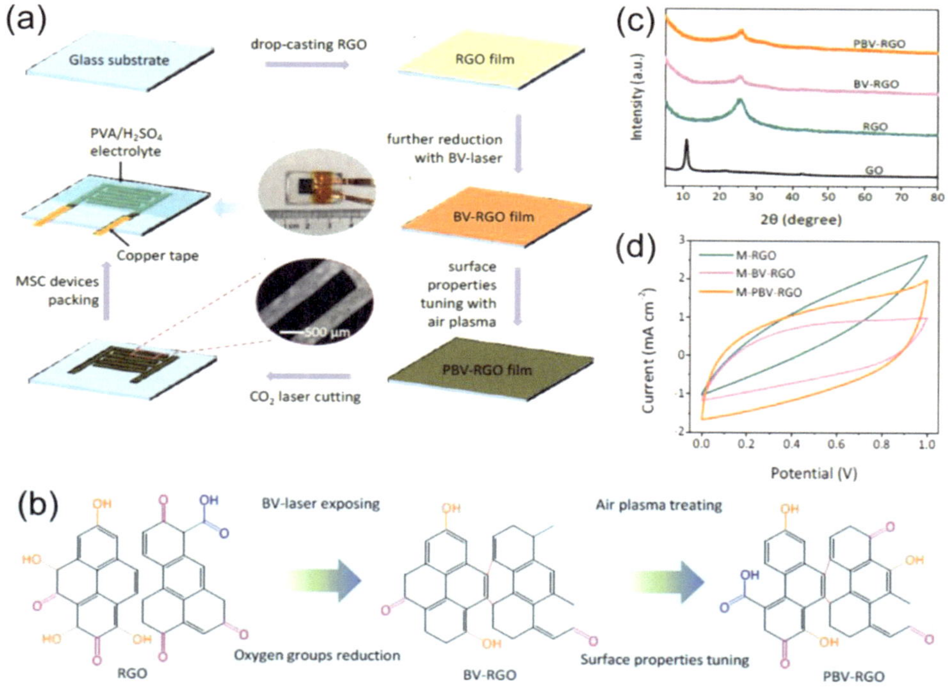

Figure 3.
(a) Fabrication steps involved in the manufacturing of in-planar PBV-RGO MSC developed by Wu et al. (b) the suggested pathway transformation of RGO on exposure with BV-laser exposing and air-plasma treatment to produce PBV-RGO electrode material (c) XRD pattern of PBV-RGO electrode material (d) CV curves of RGO (M RGO) after BV-laser treatment (M BV-RGO) followed by air-plasma treatment (M-PBV-RGO) [39].

and power density around 2.49 μWh cm^{-2} and 5 mW cm^{-2} and superior long term stability after 10,000 cycles with 99% retention, which exceeds most reported all-solid-state graphene-based MSCs. This manufacturing method is suitable for the process of micro-integrated circuit machining and has an enormous potential in the production of on-chip devices [39]. Liu *et al.* adopted a new series of inter-digital setups without any internal connection by combining techniques such as photolithography and liquid-air interface self-assembly methods. This solid-state planar on-chip MSC without internal connection shows excellent cyclic stability and outstanding energy and power density. This work demonstrates that graphene-based planar on-chip MSCs with no internal connection can integrate better with electronic components [40].

4.2 Transition metal dichalcogenids (TMDs)

Single or few layers TMDs have attracted considerable attention because of their tunable band gaps and extensive natural reserves [2]. These compounds show a typical MX2 formula, where M is an element in Group IV-VI metal, and X is a chalcogen (S, Se, or Te). In this case, Stoichiometry relies on the process and the strategy of producing a compound made up of transition metal and chalcogen elements. The layered TMDs are typically 6 to 7 Å thick and consist of an X-M-X hexagonal sandwich with a metal-atomic layer separated by both layers of chalcogen [41]. The physical properties of bulk TMD's vary from true metals such as VSe_2 and TaS_2, semi-metals such as WTe_2 and TiS_2, semiconductors such as MoS_2 and

SnS_2 and insulators such as HfS_2. Suitable electrode materials for MSCs are among these metallic TMDs with large surface area and high conductivity [2].

MoS_2 can effectively store charges over a single atomic layer utilizing an inter and intrasheet double layers. Here the central atom Mo shows an oxidation state ranging from +2 to +6 and shows a pseudocapacitive behavior with a theoretical capacitance of about 1000 F/g. But aggregation and low electrical conductivity between the atomic layers of MoS_2 hinder their extensive use in MSCs. Hybridization of TMDs with carbon material, which provides quick-electron transport and more active-sites, is one approach to solve these problems. Hence Yang *et al.* reported a solid-state MSC using MoS_2@rGO– photoresist-derived carbon-nanotube (CNT) hybrid composite by spin coating followed by photoetching and pyrolysis similar to the MoS_2@sulfonated rGO hybrid prepared by Xiao *et al.* shown in **Figure 4** [42]. This hybrid prepared by Yang *et al.* was then embedded in carbon microelectrodes, which synergistically increase the performance of the MSCs and exhibited high energy density (5.6 mWh cm^{-3}) as well as areal capacitance (13.7 mF cm^{-2}) with good capacitance retention [43]. Similarly, Haider *et al.* have reported other carbon microelectrodes based on TMD using the advantages of metallic VS_2. The high energy density (15.6 mWh cm^{-3}) and specific capacitance (86.4 Fcm^{-3}) result from the synergistic combination of VS_2 and carbon. This system exhibited excellent energy density and power density compared with the energy storage system developed by Xiao *et al.* and Yang *et al.* [44]. Besides constructing MoS_2 hybrids with conductive carbonaceous material, the phase modification in which the Mo coordination changes from the trigonal prismatic (2H) phase to the octahedral (1 T) phase is another practical approach to improve the electrochemical performance of MoS_2 [45]. Very recently, Xu *et al.* reported a femtosecond laser direct writing technique to fabricate an MSC based on 1 T MoS_2. This is the first time an MSC with excellent performance based on 1 T MoS2 has been reported. These femtosecond lasers can help achieve a submicron resolution(~800 nm), which is nearly 40 times more accurate than that achieved with traditional nanosecond lasers with a resolution of around 10-200 μM. This approach is green, facile, maskless, flexible and high vacuum environments are not required. The electrochemical performance and

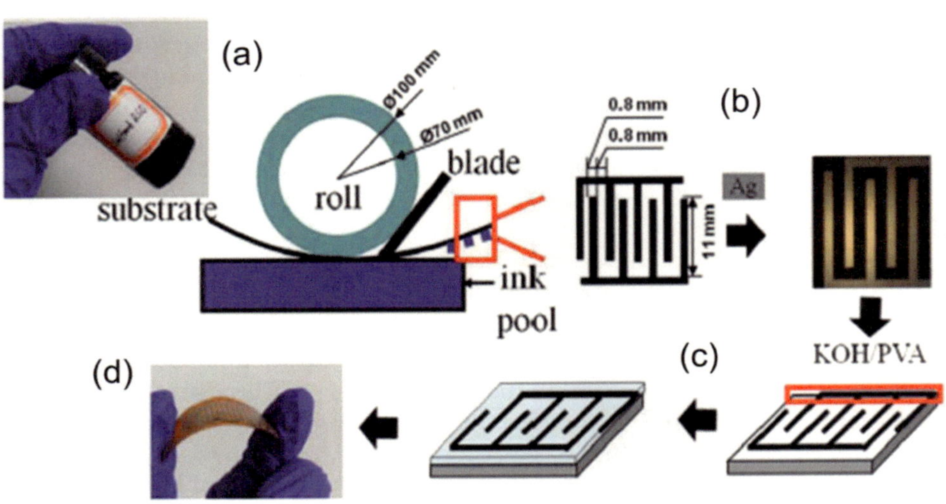

Figure 4.
(a) Schematic picture of fabrication of MoS_2@SrGO MSC using gravure printing techniques (b) photograph of prepared MSC after Ag paste painting (c) KOH-PVA gel electrolyte coated on gravure printed electrode (d) MoS_2@S r GO printed electrode in PI substrate. (Source: Reprinted from [42]).

the MSC resolution are enhanced by lowering the thermal effect to regulate the phase transition of these 1 T MoS_2 based electrode material. This MSC with in-plane configuration results in high-frequency response with ultrafast ion-diffusion rate, high specific capacitance, good cycle-life, low equivalent series resistance (ESR) value, unprecedented power density (nearly 14 kW cm^{-3}) and high energy density (15.6 mWh cm^{-3}) in PVA/H_2 SO_4 electrolyte. However, this MSC with a surface area of 100 × 100 μm^2 and a high-frequency response and time constant are suitable for AC line filters and other electronic devices demanding high power requirements [46].

The fabrication of supercapacitors with excellent energy storage capacity and flexibility in wearable smart electronics has recently attracted significant attention. Thus, a fabric supercapacitor using low-cost textile fabrics with good mechanical properties and biocompatibility coated with ternary composite poly(3,4ethylenedioxythiophene): poly(styrenesulfonate)/MoS2/poly3,4ethylenedioxythiophene) (PEDOT: PSS/MoS2/PEDOT) is manufactured by Chen and the group. This all-solid-state fabric MCS was fabricated by vapor phase polymerization (VPP) and the vapor phase deposition method exhibits an energy density of around 1.81 mWh/cm^3 and power density of around 0.82 W/cm^3. The fabric coated with this ternary composite has a 3D configuration with interconnected structure and exhibits a large surface area that enables fast electrolyte transport and provides active electrolyte accessibility. This MSC assembled in a belt-shaped device was also used by the group as transient power sources to operate the light-emitting diodes [47]. Very recently, Li *et al.* printed a MoS_2 based all-solid-state in-plane MSC using inkjet printing. This MSC printed with MoS_2 based inks has high loadings of active materials per unit area resulting in a thinner and more flexible supercapacitors than the conventional sandwich structure. PEDOT: PSS inks were first printed on PI substrate to improve the conductivity, followed by printing of MoS_2 based inks subsequently to fabricate the MSC. This scalable synthesis technique is demonstrated in **Figure 5a**. The SEM image (**Figure 5b**) shows that the layered MoS_2 formed a uniform 2D conductive network above the pre-printed PEDOT: PSS electrode, which differs from the morphology observed in the ternary composite prepared by Chen and group. They also demonstrated the relationship with the increase of electrode thickness vs. conductivity in **Figure 5c** and its practical application by powering an LED bulb by connecting MSC in series combination [48].

4.3 MXenes ($Ti_3C_2T_x$)

The overall performance of MSCs is based on the intrinsic properties of electrode materials. In many cases, carbonaceous materials such as graphene [49, 50], graphene oxide [51], CNTs [52, 53], carbide-derived carbon [54, 55] and their hybrids [56, 57] with charge storage via electric double layer, were reported in MSCs. Later, high capacity MSCs based on pseudocapacitive materials such as conductive polymers [58], transition metal oxides/hydroxides [59, 60] and sulfides (VS_2, MoS_2) [61, 62] with surface redox reactions were reported. Nevertheless, the poor electrical conductivity and lower packing density of electrode materials in these MSCs restrict the accessible volumetric and areal capacitances, the two important parameters used to indicate the performance of MSCs [63]. Recently, MoS_2 with high packing density served as a good electrode material to fabricate energy storage devices characterized by high power densities and volumetric energy [64]. A new group of layered 2D materials called MXenes, which includes transition metal nitrides, carbides, and carbonitrides, was recently reported.

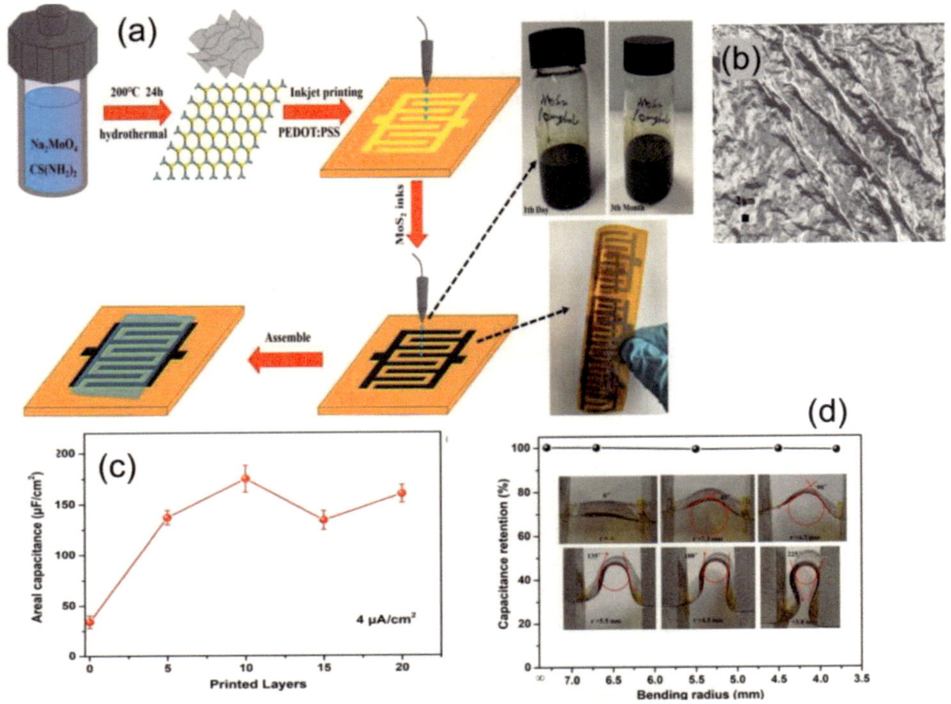

Figure 5.
(a) Steps involved in the fabrication of flexible MSC using MoS_2 inks by inkjet printing (b) SEM image of printed MSC in PI substrate (c) relationship between areal capacitance vs. number of printed layers (d) capacitance retention under different bending radius of the fabricated supercapacitor [48].

MXenes are promising layered materials derived from the precursors with the general formulae $M_{n+1}AX_n$ (M refers to Ti, Sc, Nb *etc.*; A = Al, Sn, Si, *etc.*; $n = 1, 2, 3$). In the MAX phase, the M layers are hexagonally close-packed and MXenes can be synthesized by selective etching of A element from the MAX phase using acidic-fluoride-containing aqueous solutions [65]. The presence of an aqueous medium during the synthesis can create MXene flakes with various surface functional groups such as O, F, or OH. MXenes have characteristic properties such as rich surface chemistry, hydrophilicity, layered structure, high packing densities and intrinsic electronic conductivity. The first member $Ti_3C_2T_x$ was reported in 2011, opened up an exciting research field, as revealed by the increasing number of publications on MXenes [66]. The distinctive properties and simplicity of processing have contributed to various applications such as energy storage for supercapacitors and the maximum theoretical capacity was reported to be at 615Cg^{-1} [65]. The higher pseudocapacitance and simplicity of solution processing of MXenes are highly advantageous for the designing of MSCs that are used to power wearable electronics, sensors, and micromechanical systems with low power consumption. Various methods have been adopted to fabricate MXene based MSC patterns on the submillimeter scale. These are known to be potential candidate for the design of MSCs due to the following factors: 1) MXenes have an electrical conductivity up to 6500 S cm^{-1} [67], which permit fast electron transfer and excludes the requirement for current collectors like noble metals. 2) MXenes exhibit higher gravimetric capacitances relatable to graphene, with higher packing density up to 4 g cm^{-3} [68, 69], which needs to enhance the volumetric characteristics (specific volumetric capacitance of 1000 F cm^{-3} for conventional three-electrode configuration [70], which is more than that of supercapacitors based on carbonaceous materials) [71].

Flexible MSCs are highly demanding to manufacture portable and on-chip energy storage devices because of their high security, lightweight and miniaturization [72]. Direct printing of functional inks is crucial for various applications such as smart electronic devices, healthcare, and energy storage. Nevertheless, currently available inks are distant from ideality. A low concentration of ink or the additives/surfactants are contained, which put on complexity to the fabrication and affects the printing resolution. Based on these facts, Zhang *et al.* demonstrated a direct printing method using two types of 2D MXene inks (aqueous for extrusion printing and organic for inkjet printing) to fabricate all-MXene MSCs with high resolution. The fabricated flexible MSCs displayed outstanding volumetric capacitance of 562 F cm^{-3} and a high energy density of 0.32 μW h cm^{-2}, over all other printed MSCs reported yet. The approach of direct ink printing technique plays a major role beyond energy harvesting and storage applications, including sensors, circuits, and electronics, where simple, easily integrable and cost-effective components are required. This additive-free and low-temperature ink printing technique provides many applications in sensors, antennas, smart electronics and shielding [72]. Recently, Peng *et al.* adopted a solution spray coating and laser cutting to fabricate solid-state MSC based on interdigital L-s-$Ti_3C_2T_x$ film on a glass substrate in which two layers of MXene ($Ti_3C_2T_x$) with different flake sizes were obtained. The larger MXene flakes (L-$Ti_3C_2T_x$, 3–6 μm) were stacked to form a bottom layer which act as current collectors. The top layer contains smaller MXene flakes (s-$Ti_3C_2T_x$, 1 μm) with numerous edges and defects to form an electroactive layer for energy storage. The excellent electrochemical characteristics and homogeneity in structures could offer better cyclic performance and showed excellent areal capacitance of ~ 27 mF cm^{-2} and volumetric capacitance of ~ 357 F cm^{-3} at 20 mV s^{-1}. The L-s-$Ti_3C_2T_x$ MSC showed excellent cyclic stability after 10,000 cycles without any decay of capacitance at 50 mV s^{-1}. L-s-$Ti_3C_2T_x$ film on a glass substrate was transferred onto the scotch tape substrate shows good flexibility without prominent cracks after bending up to 100 times at an angle of 60°, the areal capacitances were comparable to its original rigid structure on a glass substrate. This approach opens up different designs for the fabrication of on-chip devices using different morphologies of MXenes and their composites, flake sizes, and chemistries [73]. The integration of flexible MSCs for on-chip energy storage applications still faces some challenges due to short cycling stability, complicated manufacturing processes, and low areal energy storage. To address this, Huang *et al.* utilized spray coating of MXene ($Ti_3C_2T_x$) conductive inks for the massive preparation of paper-based flexible MSCs by using a gel-like solid-state electrode (polyvinyl alcohol and H_2SO_4) and encapsulated layer of polydimethylsiloxane. As discussed above, a highly conductive and sprayable $Ti_3C_2T_x$ interdigitated electrode served the dual role of the current collector and active materials. This flexible MSC delivers a large areal capacitance of 23.4 mF cm^{-2} and an excellent cycling capability with a capacitance retention up to 92.4% over 5000 cycles, together with exceptional flexibility [74]. The crucial obstacles in MSC applications are short discharge time, low voltage output, and low current. MSC array is known to be a solution to avoid the obstacles mentioned above. Based on the capacitors' theory, the parallel connection will increase the capacitance while the series connection will decrease the output voltage corresponding to the capacitance decrease. It means a MSC array (combination of some MSCs) can increase both capacitance and output voltage [75]. Recently, a lightweight and freestanding MXene/bacteria cellulose composite paper with outstanding electrochemical performance and mechanical stability through a facile all-solution based paper making method was fabricated by Jiao *et al*. Further, they adopted a laser-cutting kirigami patterning process for the fabrication of bendable, stretchable and twistable all-solid-state MSC arrays (**Figure 6a**). The structural design and excellent performance of MSC arrays could offer outstanding areal capacitance of 111.5 mF cm^{-2} and areal density of 0.0052 mWh cm^{-2} with electrochemical stability under mechanical deformation. The photograph of a paper crane made from MXene/BC composite paper is shown in **Figure 6b**, it is used as a conductor for lighting LED (**Figure 6c**). This technique presented a promising method for designing and manufacturing excellent mechanically deformable MSC arrays based on MXenes [76]. For practical applications, the electrodes with a 3D structure can be easily destroyed via mechanical deformation. It is possible to improve MSC's energy storage ability by fabricating them in a 3D structure [77]. In this context, Yue *et al.* developed a self-healable 3D MSC comprised of r-GO and MXene ($Ti_3C_2T_x$) composite aerogel electrode with an outer shell of self-healable

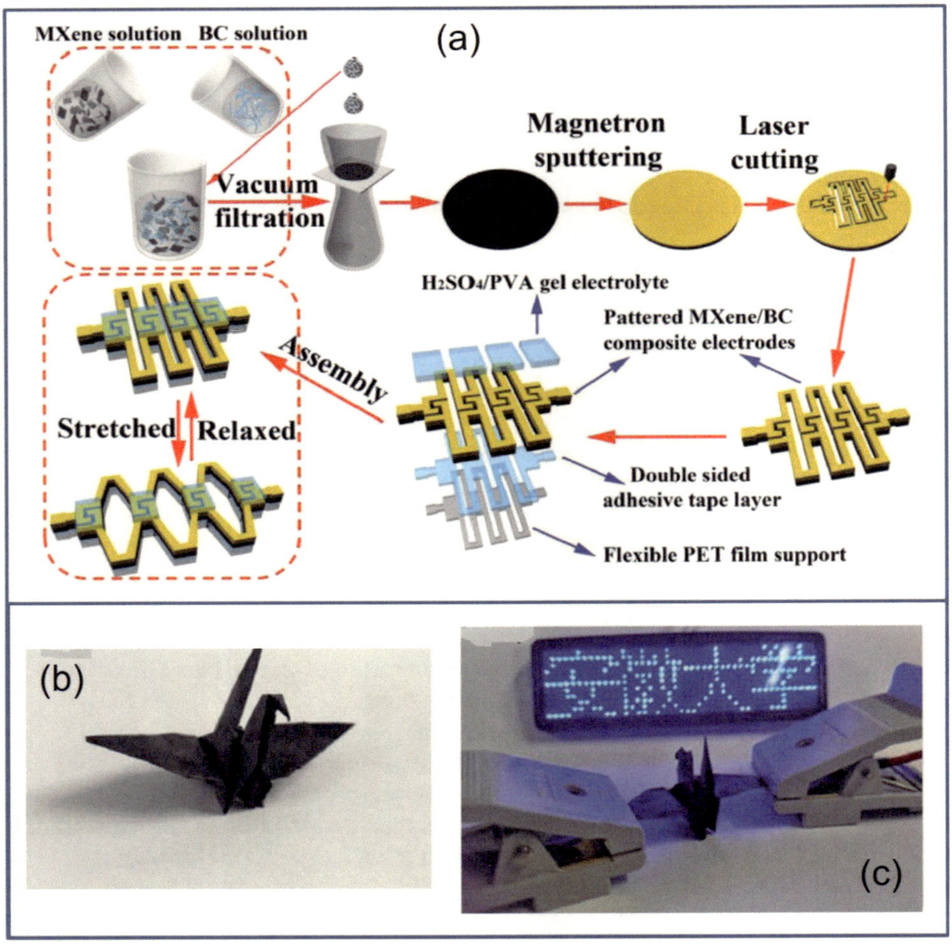

Figure 6.
(a) Schematic illustration of the manufacturing process of Mxene/bacterial composite papers and a laser-cutting kirigami patterning process for the fabrication of bendable, stretchable and twistable all-solid-state MSC arrays, (b) photograph of paper crane made from as-synthesized Mxene/BC, and (c) photograph of LED using paper crane working as conductor for lighting. (Source: Reprinted from [76]).

polyurethane through the ice-template method (**Figure 7a**). The composite aerogel electrode could resist the oxidation of MXene and prohibit the lamellar structure's restacking by combining the properties of components such as high conductivity of MXene and high surface area of r-GO. The fabricated MSc exhibited better performance with a large area-specific capacitance of 34.5 mF cm^{-2} at1mV s^{-1}, as shown in **Figure 7c-e**. The 3D-MXene-r-GO composite aerogel electrode displayed excellent cycling performance with 91% retention of capacitance after 15,000 cycles (**Figure 7f**). The 3D MSC maintained outstanding self-healing capacity (**Figure 7b**) with capacitance retention up to 81.7% after the 5th healing. The synthesis of self-healable 3D-MXene-r-GO MSC dispensed an approach for fabricating next-generation durable electronic devices with multi-functionality to meet sustainable development [77]. The performance of recently reported all-solid-state MSCs based on 2D materials are summarized in **Table 3**.

4.4 Other important 2D materials

Transition metal oxides/hydroxides (TMOs/TMHs) are electrochemical pseudocapacitor materials and widely used as electrode materials in supercapacitor applications due to their high energy density, abundance and high capacitance [87]. But their performance as supercapacitor electrode materials limited because of low intrinsic conductivity. So, 2D TMOs/TMHs have been explored in supercapacitors owing to their enhanced electronic conductivity and high specific surface area [88].

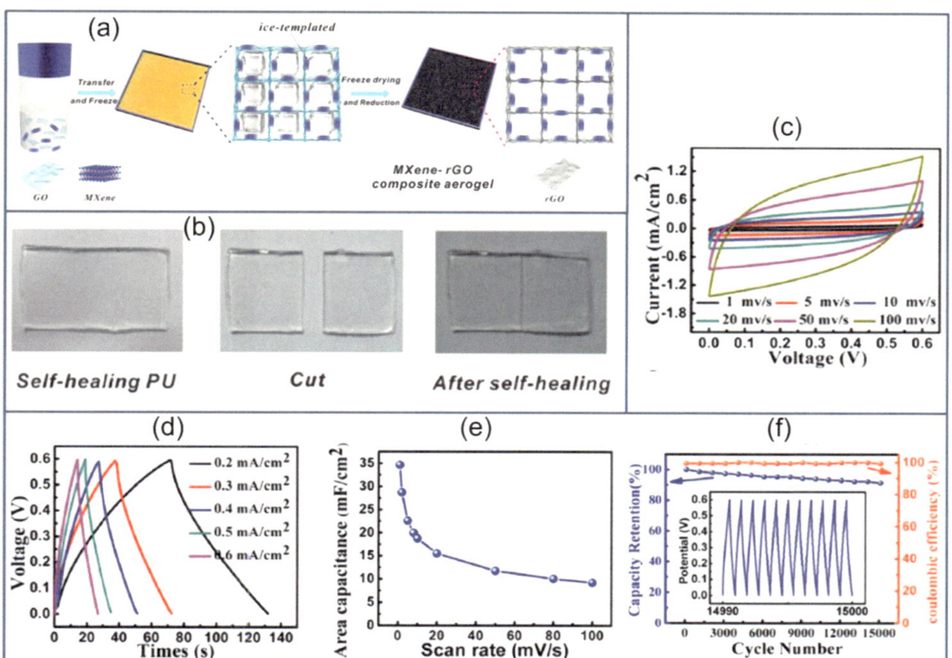

Figure 7.
(a) Schematic illustration for the fabrication process of 3DMSC based on MXene-rGO composite aerogel, (b) photographs of the self-healable carboxylated polyurethane: Left (intial), right (after the healing) and middle (after damage). Electrochemical performance of MSCs based on the MXene-rGO composite aerogel (c) CV at various scan rates, (d) GCD at different current densities, (e) the variation in areal capacitances vs *scan rate of MSC and (f) cycling stability of MSC based on MXene-rGO composite at the current density of 2 mA cm^{-2}. (Source: Reprinted from [77], with permission from, copyright@2018 ACS).*

MSC	Electrolyte	Voltage window (V)	Device performance		Specific capacitance		Cycling stability	Ref
			Power density	Energy density	Areal/mF cm^{-2}	Volumetric/ F cm^{-3}		
NOG-X-Y	PVA-H_3PO_4	0 to 1	0.23 mW cm^{-2}	2.59 μWh cm – 2	18.70	—	93% after 10,000 cycles	[28]
GMP microflakes based 3D MSC	PVA/H_2SO_4 gel	0 to 1	10 mW cm^{-2}	1 μWh cm – 2	11	—	80% after 2000 cycles	[29]
Silicon nanowire-Graphene- PANI	PVA/H_2SO_4 gel	−0.2 to 1	0.78 mW cm^{-2}	10.8 μWh cm^{-2}	77.7	—	75% after 2000 cycles	[31]
G/CA/MnO_2 based ID patterned MSC	PVA/H_2SO_4 gel	0 to 1	43.2 μW cm^{-2}	1.2 μWh cm^{-2}	8.7	—	85% after 5000 cycles	[38]
M-PBV-RGO	PVA/H_2SO_4 gel	0 to 1	5 mW cm^{-2}	2.49 μWh cm^{-2}	21.86	—	99% after 10,000 cycles	[39]
Graphene based integrated planar on-chip MCS	PVA/H_2SO_4 gel	0 to 2	68.268 mW cm^{-3}	3.792 mWh cm^{-3}	—	27.30	89% after 10,000 cycles	[40]
MoS_2@S/rGO	KOH-PVA gel		13.4 $mWcm^{-3}$	0.58 mWh cm^{-3}	6.56	—	91% after 1000 cycles	[42]
MoS_2@rGO/CNT	PVA/H_2SO_4 gel	0–1	—	5.6 mWh cm^{-3}	13.7	—	96.6% after 10,000 cycles	[43]
C/VS_2		0–1.2	2.88 Wcm^{-3}	15.6 mWh cm^{-3}	—	86.4	97.7% after 10,000 cycles	[44]
1 T MoS2 (t-lf laser)	PVA/H_2SO_4 gel	0–0.5	14 kW cm^{-3}	15.6 mWh cm^{-3}	36		93% after 5000 cycles	[46]
PEDOT: PSS / MoS2 / PEDOT	PVA/H_3PO_4 gel	−0.2 to 1	0.82 W/cm^3	1.81 mWh/cm^3	1.43	—	93.6% after 5000 cycles	[47]
Inkjet printed MSC based on MoS_2	PVA/H_2SO_4 gel	0 to 0.6	0.079 W cm^{-3}	0.215 mWh cm^{-3}		175 μF/cm^3	85.6% after 10,000 cycles	[48]

DOI: http://dx.doi.org/10.5772/intechopen.94535

MSC	Electrolyte	Voltage window (V)	Device performance		Specific capacitance		Cycling stability	Ref
			Power density	Energy density	Areal/mF cm^{-2}	Volumetric/ F cm^{-3}		
2D $Ti_3C_2T_x$/PDMS	PVA-H_2SO_4	0 to 0.6	189.9 mW cm^{-3}	1.48 mW h cm^{-3}	23.4	—	92.4% After 5000 cycles	[68]
PANI/EG/ $Ti_3C_2T_x$	PVA-H_3PO_4	0 to 0.7	159.6 mW cm^{-3} 2015 mW cm^{-3}	2.3 mWh cm^{-3} 1.3 mWh cm^{-3}	—	36.4	90% After 8000 cycles	[78]
$Ti_3C_2T_x$-PET	PVA-H_3PO_4	0 to 0.6	225 mW cm^{-3} 744 mW cm^{-3}	2.8 mWhh cm^{-3} 2.3 mWh cm^{-3}	23 25.5	—	76% after 10,000 cycles	[79]
$Ti_3C_2T_x$/polymer electrolyte (PE)	PVA-H_3PO_4	0 to 0.8	8 mW cm^{-2}	28 μWh cm^{-2}	276	—	95% After 1000 cycles	[80]
Extrusion printed MXene Inkjet printed MXene	PVA-H_2SO_4 PVA-H_2SO_4	0 to 0.5 0 to 0.5	11.4 μW cm^{-2} -	0.32μWh cm^{-2} -	43 12	- 562	97% After 10,000 cycles 100% After 10,000 cycles	[72]
$Ti_3C_2T_x$	PVA-H_2SO_4	0 to 0.6	0.7–15 W cm^{-3}	11–18 mWh cm^{-3}	27	357	100% After 10,000 cycles	[66]
3DMXene-r-GO composite aerogel	PVA-H_2SO_4	0 to 0.6	180 μW cm^{-2} 60 μW cm^{-2}	1.33 μWh cm^{-2} 2.18 μWh cm^{-2}	34.6	—	15,000(91%)	[77]
MXene/BC composite paper electrodes	PVA-H_2SO_4	0 to 0.6	—	0.00552 mWh cm^{-2}	111.5	—	5000 (72.2%)	[76]

MSC	Electrolyte	Voltage window (V)	Device performance		Specific capacitance		Cycling stability	Ref
			Power density	Energy density	Areal/mF cm^{-2}	Volumetric/ F cm^{-3}		
$Ti_3C_2T_x$@Silver-plated Nylon Fiber Electrodes	PVA-H_2SO_4	0 to 0.4	132 μW cm^{-2}	7.3 μWh cm^{-2}	328	—	10,000 (80%)	[81]
r-GO/MnO2/Ag NW-PET	SiO_2–1-butyl-3-methylimidazoliumbis(trifluoromethylsulfonyl) imide	0 to 2.5	162.0 mW. cm^{-3}	2.3 mWh. cm^{-3}	—	2.72	90.3% After 6000 cycles	[82]
VN// $Co(OH)_2$	KOH/PVA	0 to 1.6	1750 mW cm^{-3}	12.4 mWh cm^{-3}	21	39.7	84% After 10,000 cycles	[83]
LSG/Ni-Catecholate-MOF	LiCl/PVA	0 to 1.6	7 mWcm^{-2}	4.1 Wh cm^{-2}	15.2	—	87% After 5000 cycles	[84]
Asymmetric system								
FGO//FrGO	PVA/Na_2SO_4		28.3 μW cm^{-2}	2.52 μWh cm^{-2}	7.3	—	100% over 500 cycles	[34]
MXene// MXene-MoO_2-AMSCs	PVA-H_3PO_4	0 to 1.2	0.8 W cm^{-3}	9.7 mWh cm^{-3}	19	63.3	88% After 10,000 cycles	[85]
$Co(OH)_2$//erGO	PVA-KOH-KI	0 to 1.4	100.38μWh cm^{-2}	0.35μWh cm^{-2}	2.28	—	89% After 10,000 cycles	[86]

Table 3.
Summary of recently reported all-solid-state MSCs based on 2D materials.

Recently, research has been put into the fabrication of 2D TMOs/TMHs for MSC electrodes, limitations remain when using electrode based on a single material. The major disadvantages mainly rely on poor rate capacity caused by low electrical conductivity, restricted enhancement of energy density, and low capacitance, limiting their practical implementations [87]. To surpass the challenges of using single electrode materials, it is appropriate to fabricate nanoarchitectures based on composites of TMOs/TMHs. This can hone the configuration to avoid the agglomeration of 2D nanosheets and raise the performance level of various electrode materials to execute effective enhancement of supercapacitor performance [89]. Inspired from these findings, Wang *et al.* developed all-solid-state planar asymmetric MSCs based on $Co(OH)_2$/EG and porous VN nanosheets/EG as positive and negative electrodes, respectively, together with an interdigital mask placed on the Nylon membrane. The developed electrodes showed high electrical conductivity, high flexibility, and homogeneity over a large area and were acted as flexible electrodes without any need for binder, additives and metal-constituted current collectors for VN//$Co(OH)_2$ -PHMSs. The outstanding performance of the electrode was benefited from planar device geometry, synergy of $Co(OH)_2$ nanoflower (charge storing like battery) and VN nanosheets (charging storing like capacitor) based hybrid structure, and highly conducting EG nanosheets, which served as both additives and current collectors. The enhancement of capacitance in PHMss (planar hybrid MSCs) was occurred due to the porous structure of VN and nanoflower morphology of $Co(OH)_2$, these factors are suitable to enhance electrolyte ions and lessen their diffusion paths. The interdigital planar geometry of VN//$Co(OH)_2$ –PHMSs permits the ultra-fast flow of electrolyte ions between the adjacent finger electrodes with a concise diffusion pathway. This improved charge storage via the effective exploration of highly active surface area of 2D nanosheets. Consequently, the fabricated PHMss exhibited areal capacitance of 21 mF cm^{-2} and volumetric capacitance of 39.7 F cm^{-3} with a notable energy density of 12.4 mWh cm^{-3} and 84% capacitance retention over 10,000 cycles [83]. Recently, Lee *et al.* fabricated an in-plane MSCs comprised of $Co(OH)_2$ and r-GO through a cost-effective two-step fabrication method (**Figure 8a**). This method contained the fabrication of $Co(OH)_2$ and r-GO on Au electrode using photolithography and electrodeposition method. The Au metal situated at the bottom of the electrode act as a current collector and effectively transfers electrons to the active material because of its high conductivity. The electrode's large surface area promotes the reaction between the electrolyte and the active material. This electrode structure maximizes the volumetric and areal capacitance of fabricated $Co(OH)_2$//r-GO ASC (**Figure 8b**), shows a power density of 100.38 μW cm^{-2} and energy density of 0.35 μWh cm-2 for practical devices. and an excellent cycling capability with a capacitance retention up to 89% over 10,000 cycles, together with exceptional flexibility [86]. 2D MOFs are important 2D materials with tunable functionality and a designable porous structure with periodicity [90–92]. The high porosity of organic framework materials is suitable for producing electric double-layer capacitance and the heteroatoms like B, N, O and S located on the organic frameworks may show redox behavior for the pseudocapacitance [93]. The shortage of feasible microfabrication methods limits the practical implementation of MOF based electrode materials in MSCs. For the first time, a flexible symmetric MSC based on conductive Ni-catecholate-MOF possessing redox chemistry and high conductivity in the negative and positive windows was grown on 3D laser scribed graphene by Wu *et al*. The developed LSG/Ni-MOF-based MSCs showed outstanding areal capacitance of 15.2 mF cm^{-2} at 0.2 mA cm^{-2}. The π conjugation of tricatecholate ligands resulted in decent electrical conductivity. The flow of electrolytes is enhanced due to the porous 1D open channels formed by the alternative stacking of 2D layers. The fabricated MSCs displayed the

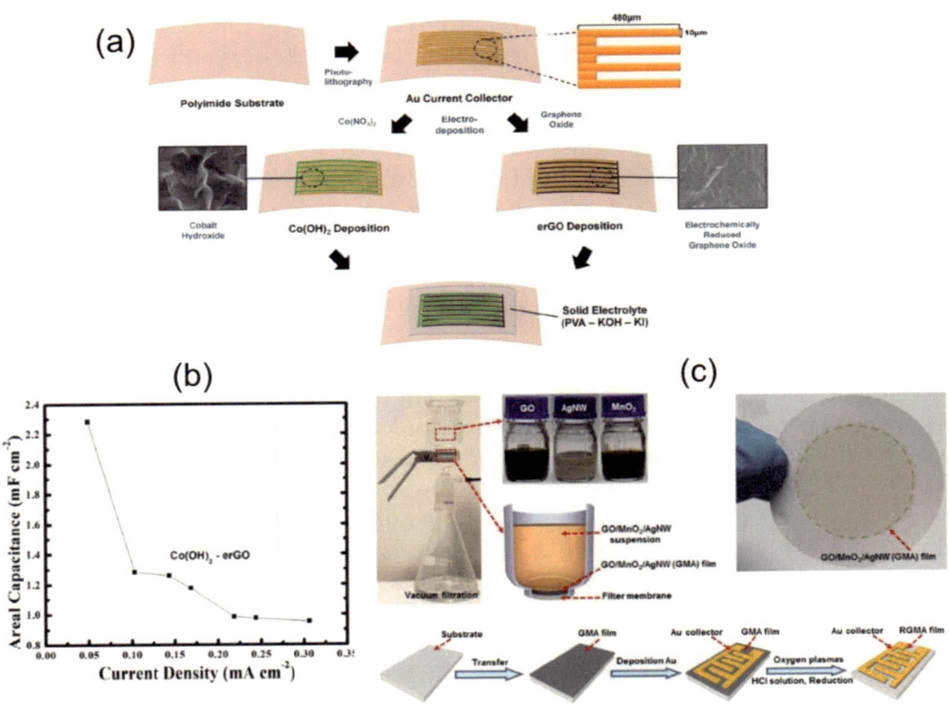

Figure 8.
(a) Schematic representation of the fabrication of Co(OH)$_2$-e-rGO, (b) the variation in areal capacitances v/s current density of Co(OH)$_2$//e-rGO MSC (Source: Reprinted from [13], with permission from RSC) and (c) schematic diagram of the fabrication of ternary hybrid film and displaying ternary hybrid film supported on cellulose acetate via vaccum filtration and fabrication process of RGO/MnO$_2$/Ag NW-MSCs on alumina (Source: Reprinted from [82], with permission from, copyright@2015, ACS).

highest power density of 7 mW cm^{-2} and a high energy density of 4.1 μW h cm^{-2}. As illustrated in fig, MOF based MSC retained 87% of its initial capacitance in gel electrolyte even after 5000 cycles. This approach may shed light on fabricating novel MOF-based MSCs and electrochemical devices [84]. Recently, Liu *et al.* presented all-solid-state MSCs based on a flexible ternary hybrid film (RGMA) of RGO/MnO$_2$/AgNW (silver nanowire) via facile vacuum filtration and thermal reduction (**Figure 8c**). It provided a great advantage to include metal oxide or metal nanoarchitectures into graphene film with strong potential for various thin-film energy storage devices. They adopted an efficient strategy to design graphene-based nanoarchitctures by incorporating the high electrical conductivity, interface integrity of the components and energy storage mechanisms (pseudocapacitance and electric double layer capacitance). Graphene is good material for the flexible energy storage devices due to its mechanical stability and MnO$_2$ served to enhance the capacitive performances and inhibited the aggregation of graphene nanosheets. The ternary hybrid film's mechanical flexibility and electrical conductivity could be enhanced by the 1D AgNW, which functioned as a conducting bridge between Needle-like MnO$_2$ and graphene nanosheets. This flexible MSC delivers a specific capacitance of 2.72 F cm^{-3} and an excellent cycling capability with a capacitance retention up to 90.3% after 6000 cycles, together with exceptional flexibility and volumetric energy density of 2.3 mWh cm^{-3} (power density of 162.0 mW cm^{-3}) in ionic liquid gel electrolyte [82]. **Figure 9** illustrates the schematic representation of fabrication and characterization of all-solid-state MSCs based on 2D materials.

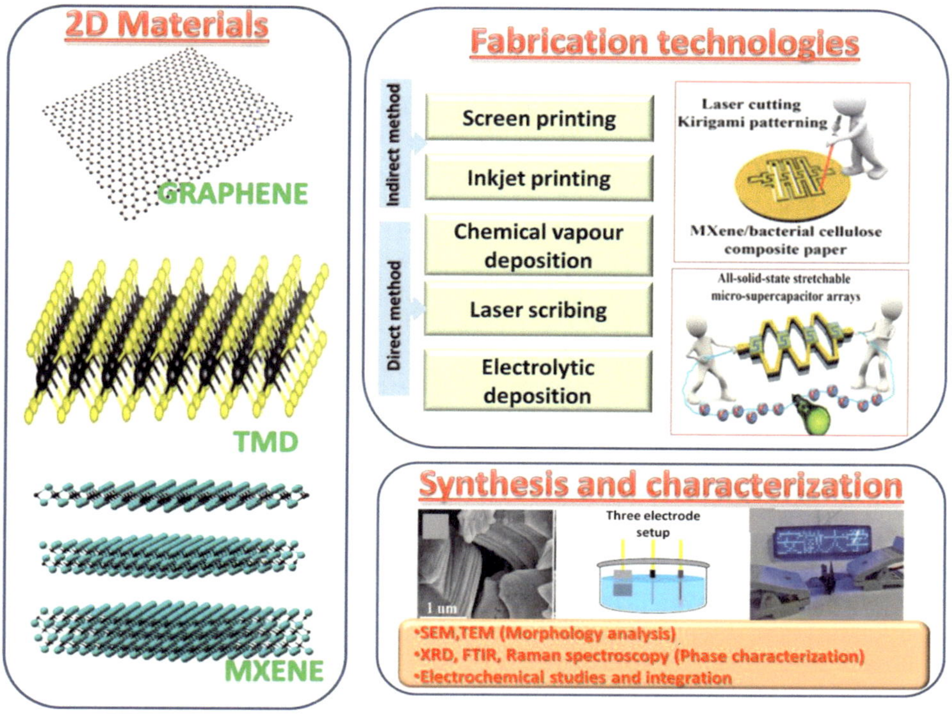

Figure 9.
Schematic of fabrication and characterization of MSC based on 2D materials. Graphene structure (Mary, 2020), TMD structure (Wikipedia contributors. (2020, October 12). Transition metal dichalcogenide monolayers. In Wikipedia, The Free Encyclopedia. Retrieved 07:20, October 17, 2020, from https://en.wikipedia.org/w/index.php?title=Transition_metal_dichalcogenide_monolayers&oldid=983161701) MXene structure (Yujuan Zhang, Ningning Zhang and Changchun Ge- First-Principles Studies of Adsorptive Remediation of Water and Air Pollutants Using Two-Dimensional MXene Material, materials 2018, 11 [11], 2281. Three electrode setup (Benjamin Hsia- Materials Synthesis and Characterization for Micro-supercapacitor Applications, Doctoral dissertion, University of California, Berkeley (2013)). Fabrication technologies and Synthesis/characterization (Source: Reprinted from [76]).

5. Conclusion

MSCs as an energy storage devices attract considerable attention due to their notable characteristics such as smaller volume and high electrochemical performance. This chapter provides a brief overview of the recent developments in the field of 2D material-based all-solid-state MSCs. A brief note on the MSC device configuration and microfabrication methods for the microelectrodes has been illustrated. 2D materials based MSCs open up new avenues for the technologically relevant real-world applications. 2D materials such as MXenes, graphene, TMDs, and 2D metal–organic framework, TMOs/TMHs materials, have been described with regard to their electrochemical properties for MSCs. It is reported that the one issue faced by 2D materials is their unavoidable aggregation or restacking owing to their intense van der Waals interactions. To overcome this, there are approaches available like expansion of interlayer space with regard to enhanced storage ability or intercalation of guest molecules to increase the active sites. Moreover, for MSCs, 2D materials with vertical orientation grown on interdigitated current collectors is favorable to attain enhanced charge transport and low interfacial resistance. Additionally, to achieve higher conductivity and large specific surface area, combining various materials with 2D hybrids is a practical approach to surpass each component material's challenges. Precisely, novel 2D materials with fascinating electrochemical properties are highly required. For example, 2D materials such as borophene, tellurene, silicene, phosphorene and germanene with higher electrical conductivity and enhanced specific surface area can be suitable candidates for high-performance MSCs. However, the coatings or surface functionalization of these 2D materials will be needed due to their chemical degradation and intrinsic surface instability under surrounding conditions. Above all, the processibility, reliability, and scalability of high-quality 2D materials are necessary not only for basic research but also for the real-world technological applications that need improved microfabrication methods such as screen printing and inkjet printing and 3D printing, *etc*.

Despite the recent advances in the design and fabrication of MSCs, MSC is still imperfect and require more developments. Some challenges limit practical implementation such as sustaining stable output voltage for wearable devices (microsystem and MSCs array just ignore these issues), current and output voltage are yet not pleased and more attempts should be assigned to design MSCs with a wider potential window. Moreover, various features, such as self-healing, hydrophobia, and stretchability, could be more developed to improve MSC performance. The device fabrication holds a significant role in technological innovation that, successively, affects the large scale production and the complexity of MSCs. It is expected that the integration of microdevices and smart functions into systems is unavoidable for facilitating the fast growth of smart electronic devices. MSCs based on 2D materials are focused on the powering of energy-consuming microdevices. Because of the complication in the smart systems' fabrication process, only limited works have been reported. So, innovative self-powered smart systems, including energy storage units, constitute a highly emerging research direction. Besides, the fabrication of a smart system with flexible, biodegradable and washable features can open the way for future independent, continuous, and intelligent daily electronics functioning. Moreover, these all-in-one self-powered systems can be used for health care applications in the future.

Acknowledgements

This work was financially supported by the Department of Science and Technology (DST)-SERB Early Career Research project (Grant No. ECR/2017/001850), DST-Nanomission (DST/NM/NT/2019/205(G), DST/TDT/SHRI-34/2018), Karnataka Science and Technology Promotion Society (KSTePS/VGST-RGS-F/2018-2019/GRD NO. 829/315).

Conflict of interest

The authors declare no conflict of interest.

Author details

Minu Mathew, Sithara Radhakrishnan and Chandra Sekhar Rout*
Centre for Nano and Material Sciences, Jain University, Jain Global Campus, Jakkasandra, Ramanagaram, Bangalore, India

*Address all correspondence to: r.chandrasekhar@jainuniversity.ac.in; csrout@gmail.com

References

[1] Wang J, Li F, Zhu F, Schmidt OG. Recent Progress in Micro-Supercapacitor Design, Integration, and Functionalization. Small Methods. 2019;3(8): 1800367.

[2] Zhang P, Wang F, Yu M, Zhuang X, Feng X. Two-dimensional materials for miniaturized energy storage devices: from individual devices to smart integrated systems. Chem Soc Rev. 2018;47(19):7426-51.

[3] Yue Y, Liu N, Ma Y, Wang S, Liu W, Luo C, et al. Highly Self-Healable 3D Microsupercapacitor with MXene–Graphene Composite Aerogel. ACS Nano. 2018 May 22;12(5):4224-32.

[4] Liu L, Zhao H, Lei Y. Advances on three-dimensional electrodes for micro-supercapacitors: A mini-review. InfoMat. 2019 Mar;1(1):74-84.

[5] Feng X, Ning J, Wang D, Zhang J, Dong J, Zhang C, et al. All-solid-state planner micro-supercapacitor based on graphene/NiOOH/Ni(OH)2 via mask-free patterning strategy. J Power Sources. 2019 Apr;418:130-7.

[6] Qi D, Liu Y, Liu Z, Zhang L, Chen X. Design of Architectures and Materials in In-Plane Micro-supercapacitors: Current Status and Future Challenges. Adv Mater. 2017 Feb;29(5):1602802.

[7] Shahrokhian S, Naderi L, Mohammadi R. High– Performance Fiber–Shaped Flexible Asymmetric Micro– Supercapacitor Based on Ni(OH)2 Nanoparticles–Decorated Porous Dendritic Ni–Cu Film/Cu Wire and RGO/Carbon Fiber Electrodes. :47.

[8] Yu D, Qian Q, Wei L, Jiang W, Goh K, Wei J, et al. Emergence of fiber supercapacitors. Chem Soc Rev. 2015;44(3):647-62.

[9] Qi et al. - 2017 - Design of Architectures and Materials in In-Plane. pdf.

[10] Liu N, Gao Y. Recent Progress in Micro-Supercapacitors with In-Plane Interdigital Electrode Architecture. Small. 2017 Dec;13(45).

[11] Beidaghi M, Gogotsi Y. Capacitive energy storage in micro-scale devices: recent advances in design and fabrication of micro-supercapacitors. Energy Environ Sci. 2014;7(3):867.

[12] Sollami Delekta S, Adolfsson KH, Benyahia Erdal N, Hakkarainen M, Östling M, Li J. Fully inkjet printed ultrathin microsupercapacitors based on graphene electrodes and a nano-graphene oxide electrolyte. Nanoscale. 2019;11(21):10172-7.

[13] Lee SC, Patil UM, Kim SJ, Ahn S, Kang S-W, Jun SC. All-solid-state flexible asymmetric micro supercapacitors based on cobalt hydroxide and reduced graphene oxide electrodes. RSC Adv. 2016;6(50):43844-54.

[14] Recent advances in micro-supercapacitors - Nanoscale (RSC Publishing) [Internet]. [cited 2020 Sep 8]. Available from: https://pubs.rsc.org/en/content/articlelanding/2019/nr/c9nr01090d#!divAbstract

[15] Qin J, Das P, Zheng S, Wu Z-S. A perspective on two-dimensional materials for planar micro-supercapacitors. APL Mater. 2019 Sep 1;7(9):090902.

[16] Gogotsi Y, Simon P. True Performance Metrics in Electrochemical Energy Storage. Science. 2011 Nov 18;334(6058):917-8.

[17] Liang J, Mondal AK, Wang D-W, Iacopi F. Graphene-Based Planar

Microsupercapacitors: Recent Advances and Future Challenges. Adv Mater Technol. 2019;4(1):1800200.

[18] Sci-Hub | 2D Materials Beyond Graphene for High-Performance Energy Storage Applications. Advanced Energy Materials, 6(23), 1600671 | 10.1002/aenm.201600671 [Internet]. [cited 2020 Sep 9]. Available from: https://sci-hub.se/10.1002/aenm.201600671

[19] Sci-Hub | Engineering 2D Architectures toward High-Performance Micro-Supercapacitors. Advanced Materials, 1802793 | 10.1002/adma.201802793 [Internet]. [cited 2020 Sep 9]. Available from: https://sci-hub.se/https://doi.org/10.1002/adma.201802793

[20] Liu C, Yu Z, Neff D, Zhamu A, Jang BZ. Graphene-Based Supercapacitor with an Ultrahigh Energy Density. Nano Lett. 2010 Dec 8;10(12):4863-8.

[21] Wu Z-S, Parvez K, Feng X, Müllen K. Graphene-based in-plane micro-supercapacitors with high power and energy densities. Nat Commun. 2013 Sep 17;4(1):2487.

[22] Xiong G, Meng C, Reifenberger RG, Irazoqui PP, Fisher TS. A Review of Graphene-Based Electrochemical Microsupercapacitors. Electroanalysis. 2014 Jan;26(1):30-51.

[23] Zhang G, Han Y, Shao C, Chen N, Sun G, Jin X, et al. Processing and manufacturing of graphene-based microsupercapacitors. Mater Chem Front. 2018 Sep 27;2(10):1750-64.

[24] Yao B, Chandrasekaran S, Zhang J, Xiao W, Qian F, Zhu C, et al. Efficient 3D Printed Pseudocapacitive Electrodes with Ultrahigh MnO2 Loading. Joule. 2019 Feb 20;3(2):459-70.

[25] Yue Y, Yang Z, Liu N, Liu W, Zhang H, Ma Y, et al. A Flexible Integrated System Containing a Microsupercapacitor, a Photodetector, and a Wireless Charging Coil. ACS Nano. 2016 Dec 27;10(12):11249-57.

[26] Peng Z, Ye R, Mann JA, Zakhidov D, Li Y, Smalley PR, et al. Flexible Boron-Doped Laser-Induced Graphene Microsupercapacitors. ACS Nano. 2015 Jun 23;9(6):5868-75.

[27] Zhou F, Huang H, Xiao C, Zheng S, Shi X, Qin J, et al. Electrochemically Scalable Production of Fluorine-Modified Graphene for Flexible and High-Energy Ionogel-Based Microsupercapacitors. J Am Chem Soc. 2018 Jul 5;140(26):8198-205.

[28] Wang Y, Zhang Y, Liu J, Wang G, Pu F, Ganesh A, et al. Boosting areal energy density of 3D printed all-solid-state flexible microsupercapacitors via tailoring graphene composition. Energy Storage Mater. 2020 Sep;30:412-9.

[29] Sollami Delekta S, Laurila M-M, Mäntysalo M, Li J. Drying-Mediated Self-Assembly of Graphene for Inkjet Printing of High-Rate Micro-supercapacitors. Nano-Micro Lett. 2020 Jan;12(1):40.

[30] Kletenik-Edelman O, Ploshnik E, Salant A, Shenhar R, Banin U, Rabani E. Drying-Mediated Hierarchical Self-Assembly of Nanoparticles: A Dynamical Coarse-Grained Approach. J Phys Chem C. 2008 Mar 1;112(12):4498-506.

[31] Van Toan N, Kim Tuoi TT, Li J, Inomata N, Ono T. Liquid and solid states on-chip micro-supercapacitors using silicon nanowire-graphene nanowall-pani electrode based on microfabrication technology. Mater Res Bull. 2020 Nov 1;131:110977.

[32] All-graphene-battery: bridging the gap between supercapacitors and lithium ion batteries | Scientific Reports [Internet]. [cited 2020 Sep 14].

Available from: https://www.nature.com/articles/srep05278

[33] Yang - 2016 - Graphene and the related conductive inks for flexi.pdf.

[34] Lu Y, Zheng Y, Zhang H, He X, Yang Q, Wu J. A high performance and flexible in-plane asymmetric micro-supercapacitor (MSC) fabricated with functional electrochemical-exfoliated graphene. J Electroanal Chem. 2020 Jun 1;866:114169.

[35] Lochmann S, Grothe J, Eckhardt K, Leistenschneider D, Borchardt L, Kaskel S. Nanoimprint lithography of nanoporous carbon materials for micro-supercapacitor architectures. Nanoscale. 2018 May 31;10(21):10109-15.

[36] Zhang C (John), Kremer MP, Seral-Ascaso A, Park S-H, McEvoy N, Anasori B, et al. Stamping of Flexible, Coplanar Micro-Supercapacitors Using MXene Inks. Adv Funct Mater. 2018;28(9):1705506.

[37] Hota MK, Jiang Q, Mashraei Y, Salama KN, Alshareef HN. Fractal Electrochemical Microsupercapacitors. Adv Electron Mater. 2017 Oct;3(10):1700185.

[38] Stamp-assisted flexible graphene-based micro-supercapacitors - ScienceDirect [Internet]. [cited 2020 Sep 15]. Available from: https://www.sciencedirect.com/science/article/abs/pii/S0378775320304699

[39] Wu Y, Zhang Y, Liu Y, Cui P, Chen S, Zhang Z, et al. Boosting the Electrochemical Performance of Graphene-Based On-Chip Micro-Supercapacitors by Regulating the Functional Groups. ACS Appl Mater Interfaces. 2020 Sep 14;acsami.0c11085.

[40] Liu F, Liu C, Li X, Zhang L, Zhao W, Zhang G. Graphene-Based Integrated Planar On-Chip Micro-Supercapacitors with No Internal Connection. Integr Ferroelectr. 2020 Mar 23;206(1):96-104.

[41] Yang S, Jiang C, Wei S. Gas sensing in 2D materials. Appl Phys Rev. 2017 Jun;4(2):021304.

[42] Gravure printing of hybrid MoS2@S-rGO interdigitated electrodes for flexible microsupercapacitors: Applied Physics Letters: Vol 107, No 1 [Internet]. [cited 2020 Sep 16]. Available from: https://aip.scitation.org/doi/10.1063/1.4926570

[43] W Y, L H, X T, M Y, H Y, X L, et al. Carbon-MEMS-Based Alternating Stacked MoS2 @rGO-CNT Micro-Supercapacitor with High Capacitance and Energy Density. Small Weinh Bergstr Ger [Internet]. 2017 May 30 [cited 2020 Sep 15];13(26). Available from: https://europepmc.org/article/med/28558128

[44] Haider WA, Tahir M, He L, Yang W, Minhas-khan A, Owusu KA, et al. Integration of VS2 nanosheets into carbon for high energy density micro-supercapacitor. J Alloys Compd. 2020 May;823:151769.

[45] Py MA, Haering RR. Structural destabilization induced by lithium intercalation in MoS_2 and related compounds. Can J Phys. 1983 Jan 1;61(1):76-84.

[46] Xu C, Jiang L, Li X, Li C, Shao C, Zuo P, et al. Miniaturized high-performance metallic 1T-Phase MoS2 micro-supercapacitors fabricated by temporally shaped femtosecond pulses. Nano Energy. 2020 Jan;67:104260.

[47] Chen Y, Zhu X, Yang D, Wangyang P, Zeng B, Sun H. A novel design of poly (3,4-ethylenedioxyt hiophene):poly (styrenesulfonate)/molybdenum disulfide/poly (3,4-ethylenedioxythiophene) nanocomposites for fabric micro-supercapacitors with favourable

performances. Electrochimica Acta. 2019 Mar;298:297-304.

[48] Li B, Liang X, Li G, Shao F, Xia T, Xu S, et al. Inkjet-Printed Ultrathin MoS_2 -Based Electrodes for Flexible In-Plane Microsupercapacitors. ACS Appl Mater Interfaces. 2020 Sep 2;12(35):39444-54.

[49] El-Kady MF, Kaner RB. Scalable fabrication of high-power graphene micro-supercapacitors for flexible and on-chip energy storage. Nat Commun. 2013 Feb 12;4(1):1475.

[50] Wu Z-S, Parvez K, Feng X, Müllen K. Photolithographic fabrication of high-performance all-solid-state graphene-based planar micro-supercapacitors with different interdigital fingers. J Mater Chem A. 2014 May 15;2(22):8288-93.

[51] Gao W, Singh N, Song L, Liu Z, Reddy ALM, Ci L, et al. Direct laser writing of micro-supercapacitors on hydrated graphite oxide films. Nat Nanotechnol. 2011 Jul 31;6(8):496-500.

[52] Hsia B, Marschewski J, Wang S, In JB, Carraro C, Poulikakos D, et al. Highly flexible, all solid-state micro-supercapacitors from vertically aligned carbon nanotubes. Nanotechnology. 2014 Feb 7;25(5):055401.

[53] Kim SK, Koo HJ, Lee A, Braun PV. Selective wetting-induced micro-electrode patterning for flexible micro-supercapacitors. Adv Mater. 2014 Aug 13;26(30):5108-12.

[54] Huang P, Lethien C, Pinaud S, Brousse K, Laloo R, Turq V, et al. On-chip and freestanding elastic carbon films for micro-supercapacitors. Science. 2016 Feb 12;351(6274):691-5.

[55] Chmiola J, Largeot C, Taberna P-L, Simon P, Gogotsi Y. Monolithic carbide-derived carbon films for micro-supercapacitors. Science. 2010 Apr 23;328(5977):480-3.

[56] Beidaghi M, Wang C. Micro-Supercapacitors Based on Interdigital Electrodes of Reduced Graphene Oxide and Carbon Nanotube Composites with Ultrahigh Power Handling Performance. Adv Funct Mater. 2012;22(21):4501-10.

[57] Song B, Li L, Lin Z, Wu Z-K, Moon K, Wong C-P. Water-dispersible graphene/polyaniline composites for flexible micro-supercapacitors with high energy densities. Nano Energy. 2015 Sep 1;16:470-8.

[58] Kurra N, Jiang Q, Alshareef HN. A general strategy for the fabrication of high performance microsupercapacitors. Nano Energy. 2015 Sep 1;16:1-9.

[59] Kurra N, Alhebshi NA, Alshareef HN. Microfabricated Pseudocapacitors Using Ni(OH)2 Electrodes Exhibit Remarkable Volumetric Capacitance and Energy Density. Adv Energy Mater. 2015;5(2):1401303.

[60] Si W, Yan C, Chen Y, Oswald S, Han L, Schmidt OG. On chip, all solid-state and flexible micro-supercapacitors with high performance based on MnOx/Au multilayers. Energy Environ Sci. 2013 Oct 18;6(11):3218-23.

[61] Tahir M, He L, Yang W, Hong X, Haider WA, Tang H, et al. Boosting the electrochemical performance and reliability of conducting polymer microelectrode via intermediate graphene for on-chip asymmetric micro-supercapacitor. J Energy Chem. 2020 Oct 1;49:224-32.

[62] Cao L, Yang S, Gao W, Liu Z, Gong Y, Ma L, et al. Direct Laser-Patterned Micro-Supercapacitors from Paintable MoS2 Films. Small. 2013;9(17):2905-10.

[63] Beidaghi M, Gogotsi Y. Capacitive energy storage in micro-scale devices: recent advances in design and fabrication of micro-supercapacitors.

Energy Environ Sci. 2014 Feb 20;7(3):867-84.

[64] Acerce M, Voiry D, Chhowalla M. Metallic 1T phase MoS2 nanosheets as supercapacitor electrode materials. Nat Nanotechnol. 2015 Apr;10(4):313-8.

[65] Zhang P, Wang F, Yu M, Zhuang X, Feng X. Two-dimensional materials for miniaturized energy storage devices: from individual devices to smart integrated systems. Chem Soc Rev. 2018 Oct 1;47(19):7426-51.

[66] Peng Y-Y, Akuzum B, Kurra N, Zhao M-Q, Alhabeb M, Anasori B, et al. All-MXene (2D titanium carbide) solid-state microsupercapacitors for on-chip energy storage. Energy Environ Sci. 2016 Aug 31;9(9):2847-54.

[67] Dillon AD, Ghidiu MJ, Krick AL, Griggs J, May SJ, Gogotsi Y, et al. Highly Conductive Optical Quality Solution-Processed Films of 2D Titanium Carbide. Adv Funct Mater. 2016;26(23):4162-8.

[68] Naguib M, Mochalin VN, BarsoumMW,GogotsiY.Two-Dimensional Materials: 25th Anniversary Article: MXenes: A New Family of Two-Dimensional Materials (Adv. Mater. 7/2014). Adv Mater. 2014;26(7):982-982.

[69] Ghidiu M, Lukatskaya MR, Zhao M-Q, Gogotsi Y, Barsoum MW. Conductive two-dimensional titanium carbide 'clay' with high volumetric capacitance. Nature. 2014 Dec;516(7529):78-81.

[70] Ghidiu M, Lukatskaya MR, Zhao M-Q, Gogotsi Y, Barsoum MW. Conductive two-dimensional titanium carbide 'clay' with high volumetric capacitance. Nature. 2014 Dec;516(7529):78-81.

[71] Yan J, Wang Q, Wei T, Jiang L, Zhang M, Jing X, et al. Template-assisted low temperature synthesis of functionalized graphene for ultrahigh volumetric performance supercapacitors. ACS Nano. 2014 May 27;8(5):4720-9.

[72] Zhang C (John), McKeon L, Kremer MP, Park S-H, Ronan O, Seral-Ascaso A, et al. Additive-free MXene inks and direct printing of micro-supercapacitors. Nat Commun. 2019 Apr 17;10(1):1795.

[73] Peng Y-Y, Akuzum B, Kurra N, Zhao M-Q, Alhabeb M, Anasori B, et al. All-MXene (2D titanium carbide) solid-state microsupercapacitors for on-chip energy storage. Energy Environ Sci. 2016 Aug 31;9(9):2847-54.

[74] Huang H, Chu X, Su H, Zhang H, Xie Y, Deng W, et al. Massively manufactured paper-based all-solid-state flexible micro-supercapacitors with sprayable MXene conductive inks. J Power Sources. 2019 Mar 1;415:1-7.

[75] Zhang H, Cao Y, Chee MOL, Dong P, Ye M, Shen J. Recent advances in micro-supercapacitors. Nanoscale. 2019 Mar 28;11(13):5807-21.

[76] Jiao S, Zhou A, Wu M, Hu H. Kirigami Patterning of MXene/Bacterial Cellulose Composite Paper for All-Solid-State Stretchable Micro-Supercapacitor Arrays. Adv Sci. 2019;6(12):1900529.

[77] Yue Y, Liu N, Ma Y, Wang S, Liu W, Luo C, et al. Highly Self-Healable 3D Microsupercapacitor with MXene–Graphene Composite Aerogel. ACS Nano. 2018 May 22;12(5):4224-32.

[78] Li P, Shi W, Liu W, Chen Y, Xu X, Ye S, et al. Fabrication of high-performance MXene-based all-solid-state flexible microsupercapacitor based on a facile scratch method. Nanotechnology. 2018 Sep;29(44):445401.

[79] Jiang Q, Wu C, Wang Z, Wang AC, He J-H, Wang ZL, et al. MXene

electrochemical microsupercapacitor integrated with triboelectric nanogenerator as a wearable self-charging power unit. Nano Energy. 2018 Mar 1;45:266-72.

[80] Xu S, Liu W, Liu X A MXene based all-solid-state microsupercapacitor with 3D interdigital electrode, 19th international conference on solid state sensors, Actuators and Sensors. 2017 Jun 18;706-709.

[81] Hu M, Li Z, Li G, Hu T, Zhang C, Wang X. All-Solid-State Flexible Fiber-Based MXene Supercapacitors. Adv Mater Technol. 2017;2(10):1700143.

[82] Liu W, Lu C, Wang X, Tay RY, Tay BK. High-Performance Microsupercapacitors Based on Two-Dimensional Graphene/Manganese Dioxide/Silver Nanowire Ternary Hybrid Film. ACS Nano. 2015 Feb 24;9(2):1528-42.

[83] Wang S, Wu Z-S, Zhou F, Shi X, Zheng S, Qin J, et al. All-solid-state high-energy planar hybrid micro-supercapacitors based on 2D VN nanosheets and Co(OH) 2 nanoflowers. Npj 2D Mater Appl. 2018 Mar 26;2(1):1-8.

[84] Wu H, Zhang W, Kandambeth S, Shekhah O, Eddaoudi M, Alshareef HN. Conductive Metal–Organic Frameworks Selectively Grown on Laser-Scribed Graphene for Electrochemical Microsupercapacitors. Adv Energy Mater. 2019;9(21):1900482.

[85] Zhang L, Yang G, Chen Z, Liu D, Wang J, Qian Y, et al. MXene coupled with molybdenum dioxide nanoparticles as 2D-0D pseudocapacitive electrode for high performance flexible asymmetric micro-supercapacitors. J Materiomics. 2020 Mar 1;6(1):138-44.

[86] Lee SC, Patil UM, Kim SJ, Ahn S, Kang S-W, Jun SC. All-solid-state flexible asymmetric micro supercapacitors based on cobalt hydroxide and reduced graphene oxide electrodes. RSC Adv. 2016 May 3;6(50):43844-54.

[87] Jiang J, Li Y, Liu J, Huang X, Yuan C, Lou XWD. Recent advances in metal oxide-based electrode architecture design for electrochemical energy storage. Adv Mater Deerfield Beach Fla. 2012 Oct 2;24(38):5166-80.

[88] Gao S, Sun Y, Lei F, Liang L, Liu J, Bi W, et al. Ultrahigh Energy Density Realized by a Single-Layer β-Co(OH)2 All-Solid-State Asymmetric Supercapacitor. Angew Chem Int Ed. 2014;53(47):12789-93.

[89] Guan M, Wang Q, Zhang X, Bao J, Gong X, Liu Y. Two-Dimensional Transition Metal Oxide and Hydroxide-Based Hierarchical Architectures for Advanced Supercapacitor Materials. Front Chem [Internet]. 2020 [cited 2020 Sep 19];8. Available from: https://www.frontiersin.org/articles/10.3389/fchem.2020.00390/full

[90] Dang S, Zhu Q-L, Xu Q. Nanomaterials derived from metal–organic frameworks. Nat Rev Mater. 2017 Dec 5;3(1):1-14.

[91] Mandal AK, Mahmood J, Baek J-B. Two-Dimensional Covalent Organic Frameworks for Optoelectronics and Energy Storage. ChemNanoMat. 2017;3(6):373-91.

[92] Liu J, Wöll C. Surface-supported metal–organic framework thin films: fabrication methods, applications, and challenges. Chem Soc Rev. 2017 Oct 2;46(19):5730-70.

[93] Zhang P, Wang F, Yu M, Zhuang X, Feng X. Two-dimensional materials for miniaturized energy storage devices: from individual devices to smart integrated systems. Chem Soc Rev. 2018 Oct 1;47(19):7426-51.

Chapter 14

Pulsed Electrochemical Micromachining in Stainless Steel

Pablo Rodríguez, Daniel Hidalgo and Julio Eduardo Labarga

Abstract

This chapter presents research on pulsed electrochemical micromachining of stainless steel. Suitable equipment to study the process is described as well as a fitting procedure to machine and measure the variables involved. The pulse on-time must be maintained in the order of ns to achieve a good current confinement since the tool is active. Some experiments were carried out to assess the most important variables of the process: current confinement, surface roughness, material removal rate and efficiency. The current confinement has been observed to worsen when the pulse on-time increases, as well as the surface roughness. The material removal rate and efficiency increase with the voltage amplitude and the pulse on-time. The voltage amplitude must be higher than 12 V so that the phenomenon of passivation does not affect the process. There is a compromise in the choice of the variables, so a suitable combination of parameters is determined in order to achieve a good material removal rate with an acceptable result.

Keywords: pulsed electrochemical micromachining, current confinement, material removal rate, efficiency

1. Introduction

Microfabrication consists of obtaining products or parts with features at micro- or submicroscale, therefore requiring very narrowly controlled material removal. Microfabrication has been widely used for the manufacture of holes in injectors, fluidic microchemical reactors requiring microscale pumps, micromoulds and many more applications, as described by Brousseau et al. [1]. Microfabrication plays an increasingly important role in the miniaturisation of components from biomedical applications to manufacturing sensors. Surfaces to be obtained are slots, complex surfaces, microholes, etc. Combinations of those features must frequently be achieved in the industry of microelectronics. These parts are very often manufactured by conventional processes with all the limitations and problems involved, such as tool wear, inaccuracy due to low rigidity of the tool, heat generated by the process, etc. With the development of MEMS and multiple benefits of the microsystems, microproducts are widely accepted in various fields of applications like aerospace, automotive, biomedicine, etc. [2]. In this context, non-conventional processes, and especially electrochemical micromachining, acquire greater significance due to their specific characteristics to avoid the problems of conventional processes.

Since the first years of developments in electrochemistry, electrochemical methods have played an important role in precision technologies to machine structures and parts. In the 1950s, electrochemical machining arose as the most widely used technique to manufacture complex geometries, such as turbine blades, generally in dense materials. The ease of application of this technology along with the inherent advantages of the process, such as good surface roughness, promoted its application to more advanced processes in the field of micromechanics, microelectronics and micro-systems [3]. Electrochemical deposition techniques were used as standard technology to deposit copper to obtain connections in high performance circuits while lithographic techniques, LIGA, are used to manufacture micromoulds [4, 5].

Electrochemical micromachining is a highly specialised process used in the aerospace industry. Today, it is starting to be used in other industries, where difficult-to-manufacture parts, complex surfaces and components of a microscopic scale need to be obtained. Electrochemical micromachining is today widely used for manufacturing semiconductor elements and thin metallic films [6]. In addition, electrochemical micromachining can be easily hybridised with other processes to broaden the process capabilities and material processing window [7].

Analogous to conventional electrochemical machining (ECM), pulsed electrochemical micromachining (PECMM) is a controlled process of anodic dissolution to remove the material with current densities in the order of 10^5 A/m^2 between the tool (cathode) and the workpiece (anode) through the electrolyte [8]. PECMM uses a pulsed voltage signal and must be analysed per pulse according to the structure of the Helmholtz/Gouy-Chapman/Stern double layer [9], which can be modelled as a resistance in parallel with a capacitor. This model has provided good results in experiments and indicates that the current is used at first to charge the capacitor (capacitive current). When the charge is high enough, that is, when its voltage is high enough, some current will flow to be used in the anodic dissolution process (faradaic current) since the polarisation or overpotential will have a significant value. Therefore, two stages can be distinguished in each pulse. The first part of the pulse on-time is a transient period in which the current is used in the polarisation of the double layer, which has to be high to achieve fast polarisation. The second stage is a steady period in which the current is used mainly for the anodic dissolution. In this context, what seems most fitting is that the transient process (non-faradaic) should be very short and the steady process (faradaic) very long. In addition, the intermittent supply of voltage provides idle time to flush the hydrogen bubbles and sludge from the machining zone and also increases control over the dissolution process [10]. However, a long steady period decreases the accuracy of the process as the current confinement under the tool tip worsens when this period is lengthened. Therefore, a compromise in the time of the steady-state period is required. By solving the differential equation of the equivalent circuit, the expression of the current as a function of time is obtained. The resulting time constant is the product of the electrolyte resistivity, the capacity of the double layer and the distance between the interelectrode gap (IEG) [8].

$$\tau = \rho \cdot \mathrm{IEG} \cdot c_{DL} \qquad (1)$$

A high value of the constant time will cause the current lines to spread over a broad area from the tool tip, thus reducing the accuracy of the process. Therefore, a low pulse-on time must be chosen to achieve accuracy.

By causing the tool to move towards the workpiece, the material is removed under its tip, since the current density is higher at a lower distance between the tool and workpiece, and thus, the geometry of the tool is copied as a cavity in the workpiece. As compared with other processes, PECMM is a high-precision technique to obtain holes of a small diameter or to obtain crack-free microcomponents without any residual stress. There are two methods of achieving accuracy with electrochemical micromachining. One of them is to use a tool in which all the surface is isolated except for the tip. This method ensures that all the current flows from the tip and that the cavity obtained is equal to that tip, since this current is responsible for the anodic dissolution of the material. Another method is the use of ultrashort voltage pulses, usually shorter than 100 ns. This method achieves high accuracy by confining the faradaic current density under the tool due to the incomplete charge of the double layer in areas far from the part through which a very low current will flow.

An important phenomenon which affects the process is the formation of a passive oxide layer that hinders the anodic dissolution [11]. The characterisation of this phenomenon is very important, since some processes like electropolishing are performed more adequately in passivation conditions [12]. When this takes place, the voltage applied has to be above a threshold value to cause effective machining [13]. It can also be avoided by adding acid to the electrolyte, such as HCl or H_2SO_4, which dissolves the passive layer. This layer can be considered an additional electrical resistance in the equivalent circuit, which prevents the current from being confined under the tool tip [14]. According to this explanation, the current which flows from the sides of the tool finds a similar resistance to that which flows from the tool tip and therefore the current is spread over a broad surface.

Significant advances have been made in the research of this process on many materials such as aluminium, titanium, steel and copper [15, 16]. Stainless steel is a very important material to be used in any type of microcomponent, but dissolution is difficult since its chemical properties are not very suitable for this process. Some of the existing studies were performed specifically on stainless steel [17–20]. Nevertheless, the pulse on-time used in those cases is too high to obtain a good confinement of the current. Furthermore, there are few studies in which the size of the tool is as small as a few microns. Though some work has been done in order to control the process by varying the main parameters [21], there is a huge amount of work to be done to characterise this process correctly as regards the values of the parameters in order to obtain a good result in terms of current confinement, surface roughness, material removal rate (MRR) and experimental set-up. In this work, a broad study has been made of the results of PECMM in stainless steel with pulse on-time values in the order of ns as a function of the main variables.

2. Experimental set-up

The experiments performed for the study were made by means of equipment that allows accuracy and ease of handling of tools and parts to be achieved. **Figure 1** shows a sketch of this equipment.

The equipment for the experiments rests on an anti-vibrations Table TMC, which provides a floating bench that prevents the tool and the part from oscillating. The position of the recipient is controlled by a three-dimensional (3D) nanometric positioning system based on a piezoelectric technology and with a resolution of 1 nm. There is a system of recirculation for the electrolyte, which flows constantly through the cell to a tank from which it is pumped to the cell after passing through a filter. Thus, the particles that appear in the cell are constantly being removed from the electrolyte. Experiments were performed in a solution of $NaNO_3$ at 2% in weight as the electrolyte.

The material of the workpiece is AISI 304 stainless steel and the tool is made of Tungsten, 99.7% high purity. The tools are pins with a very small tip, measuring about 5 μm in diameter. The tool tip is sharpened by means of anodic dissolution in which the tungsten pin is used as the anode and the sheet of stainless steel as the cathode. The electrolyte used for this process is a solution of KOH at 5% in weight. **Figure 2** shows a picture of the equipment used for the process. In **Figure 3**, a microtool used for the process is shown.

In order to apply the voltage pulses to the system, a Function Generator Agilent 33,250 A is used, which generates voltage signals of several types and a broad range of frequency, up to 100 MHz, which corresponds to a width of 10 ns in the

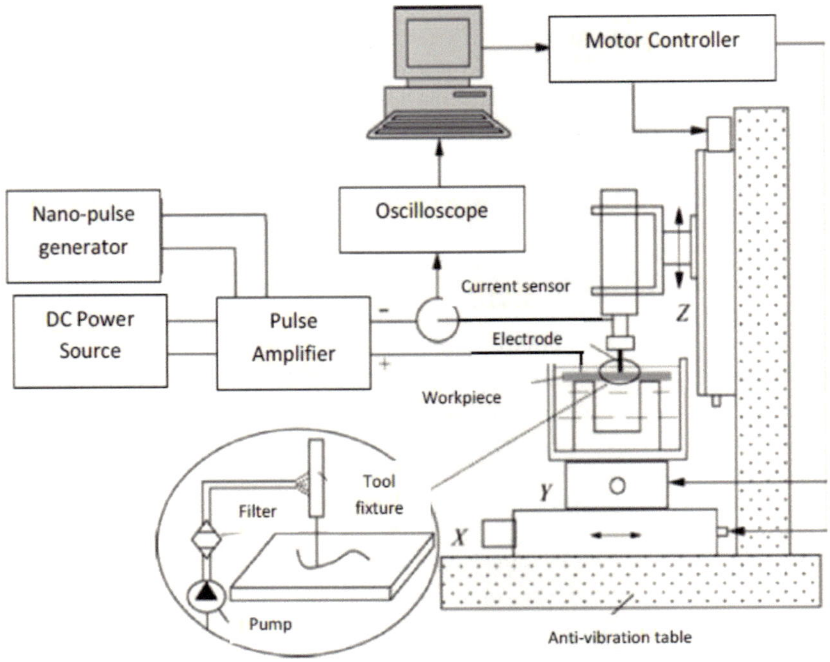

Figure 1.
Sketch of the equipment used for experiments.

Figure 2.
Equipment for electrochemical micromachining.

voltage pulses. The signal applied by the generator passes through a developed pulse amplifier that provides the necessary current for the process corresponding to the voltage amplitude. The amplifier is fed by a DC power source Keytheley 2220G-30-1 which provides a current limiting system, so that the amplifier is not overloaded. The graphs of voltage and current between electrodes for a machining process are shown in **Figure 4**.

In this graph, the current rises from zero to the stationary value going through a transient period of about 50 ns. Taking the criterion that the time constant is the time taken by the system to reach 63% of the total amount of change, a value of 25 ns for the time constant is deduced for a value of IEG = 1 μm and the other conditions described above. Therefore, values of on-time pulse above 50 ns are adequate for the process.

The electrochemical process is observed by means of a Supereyes USB Portable Digital Microscope B008 connected to a computer in which the amplified image

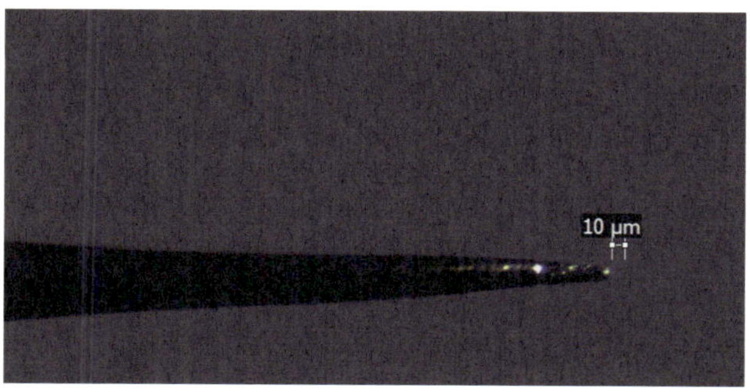

Figure 3.
Microtool used for the process.

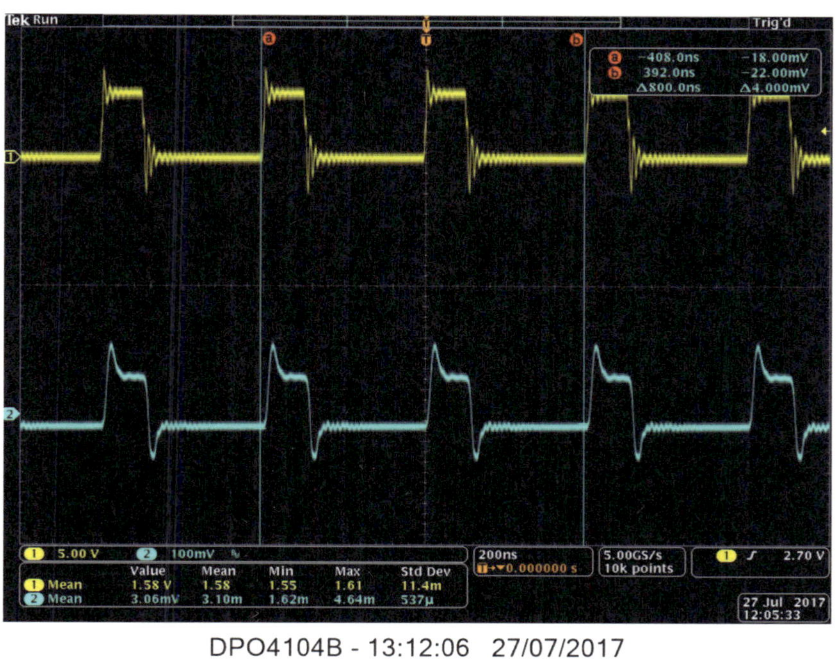

Figure 4.
Signals of voltage and current between the electrodes in the machining process. Signal 1: voltage (V) and signal 2: current (mA).

of the tool and the area of the part being machined can be seen. This microscope is also helpful to set the approach of the tool to the workpiece in order to establish the reference of distance.

The voltage applied to the cell as well as the current passing through it is measured by means of a digital oscilloscope Tektronic DPO 4104, which allows several signals with up to 3 GHz to be visualised by using a maximum sample rate of 5 Gs/s. It also permits mean signal values to be measured, applying filters and making mathematical operations with signals, such us obtaining Fourier Transforms.

In order to observe and measure the dimensions of the features machined, as well as the tip of the tools, a scanning electron microscope and an optic microscope were used.

The reference for the position of the tool is taken as the point of value 0 for the IEG. That position was found by electrical contact between the tool and the workpiece. It is observed that, when using an active tool, the current does not change significantly as IEG decreases. However, when there is electrical contact, the current increases suddenly to a very high value. This phenomenon allows the reference to be found with a very slow movement of the tool and, therefore, a brusque impact is avoided, which could damage the tool tip.

3. Results and discussion

PECMM works on the principle of Faraday's laws of electrolysis. The process consists of applying a potential difference between the tool and the workpiece so that an electrochemical reaction arises, which removes material from the workpiece. The metal is detached atom by atom from the anode surface and appears in the electrolyte as ions (Fe^{2+}). These ions result in the precipitation of ferrous hydroxide $Fe(OH)_2$. Simultaneously, hydrolysis causes the water molecules to gain electrons from the cathode and they separate into free hydrogen gas and hydroxyl ion [22]. The reactions can be summarised in the following equations:

$$Fe \rightarrow Fe^{2+} + 2e^{-} \tag{2}$$

$$H_2O + 2e^{-} \rightarrow H_2 \uparrow + 2OH^{-} \tag{3}$$

$$Fe^{2+} + 2OH^{-} \rightarrow Fe(OH)_2 \tag{4}$$

3.1 Current confinement and surface roughness

In order to achieve precision in the machining the process must take place only under the tool tip, so that the cavity obtained in the workpiece is exactly the one determined by the profile of the tool. Therefore, current through the sides of the tool must be avoided, since it would remove material from other areas far from the tool tip. There are two methods of attaining this goal. The first one is isolating the side surface of the tool and using DC voltage as the process signal. The other one is using ultrashort voltage pulses and a very low IEG. The second method is used by several researchers [8, 11, 12] due to its ease of use if a function generator is available.

The confinement of the current can be assessed by observing the edge of the hole machined. If there is confinement, the contour of the hole will be sharp; otherwise the edge will be rounded. This phenomenon was studied by machining slots with several values of pulse on-time and maintaining a constant voltage and period. By observing the size of the machined area, an assessment of the confinement of current can be achieved. The conditions for the experiments are shown in **Table 1**.

The average current is seen to decrease as the pulse on-time becomes lower, since the current only flows in the voltage pulse periods. A photograph of the holes machined in experiments R-01 to R-04 described in **Table 1** is presented in **Figure 5**. All the slots were machined with the same tool, which had a tip diameter of 10 μm. However, the width of the slot decreases with the pulse on-time from 150 to 70 μm approximately, as can be seen in the image. This is a consequence of the spreading of the current, which will be higher with an increased pulse on-time. In addition, the roundness of the edges can clearly be seen to be higher when the pulse on-time is augmented. A bright area can be observed around the slots, which suggests that the current also spread outside the hole and hence some material was removed from that area. It can therefore be deduced that when an active tool is used there is always a spreading of current outside the area under the tool tip, even if the edge of the hole is sharp. The holes machined with an isolated tool are shown in **Figure 6**. The current has clearly spread over a much smaller area since the diameter of the holes is much lower and there is no bright area around them.

Regarding surface roughness, the relationship between conditions and results are similar to those in mechanical machining, since a high MRR produces high surface roughness and vice versa. Therefore, a compromise must be achieved between surface roughness and process speed.

Experiment	IEG (μm)	Voltage (V)	Pulse width (ns)	Period (ns)	Average current (mA)
R-01	1	16	120	370	**26**
R-02	1	16	110	370	**22**
R-03	1	16	100	370	**16.1**
R-04	1	16	90	370	**10.3**

Table 1.
Conditions of the experiments for assessing current confinement and surface roughness.

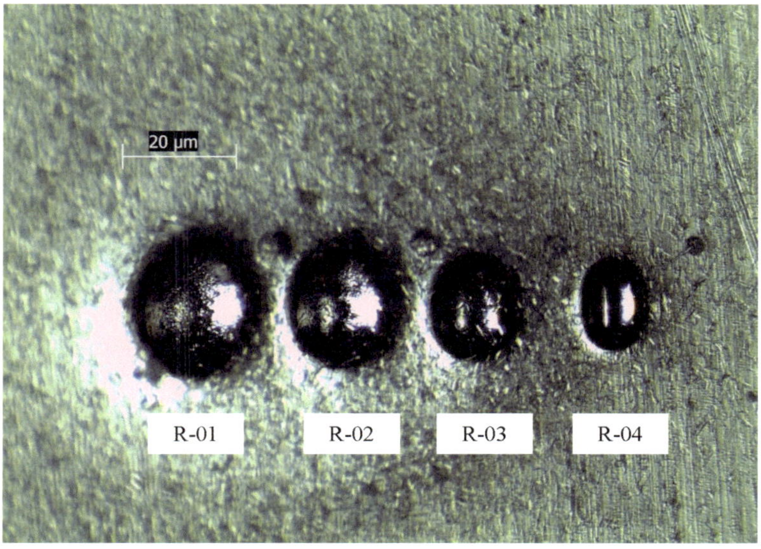

Figure 5.
Slots machined in experiments R-01 to R-04.

Figure 6.
Holes machined with an isolated tool.

Electrochemical machining has been observed to cause tiny craters in the workpiece surface, as a result of the localised current flowing through the electrolyte at the points of least electrical resistance. Therefore, if the current intensity is lower, the craters will be less deep and the resultant surface will be smoother. This can be seen clearly in **Figure 4**, which shows that the roughness is increasingly higher in the holes corresponding to R-04, R-03, R-02 and R-01, that is, when the pulse on-time grows. Therefore, it can be concluded that a good result is achieved by applying a voltage of 16 V and a pulse on-time of 80 ns and both confinement and surface roughness worsen when more aggressive values are used.

3.2 Material removal rate (MRR)

Material removal rate is a crucial variable in machining, since it determines the productivity of the process. This variable depends on the overpotential η, according to the principles of electrochemistry [9]. Therefore, the amplitude of the voltage signal determines the current intensity. Nevertheless, as the voltage signal applied to the cell consists of pulses, what determines MRR is the mean value of the current, according to Faraday's law of electrolysis:

$$MRR = \dot{m} = \frac{A \cdot I}{Z \cdot F} \tag{5}$$

where A is the gram atomic weight, Z is the valence of dissolution, F is Faraday's constant and I is the average current. In turn, the average current depends on the ratio between the period and the pulse on-time of the signal. Therefore, the main parameters which determine MRR are the pulse amplitude and the ratio between pulse on-time and period.

In order to determine the value of the parameters to attain a maximum of MRR several experiments were performed, setting the combination of parameters by means of an experiment design in which the voltage varied between 7 and 16 V and the pulse on-time from 50 to 80 ns, keeping the period constant at 370 ns. The output variable considered was the current density, which provides more information regarding the performance of the process than the current intensity, as it takes the tool tip size into account. The results can be observed in **Figure 7**, which shows the variation of the average current density as a function of the voltage for every value of the pulse on-time.

As can be seen in the graph, from 7 to 11 V, there is a decrease in current density as the voltage increases. This is due to the passivation phenomenon which occurs on the stainless steel surface. At a value of 12 V, the current density increases dramatically and then remains approximately constant. The range beyond 12 V is therefore the transpassive area, in which the voltage of the tool is enough to dissolve the passive layer under the tool tip and to remove material locally. As the current density was calculated by dividing the total current by the area of the tool tip, the sudden increase in the current density in that area does not involve a significant increase in the current as whole. Therefore, the average current grows in a constant manner as the voltage increases, as can be seen in **Figure 8**.

This effect determines that, in order to achieve good machining without dispersion of the current, the voltage value must be high beyond the passive area of the stainless steel so that the MRR is maximum.

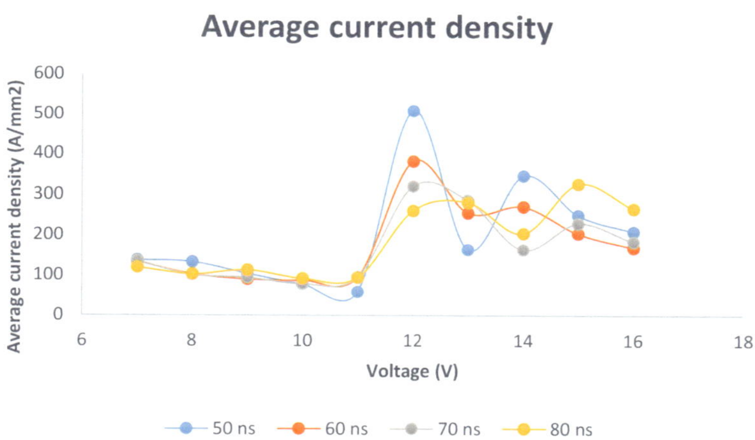

Figure 7.
Average current density as a function of voltage amplitude and pulse on-time.

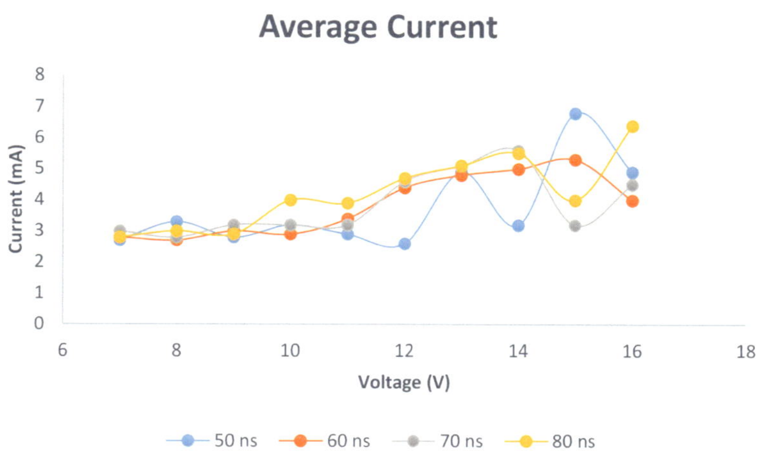

Figure 8.
Average current as a function of voltage amplitude and pulse on-time.

The method of assessing the real MRR is to observe the volume of material removed, which is determined by the geometry of the hole made in every experiment. The volume removed can be represented as a function of the voltage applied and the pulse on-time. These graphs are shown in **Figure 9**.

The graph shows that the volume removed increased with the voltage applied for every value of the pulse on-time. According to this tendency, the best value of the voltage to achieve a good MRR is the highest possible one. On the other hand, the increase is observed to be faster when the pulse on-time grows, so the value of this parameter should be as high as possible while maintaining the conditions of confinement. This graph can be presented in a 3D format in order to show the join effect of voltage and pulse on-time, as it can be seen in **Figure 10**.

3.3 Efficiency

The efficiency of the electrochemical machining can be obtained by comparing the theoretical value of material removed with the real one. The theoretical value is given by Faraday's law (5) and can be calculated from the current in the process.

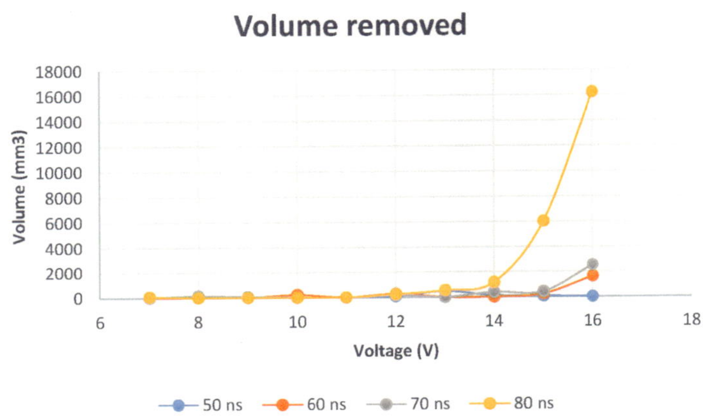

Figure 9.
Volume of material removed as a function of voltage amplitude and pulse on-time.

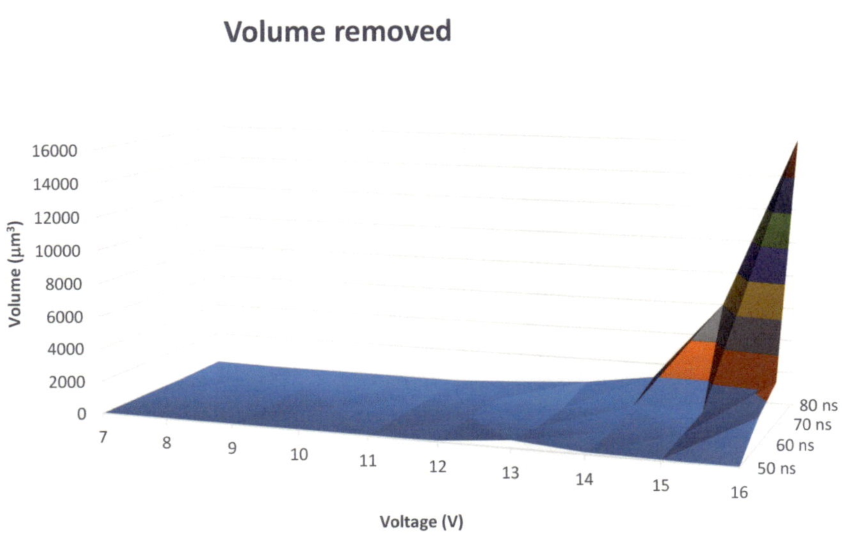

Figure 10.
3D representation of the volume removed as a function of voltage amplitude and pulse on-time.

The real value can be calculated from the geometry of the machined feature, as explained in Section 3.2.

This characteristic of the process is of great significance to the cost of the process, most of all, at an industrial level, and should be optimised by choosing the appropriate parameters.

In order to assess the efficiency of the process the results of the experiments made for observing the MRR were used. The ratio between the volume removed and the theoretical volume corresponding to the current was obtained and represented in **Figure 11**.

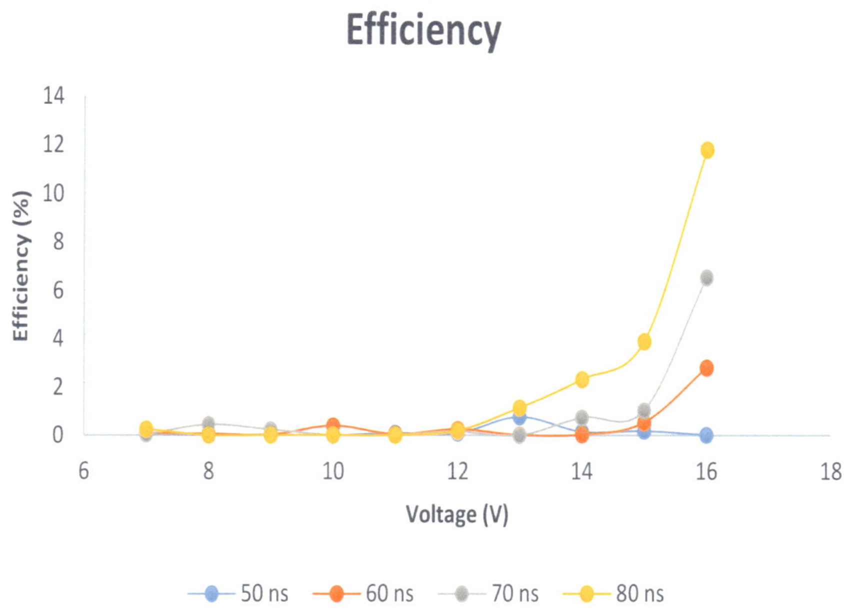

Figure 11.
Efficiency of the process as a function of voltage amplitude and pulse on-time.

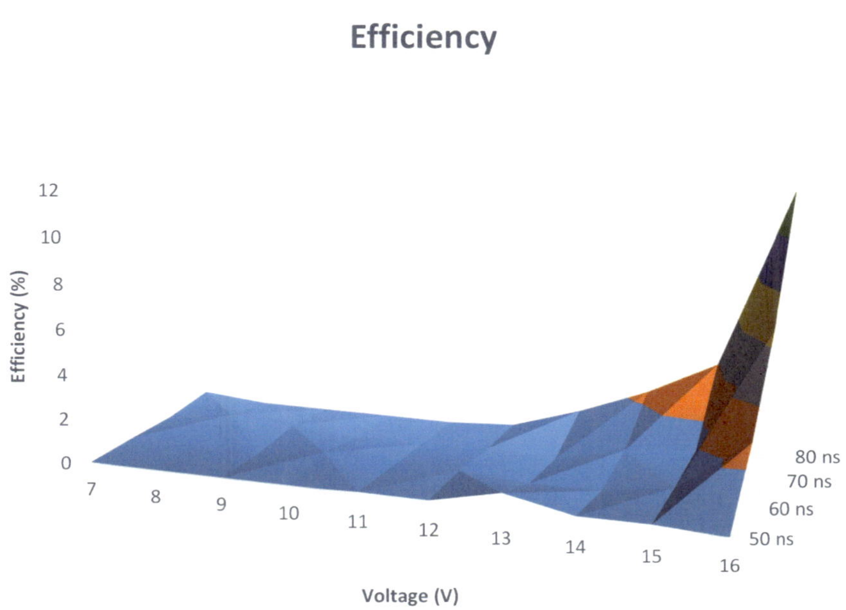

Figure 12.
3D representation of the efficiency of the process as a function of voltage amplitude and pulse on-time.

In these graphs, very low values of efficiency can be seen, because the maximum efficiency is lower than 12%. This is a consequence of the dispersion of the current, as the results presented in Section 3.1 show. According to the graphs, the higher the voltage and pulse on-time, the better the confinement of the current in the area under the tool tip and hence the higher the efficiency. The join effect of voltage and pulse on-time in efficiency is shown in the 3D of **Figure 12**.

These results can be analysed along with those presented in Section 3.2. Observing those graphs and **Figure 5** it can be deduced that the reason why the volume removed increases so drastically for voltage values higher than 14 V is not the increased current, but the clear increase in the efficiency of the process for those values. So, in order to achieve the best efficiency along with a good value of material removed, the highest possible value of voltage amplitude must be used along with the widest pulse maintaining the confinement and surface roughness within acceptable values.

4. Conclusions

A study of the optimum conditions for pulsed electrochemical micromachining of stainless steel has been presented. The equipment and the conditions for the process have been described. In order to find the optimum parameters for the process, the most important variables for the performance of the process have been taken into account. These variables were confinement, surface roughness, material removal rate and efficiency. Observing the results of the experiments, it can be stated that surface roughness increases with the pulse on-time of the voltage signal, whereas the confinement is better when the pulse on-time is lower. The passivation phenomenon takes place at voltage amplitude values lower than 12 V and disappears at higher voltages. The material removal rate is higher when both voltage amplitude and pulse on-time grow. The efficiency of the process is an important variable which increases with voltage amplitude and pulse on-time. Nevertheless, these variables must not be chosen beyond the limits of acceptable surface roughness and confinement. These limits have been set at 16 V and 80 ns, respectively, in this study.

Author details

Pablo Rodríguez*, Daniel Hidalgo and Julio Eduardo Labarga
Department of Manufacturing Engineering, Industrial Engineering Faculty, University of Leon, León, Spain

*Address all correspondence to: pablo.rodriguez@unileon.es

References

[1] Brousseau EB, Dimov SS, Pham DT. Some recent advances in multi-material micro- and nano-manufacturing. International Journal of Advanced Manufacturing Technology. 2010;**47**:161-180. DOI: 10.1007/s00170-009-2214-5

[2] Rathod V, Doloi B, Bhattacharyya B. Fabrication of microgrooves with varied cross-sections by electrochemical micromachining. International Journal of Advanced Manufacturing Technology. 2017;**92**:505-518. DOI: 10.1007/s00170-017-0167-7

[3] Schultze JW, Bressel A. Principles of electrochemical micro- and nano-system technologies. Electrochimica Acta. 2001;**47**:3-21. DOI: 10.1016/S0013-4686(01)00584-9

[4] Dunkel K, Bauer H-D, Ehrfeld W, Hoßfeld J, Weber L, Hörcher G, et al. Injection-moulded fibre ribbon connectors for parallel optical links fabricated by the LIGA technique. Journal of Micromechanics and Microengineering. 1998;**8**(4):301-306

[5] Datta M, Landolt D. Fundamental aspects and applications of electrochemical microfabrication. Electrochimica Acta. 2000;**45**:2535-2558. DOI: 10.1016/S0013-4686(00)00350-9

[6] Datta M, Shenoy RV, Romankiw LT. Recent advances in the study of electrochemical micromachining. Journal of Industrial and Engineering - Transactions of the ASME. 1996;**118**:29-36

[7] Saxena KK, Qian J, Reynaerts D. A review on process capabilities of electrochemical micromachining and its hybrid variants. International Journal of Machine Tools and Manufacture. 2018;**127**:28-56. DOI: 10.1016/j.ijmachtools.2018.01.004

[8] Hotoiu EL, Van Damme S, Albu C, Deconinck D, Demeter JD. Simulation of nano-second pulsed phenomena in electrochemical micromachining processes-effects of the signal and double layer properties. Electrochimica Acta. 2013;**93**:8-16. DOI: 10.1016/j.electacta.2013.01.093

[9] Bard AJ, Faulkner LR. Electrochemical Methods: Fundamentals and Applications. New York: John Wiley & Sons, Inc; 2001. DOI: 10.1146/annurev.matsci.30.1.117

[10] Patel DS, Sharma V, Jain VK, Ramkumar J. Reducing overcut in electrochemical micromachining process by altering the energy of voltage pulse using sinusoidal and triangular waveform. International Journal of Machine Tools and Manufacture. 2020;**151**:103526. DOI: 10.1016/j.ijmachtools.2020.103526

[11] Kock M, Kirchner V, Schuster R. Electrochemical micromachining with ultrashort voltage pulses-a versatile method with lithographical precision. Electrochimica Acta. 2003;**48**:3213-3219. DOI: 10.1016/S0013-4686(03)00374-8

[12] Labarga JE, Bastidas JM, Feliu S. A contribution to the study on electropolishing of mild steel and aluminium using alternating current. Electrochimica Acta. 1991;**36**:93-95

[13] Anasane SS, Bhattacharyya B. Experimental investigation into fabrication of microfeatures on titanium by electrochemical micromachining. Advanced Manufacturing. 2016;**4**:167-177. DOI: 10.1007/s40436-016-0145-6

[14] Sueptitz R, Dunne P, Tschulik K, Uhlemann M, Eckert J, Gebert A. Electrochemical micromachining of passive electrodes. Electrochimica Acta. 2013;**109**:562-569. DOI: 10.1016/j.electacta.2013.07.139

[15] Schuster R, Kirchner V, Allongue P, Ertl G. Electrochemical micromachining. Science. 2000;**289**:98-101. DOI: 10.1126/science.289.5476.98

[16] Bhattacharyya B. Electrochemical Micromachining for Nanofabrication, MEMS and Nanotechnology. Oxford, UK: Elsevier; 2015

[17] Yong L, Yunfei Z, Guang Y, Liangqiang P. Localized electrochemical micromachining with gap control. Sensors and Actuators, A: Physical. 2003;**108**:144-148. DOI: 10.1016/S0924-4247(03)00371-6

[18] Ahn SH, Ryu SH, Choi DK, Chu CN. Electro-chemical micro drilling using ultra short pulses. Precision Engineering. 2004;**28**:129-134. DOI: 10.1016/j.precisioneng.2003.07.004

[19] Kim BH, Na CW, Lee YS, Choi DK, Chu CN. Micro electrochemical machining of 3D micro structure using dilute sulfuric acid. CIRP Annals - Manufacturing Technology. 2005;**54**:191-194. DOI: 10.1016/S0007-8506(07)60081-X

[20] Mingcheng G, Yongbin Z, Lingchao M. Electrochemical micromachining of square holes in stainless steel in H2SO4. International Journal of Electrochemical Science. 2019;**14**:414-426. DOI: 10.20964/2019.01.40

[21] Xu L, Ning J, Zhao C. Electrochemical micromachining based on time constant control. Mechanical Systems and Signal Processing. 2020;**145**:106920. DOI: 10.1016/j.ymssp.2020.106920

[22] Davim JP. Machining. Fundamentals and Recent Advances. New York, London: Springer; 2008

Chapter 15

RF MEMS Switch Fabrication and Packaging

Lakshmi Swaminathan

Abstract

RF (Radio Frequency) MEMS (Micro Electro Mechanical Systems) technology is the application of micromachined mechanical structures, controlled by electrical signals and interacting with signals in the RF range. The applications of these devices range from switching networks for satellite communication systems to high performance resonators and tuners. RF MEMS switches are the first and foremost MEMS devices designed for RF technology. A specialized method for fabricating microsturctures called surface micromachining process is used for fabricating the RF MEMS switches. Die level packaging using available surface mount style RF packages. The packaging process involved the design of RF feed throughs on the Alumina substrates to the die attachment, wire bonding and hermetic sealing using low temperature processes.

Keywords: RF MEMS switches, surface micromachining, low temperature packaging

1. Introduction

Due to the reduced size, cost and low power consumption as well as very high precision, MEMS applications have extended from mere pressure and temperature sensors to vast array of applications viz., Aerospace, Automobile, Biotechnology, Consumer products, Defense and the most important and pertinent Telecommunications [1]. Hence RF MEMS devices have the advantage of increased functionality, substantial performance improvements, high agility, modularity and reconfigurability [2]. These devices are applicable to high performance communication systems such as satellite communication and m applications [3].

RF MEMS switches are the first and foremost MEMS devices designed for RF technology. RF MEMS switches compared to their semiconductor counterparts such as FET and PIN diodes show far superior performance. The current–voltage non-linearity that is the bane of semiconductor devices is non-existent in the case of RF MEMS switches. The power consumed by these switches is far less since most of the switches using electrostatic and piezoelectric actuation require negligible power requirements. They are also not plagued by issues of harmonics and intermodulation of signals. They exhibit very low insertion loss in the range − 0.05 to −0.2 dB at a frequency of 40 GHz. They also possess very high isolation in the range of −40 dB at 40 GHz [4, 5]. The only drawback is that their switching speed is far inferior compared to their semiconductor counterparts. However, there are several high performance communication circuits such as in defense and satellite systems where speed may not be the criteria whereas low power consumption and high RF performance would be the key features required. Due to these features they improve the overall performance of the systems into which they are integrated. Hence, the focus of this work is on RF MEMS switches which are a superior alternative to existing semiconductor switches.

MEMS devices are fabricated by the use of special techniques called micromachining. Micro fabrication or micromachining or micro manufacturing is the use of a set of manufacturing tools based on thin and thick film fabrication techniques commonly used in the electronics industry. It is also a technology for creating small three dimensional structures with dimensions ranging from sub centimeters to sub micrometers. A vast majority of MEMS structures are fabricated using bulk micromachining process. This involves etching of bulk wafer leading to three dimensional structures such as beams, cantilevers and cavities. These processes can be realized on substrates such as Silicon, Glass and Gallium Arsenide etc. The thickness of the structures can range from a few micrometers to 200 mm. The resulting dimensions of microstructures are much larger compared to surface micromachining process. Surface micromachining is a process based on building up of material layers and then selectively retaining or etching by continued processing. The bulk of the substrate remains untouched. LIGA processes combine IC lithography and electroplating and molding to obtain depth. Patterns are created in a substrate and then electroplated to create 3D molds. These molds can be used as the final product, or various materials can be injected into them. This process has two advantages. Materials other than Silicon can be used e.g. metal, plastic and devices with very high aspect ratios can be built [6].

This chapter provides the complete details of the unit step processes used for the fabrication and packaging of RF MEMS switches. The focus is on fabrication of low actuation voltage RF MEMS switches [7–10]. There are several challenges involved in the fabrication of MEMS switches such as, structural deformation, residual stress, non-release of structural layer to name a few. These challenges are overcome and addressed throughout the fabrication process by optimization of several unit processes. The unit processes used is discussed in each section of this chapter.

1.1 Fabrication process steps

Surface micromachining process is used for fabricating the switches. In the present work, fabrication costs were brought down by

- low resistivity Silicon wafers as substrate
- Use of only four masks for fabrication [11]

The sections below give the detailed description of the fabrication steps followed for successful fabrication of RF MEMS shunt switches.

The test wafers used in this work is P-type {100} low resistivity 4″ wafers with resistance ranging from 1 to 100 Ω. Using low resistivity wafers to fabricate RF MEMS switches has the advantage that integration with CMOS circuits is easier. However, use of low resistivity Silicon wafer leads to higher insertion loss due to inherent parasitics.

The following are the process steps used for fabrication:

i. Cleaning of test wafer: Using RCA-1 and RCA-2 processes.

ii. Oxidation of the test wafer: Using wet oxide process

iii. CPW metal layer patterning: Using sputtering and lithography steps

iv. Dielectric deposition and layer patterning: Using PECVD for Silicon Nitride depostion followed by lithography steps.

v. Sacrificial layer deposition and patterning: Using Photoresists and lithography steps

vi. Top layer deposition and patterning: Using sputtering and lithography steps.

vii. Top layer release: Using Critical point dryer.

Figure 1 gives pictorial representation of the process steps followed for fabrication of the RF MEMS shunt switches.

1.1.1 Cleaning of test wafer

The cleaning of the Silicon wafer is the first process employed to removing any organic residue or films on the Silicon wafers. The cleaning process is performed in two parts [12]. The first part of the cleaning process is the famous RCA-1 named after the laboratory at which it was developed. In this process five parts of water is mixed with one part of Ammonium Hydroxide (NH_4OH) and one part of Hydrogen Peroxide (H_2O_2). This mixture is then heated to 75°C on a hot plate. Once the solution bubbles vigorously the Silicon wafer is soaked in this solution for 15 minutes. The wafer is then dipped in a solution made of one part of Hydrofluoric acid (HF) and 50 parts of water for 30 seconds. This solution serves the purpose of etching out the thin oxide layer developed on the wafer. The wafer is again washed with DI water. The next step also called RCA-2 involves the use of Hydrochloric (HCl) acid, Hydrogen Peroxide (H_2O_2) and DI water in the ratio of 1:1:6. This solution is then heated to a temperature of 75°C for 15 minutes after which the Silicon wafer is placed in this solution. RCA-2 completely removes the traced of ionic contaminants from the wafer surface.

1.1.2 Oxidation of test wafer

The oxidation of Silicon wafer leads to the formation of a layer of native oxide i.e., Silicon Dioxide on the wafer surface. It is seen that only Silicon material has the

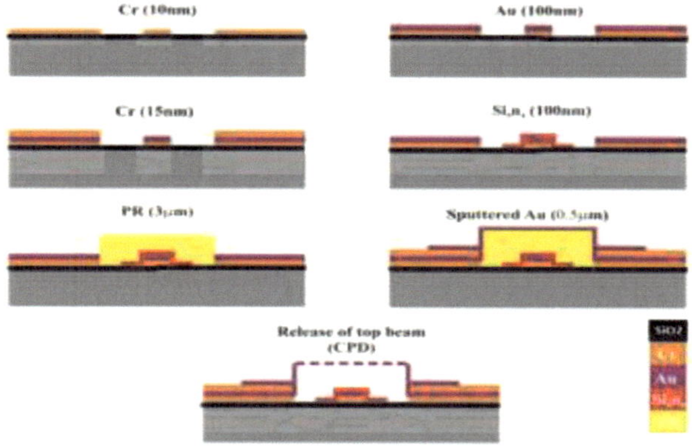

Figure 1.
Steps involved in fabrication of capacitive shunt switches.

ability to form a native oxide which has led to its wide usage in the IC industry. This layer serves a number of purposes. It acts as a surface passivation layer by protecting the surface from moisture and other atmospheric contaminants.

The main aim of using Silicon dioxide for RF MEMS switches is for the need for isolation and insulation from the low resistivity silicon wafer used as the substrate. By using Silicon Dioxide it is seen that the parasitics between the Co-Planar Wave-guide (CPW) layer and the silicon substrate underneath are drastically reduced. This approach leads the application of silicon substrate for RF circuits and wireless communication systems [13–16]. The formation of oxide layer in this work is through the wet oxidation process since the requirement is only for passivation.

The wafer was placed in a Nano pyrogenic furnace as shown in **Figure 2** to obtain a Silicon Dioxide layer of 1 μm thickness. The following steps were followed to oxidize the wafers. The time required for the Silicon Dioxide thickness of 1 μm was calculated to be approximately 4 hours, 30 minutes.

1. The furnace temperature is ramped to 500°C with Nitrogen gas flow at 5 liters/min. The furnace temperature is then raised to a temperature of 1100°C. This process of heating up takes 1–2 hours.

2. Once the set point temperature is reached, the wafers are put into a Quartz boat and loaded into the tube utilizing a furnace loader.

3. During the heating up process, pure oxygen and hydrogen flows through the water bubbler for 4 hrs and 30 minutes resulting in gas saturation with water vapor.

4. The wafers were then annealed using Nitrogen gas with the gas allowed to flow at 5 litre/min for 10 minutes.

5. The wafers are then cooled for ten minutes and checked for oxide thickness.

The thickness of the oxide layer was measured using an ellipsometer and was found to be around 1.063 μm.

1.1.3 CPW metal layer patterning

The proposed RF MEMS capacitive shunt switches have been integrated with a CPW line. The fabrication of CPW lines is easily integratable with the fabrication

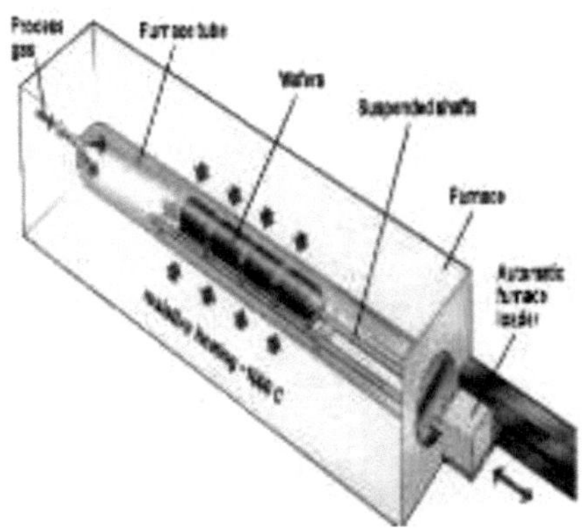

Figure 2.
Details of oxidation furnace at CeNSe, IISc.

steps required for the RF MEMS switch, which justifies the choosing of CPW lines over microstrip lines. This section gives fabrication steps for the CPW layer formation on the Silicon wafer.

a. **Sputtering of Gold layer**: The sputtering of gold layer depends on various parameters such as temperature, target distance, deposition pressure and Argon flow rate [17]. TECPORT sputter coater is used for obtaining the Chrome/Gold layer as shown in **Figure 3**. The process parameters of the sputter coater were set at a base pressure of $5x10^{-6}$ Torr, deposition pressure of $6.5x10^{-3}$ Torr, target to substrate distance set at 7.5 cm, with the Argon flow rate at 250 Scc/m. A seed layer of 10 nm is sputtered using a DC power of 100 W, a pre-sputtering time of 600 seconds and a deposition time of 22 seconds. For Gold DC Power was set at 25 W with a pre sputtering time of 30 seconds followed by a deposition time of 220 seconds with the deposition rate at 5 ^{0}A/sec. This was followed by Chrome sputtering to form a layer of 15 nm thickness. This process step would ensure good adhesion of the anchors of the top Gold beam with the bottom layer.

b. **Lithography for CPW layer**: The first photolithography step is used to pattern the CPW lines. A positive Photoresist (PR) AZ5214E is spin coated at speed of 4000 rpm using the spin coater for 40 seconds. It is then soft baked at 110°C for 1 minute. The wafer is then loaded into the EVG Mask aligner for PR exposure as shown in **Figure 4**. The proximity of the mask aligner is set at 30 μm and the energy for UV rays is set at 15 mJ. The mask used for this layer is as shown in **Figure 4**. The wafer is then post baked at 110°C for 1 minute and flood exposed using 75 mJ. The wafer is then immersed in the developer MF 26 A for around 20–30 seconds. The wafer is then subjected to a hard bake at 110°C for 3 minutes. The wafer is then inspected under the microscope to ascertain that the PR has developed.

c. **Gold/Chromium etch:** The etching of Gold (Au)/Chromium (Cr) is achieved by Potassium Iodide and Iodine (KI/I_2) solution in a ratio of KI: I_2: H_2O = 4 g: 1 g: 40 ml. At room temperature etch rate is approximately 1 μm/min for Chrome/Gold. For the Cr/Au/Cr thicknesses of 10 nm/100 nm/15 nm respectively the time is set to 10 to 20 sec for Cr etch, 60 to 120 sec for Au

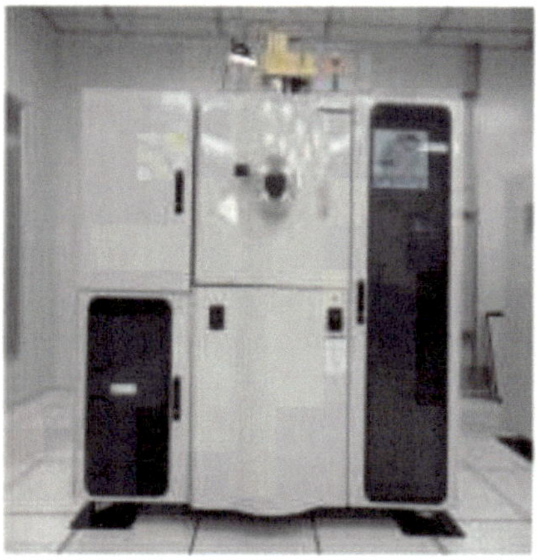

Figure 3.
TECPORT sputter coater.

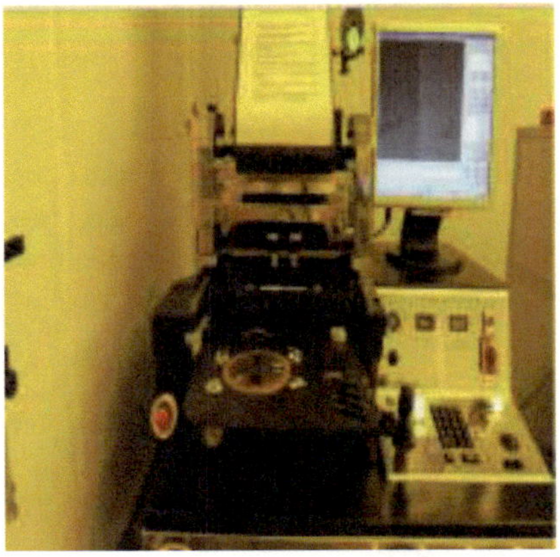

Figure 4.
EVG mask aligner at CeNSE, IISc.

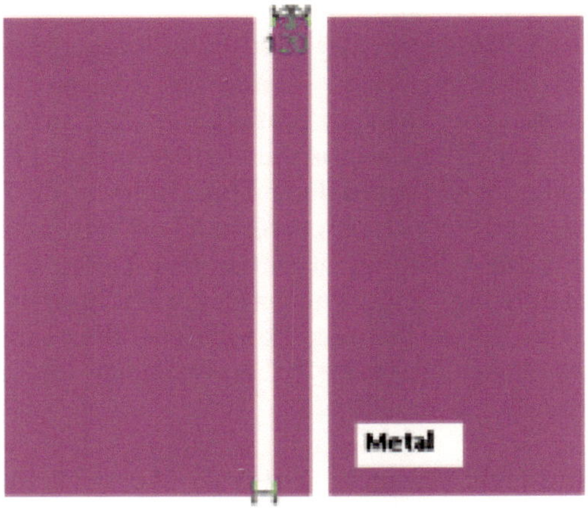

Figure 5.
Mask 1 for CPW layer patterning.

etch and 10 to 20 sec for Cr etch. **Figures 5** and **6** represent the mask for patterning and the resulting CPW layer respectively.

1.1.4 Dielectric layer deposition and patterning

The following process steps were followed for the deposition and patterning of dielectric Silicon Nitride (Si_3N_4) on the central signal line of the CPW layer.

a. **Deposition of Si_3N_4:** This layer provides the dc isolation between the signal line and the ground line when the switch is actuated to the down-state position. A thinner layer of Si_3N_4 will result in a higher capacitance in the downstate but would lead to pinhole problems which occur in thin dielectric layers. Also, the thin dielectric layer must be able to withstand the actuation voltage without breakdown.

Oxford Instruments Plasma technology Plasma Enhanced Chemical Vapor Deposition (PECVD) system is used for deposition of Si_3N_4 as shown in **Figure** 7. PECVD is a process by which thin films are deposited from the conversion of

Figure 6.
Optical microscope image.

Figure 7.
Oxford PECVD for Si_3N_4 deposition at CeNSE.

gaseous materials into solid state, due to a chemical reaction occurring in the presence of plasma. PECVD uses electrical energy to generate the plasma. Due to the presence of plasma, the gas mixture is transformed into highly reactive ions and molecules, which leads to low temperature requirements as compared to CVD processes. PECVD processes results in high quality films which have good adhesion, uniformity and good step coverage [18].

Silane (SiH_4) is usually supplied along with an inert gas like Nitrogen, Argon or Helium. Silane reacts with Ammonia (NH_3) to produce Si_3N_4 and a by-product Hydrogen. This reaction is as depicted by the chemical reaction as given below.

$$3SiH_4 + 4\,NH_3 \rightarrow Si_3N_4 + 12\,H_2$$

b. **Lithography for Si_3N_4:** The patterning of Si_3N_4 is achieved by first depositing a positive photoresist AZ4562 by placing it on a spin coater. The spin coater rotates at 4000 rpm for 40 sec. After soft baking at 110°C for 1 minute, the PR is exposed to UV rays through a mask aligner at proximity of

30 μm and energy of 110 mJ. The PR is then developed using the developer AZ 351B for 45–60 seconds. Next, the wafer is hard baked on an oven at 110°C for 3 minutes.

c. **Etching of Si_3N_4:** The etching of Si_3N_4 is performed using a dry etch process called Reactive Ion Etch (RIE). Reactive Ion etching is a process wherein the reactive species react with the material to be etched only when the surfaces of the material are activated by the collision of incident ions from the plasma. The etching characteristics like etch rate, etch profile, etch uniformity, etch selectivity can be controlled very precisely by selecting the right combination of recipes of chamber pressure, flow rate of gases, applied RF power and electrode bias. The etch rates are slow typically about 10 nm/ min up to 50 nm/min.

The RIE-F equipment used at CeNSe, IISc is as shown in **Figure 8**. For etching of Si_3N_4 the chamber pressure is set at 10 mTorr, RF power at 50 W with the main power at 2000 W. The flow rate of Sulfur Hexa Flouride (SF_6) is set at 45 scc/m with the temperature at 5°C. For etching out 100 nm of Si_3N_4 the required time was 12 seconds. The mask used for the patterning of the Si_3N_4 layer is as shown in **Figure 9**.

d. **Photoresist strip:** This is followed by the wet etching of the photoresist by dipping the wafer in acetone for 5 minutes followed by immediate cleaning

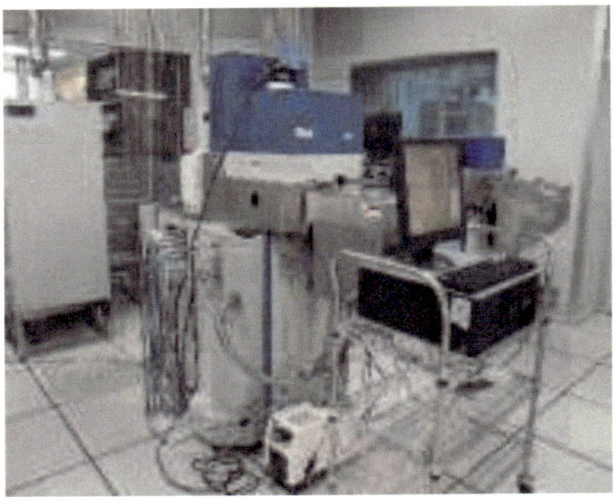

Figure 8.
RIE F CeNSe, IISc.

Figure 9.
Mask 2 for silicon nitride.

using Isopropyl Alchohol (IPA). This is to prevent the re-deposition of stripped photoresist on the substrate since Acetone has high vapor pressure. This is followed by cleaning with Ultrsonicate Acetone for 3 minutes. **Figure 10** shows the patterned silicon nitride layer.

1.1.5 Sacrificial layer deposition and patterning

The sacrificial layer is the layer which will be etched out to release the top metal layer. The topography and planarity of the top membrane is defined by the sacrificial layer planarity. Several materials like metals, dielectrics and photoresists have been used as the sacrificial layer. The choice of the sacrificial layer is based on the processing steps that follow the deposition of this layer, the temperature range, the required planarity and profile of surface. Here, a positive Photoresist (PR) S1813 is used as the sacrificial layer. This PR has to be deposited with utmost accuracy in order to define the gap between the top electrode and bottom electrode of the RF MEMS switch. The complete process of sacrificial layer deposition and patterning can be explained by the following steps:

Sacrificial layer Optimization: The PR S1813 is a positive photoresist which has excellent adhesion, excellent coating uniformity with effective broadband exposure. This PR is used for a wide variety of process flow requirements such as lift-off, dry etch, wet etch, the thickness of the PR to name a few. The plot in **Figure 11** gives the resist thickness versus spin for the Shipley family of PRs. Thick PR layers can be achieved in one step, however they have the disadvantage of being non-uniform over the wafer surface. In order to achieve uniform and thick PR coating, the coating process is performed in three steps. In the first step, the spin coater is run at low speeds of 500 rpm for 30 sec. This low spin speed and reduced spin time

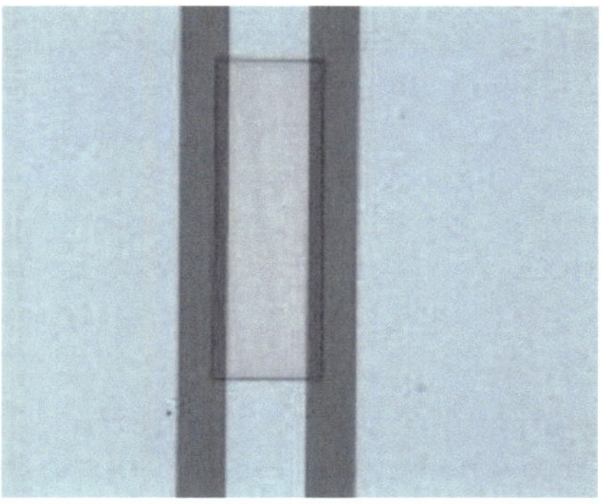

Figure 10.
Optical microscope image of silicon nitride layer formed.

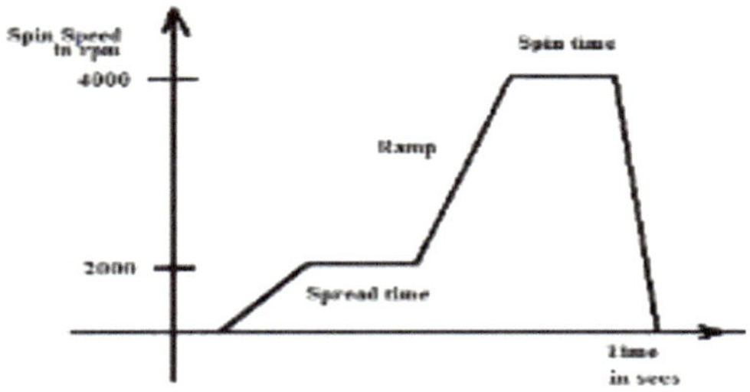

Figure 11.
PR deposition using multiple step method.

will result in uniform coating of thick resist on the wafer. In the second step the speed is ramped upto 1000 rpm within a time of 30 sec. A solid film of the photoresist is formed with the complete evaporation of the solvent. This step decides the thickness and uniformity of the photoresist. The third step consists of the spin coater speed set at 2000 rpm for 40 sec. This last step ensures that any leftover solvent is completely evaporated. The complete cycle of spin coating is as shown in **Figure 11**. Using a Dektak optical profiler the thickness of this layer was confirmed to be 3 μm.

Sacrificial layer patterning: The patterning of the sacrificial layer photoresist is processes by first depositing one more layer of positive PR S1813 on this layer. This was achieved by the spin coater speed set to 500 rpm for 30 seconds, followed by a ramp up of 1000 rpm for 30 sec and 200 rpm for 40 sec. After soft baking the PR is exposed to UV rays through a mask aligner at a proximity of 30 μm and energy of 75 mJ. The mask used for generating the pattern for this layer is as shown in **Figure 12**.

The PR is then developed using the developer AZ 351B for 30–60 seconds. Next, the wafer is hard baked on an oven at 90°C for 30 minutes. The PR layer thickness shrunk from 3 μm to 2.09 μm after development and baking.

1.1.6 Top layer deposition and patterning

The top layer or beam formation defines the performance of the RF MEMS switch. The top layer designs were simulated using Coventorware™. These designs have been chosen due to their lower pull-in voltages. Gold is the choice for the top layer due to its favorable characteristics such as, its high conductivity, non-tarnishing property, high Young's Modulus and compatibility with micromachining processes. The top metal layer deposition and patterning is described in the following sections.

a. **Gold layer deposition:** The deposition of this layer was carried out using the TECPORT sputtering equipment. It may be recalled that the bottom layer has the composition of Cr/Au/Cr. This composition would lead to excellent adhesion of the top layer anchors with the previously deposited Chrome layer. Several Iterations were carried out in order to sputter the top Gold layer without residual stress. Several parameters such as temperature, rate of deposition were optimized in order to arrive at top layers without buckling after release process.

Finally, with the optimized parameters setting temperature and rate of deposition a stress free top layer was arrived at. The stress free top layer is of critical importance for reduction in actuation voltage. The process parameters of the sputter coater were set at a base pressure of $5x10^{-6}$ Torr, deposition

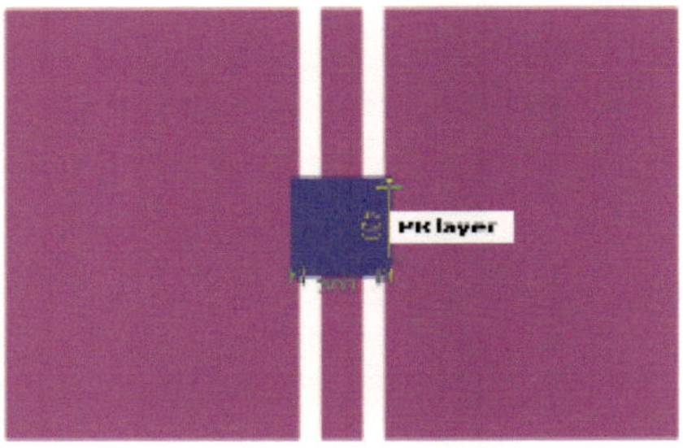

Figure 12.
Mask 3 for PR layer.

pressure of $6.5x10^{-3}$ Torr, target to substrate distance set at 7.5 cm, with the Argon flow rate at 250 sccm. The DC Power was set at 25 W with a pre sputtering time of 30 seconds followed by a deposition time of 1100 seconds with the deposition rate set at 5^0A/sec.

b. **Gold layer patterning:** The four switch designs chosen for the top Gold layer are shown as four respective masks in **Figure 13**. The lithography involved the use of AZ5412E positive PR. This was spin coated at 4000 rpm for 40 sec. The wafer was then soft baked aligner at a proximity of 10 μm and energy of 50 mJ The PR is then developed using the developer MF 26A with the wafer dipped in the developer of 20–140 sec. Next, the wafer is hard baked on an oven at 90°C for 30 minutes. For Gold etch, freshly made Potassium Iodide and Iodine (KI/I_2) solution in a ratio of $KI:I_2:H2O$ = 4 g:1 g:40 ml is used. The unwanted Chrome deposition on the bottom layer is also etched out using a Chrome etchant for 5 to 10 seconds.

1.1.7 Top layer release

The release of the top switch membrane is the most crucial step in the whole fabrication process. There are many methods to release the top layer without deformation and stiction. The first step in the top layer release is to etch the sacrificial layer. This could be achieved by using dry etching or wet etching. In wet etching, conventional liquid solvents are used to completely remove the sacrificial layer followed by drying. The drying could be through the process of air drying or through critical point drying.

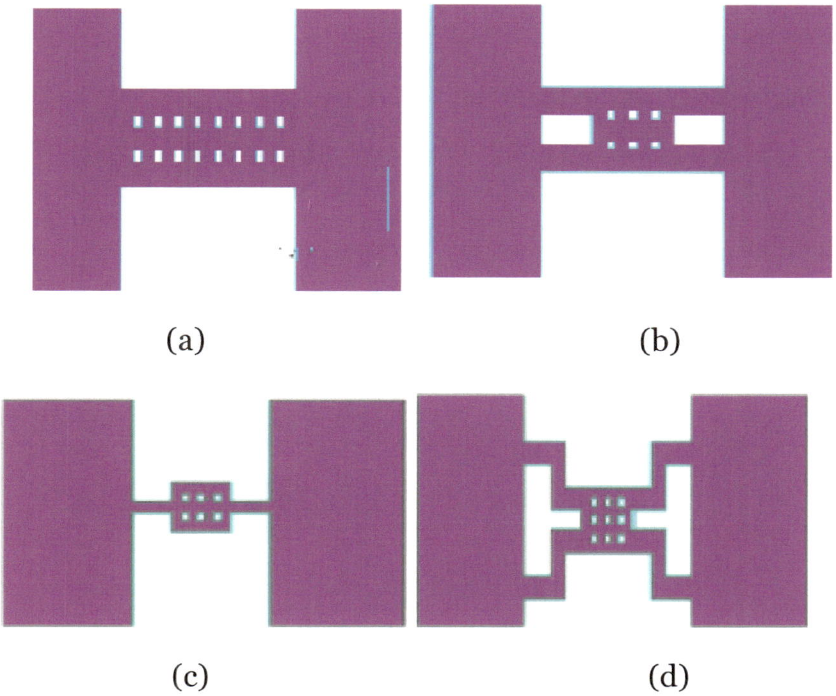

Figure 13.
Four top layer designs for RF MEMS switch. (a) Fixed-fixed beam switch. (b) Fixed-fixed Flexure switch. (c) Fixed-Fixed Single Flexure switch. (d) Crab leg Flexure switch.

Critical Point Drying (CPD) was found to be the best method for MEMS devices [19]. In this work the wet etch was followed by CPD to release the top layer. PR layer first stripped by using Piranha solution. The Piranha solution is prepared by mixing Sulfuric Acid and Hydrogen Peroxide in the ratio of 3:1. This is an extremely strong oxidizing agent which removes organic residues and especially PRs from the substrate.

Critical point drying.

There was the requirement of a drying technique wherein surface tension could be reduced to zero and a continuity of state of the liquid could be obtained. It was found that if the temperature of the liquefied gas is increased the resulting pattern of the meniscus is flat indicating a reduction in surface tension. This results a very low surface area of the liquid which in turns leads to the evaporation of the liquid. This is called the critical point of the liquid. The critical phenomena can be utilized as a drying technique as it achieves a phase change from liquid to dry gas without the effects of surface tension and is therefore suitable for delicate biological specimens. MEMS devices. Of all the gases that were tested for the critical point, Carbon Dioxide (CO_2) remains the most common medium for the CPD procedure and is termed the 'Transitional Fluid'. However, CO_2 is not miscible with water and therefore water has to be replaced in the specimen with another fluid which is miscible with CO2, this is termed the 'Intermediate Fluid'. IPA is solvable in CO_2 and hence most of the MEMS devices are place in this liquid for CPD process.

The critical point dryer used in this work was the Tousimis Samdri® line of Supercritical Point Drying machine as shown in **Figure 14**. The wafer after the Piranha dip was placed with great care in a petri dish containing IPA. This was then carefully transferred to the CPD equipment. Once the release cycle was finished, the Switches were inspected under a microscope and then using Scanning Electron Microscope (SEM) and were found to be free of residual stress on the top beam. Also, the gap between the top membrane and the bottom electrode was clearly visible without any PR residues.

1.2 Packaging of RF MEMS shunt capacitive switches

The main objectives of packaging of MEMS devices are to protect the actual functioning of the device from external environmental influences like chemicals, temperature, electromagnetic influences. The packaging forms a foundation on which the actual device is mounted thus giving much needed mechanical support. Packaging also helps in routing of interconnections of the chip with the outside world.

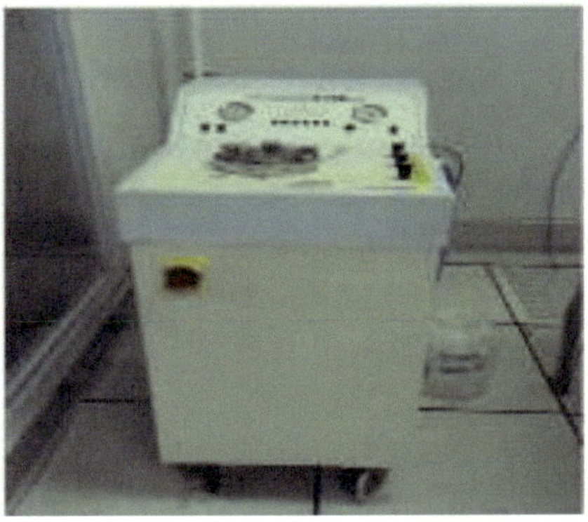

Figure 14.
Tousimis Samdri critical point dryer at CeNSE, IISc, Bangalore.

The most critical factor for the successful commercialization of micro level devices is packaging. With the maturity gained in IC (integrated circuits) fabrication over the past many years, the packaging of ICs also has gained great maturity and sophistication. The same cannot be said about MEMS packaging. Although some of the advancements of IC packaging can be applied to meet the requirements of MEMS devices, some specialized techniques are required for MEMS packaging. Packaging of MEMS devices is much more complex and expensive than conventional IC packaging. This is because MEMS devices usually consist of three dimensional structures with free movement. This leads to the requirement of encapsulated cavities. Microsystem packaging also involves, bonding, interconnecting, and assembly of micro scale component to form a microsystem product. Packaging is the last and crucial step in the lifecycle of MEMS devices and may cost anywhere between 20–90% of the total device cost. Important functions of packaging are listed below:

- Mechanical reinforcement and ruggedness

- Environment invulnerability against temperature, electromagnetic aberrations, chemical reactions

- Interfacing with outside world

- Hermetic sealing

- Assimilation of multiple chips to form a multifunctional system

In the case of MEMS devices the requirement of hermetic sealing may vary from device to device since some of the MEMS devices need an exposure to the environment in which they work and some other devices do not. It is also necessary to note that the packaging needs are special and case specific due to the micro mechanical structures. MEMS packaging involves key design and packaging considerations such as wafer thickness, wafer dicing, thermal issues, stress effects, isolation, protective coatings and hermetic sealing.

The packaging for RF MEMS devices has to meet more stringent specifications due to the high frequency range of interest. Also, the demand is for high performance, low cost strategies which is usually a challenge. Furthermore, apart from the general MEMS packaging issues, the packaging of RF-MEMS devices has the following concerns.

1. Hermiticity of the packages should be ensured to provide high reliability RF MEMS devices since their operation depends on the ambient conditions under which they perform.

2. Interconnects, package substrates and passivation layers through the package to the outside world should offer low loss and low intermodulation.

3. Footprint of the total packaged device must be small, keeping with the requirement of miniaturization and high component densities especially for satellite and wireless communication systems.

The packaging of RF MEMS devices can be classified into two broad categories, one, wafer level packaging and the other, die level packaging. This work focuses on die level packaging hence the following paragraphs will focus on this.

1.2.1 Die level packaging

This is a type of packaging used for low volume requirements. Die level packaging is also called 1-level of packaging. The 1-level package usually consists of a pre-fabricated metal can/ceramic/plastic package with leads for connecting to the outside circuits or systems. These packages come with the base as well as the lid. For both ceramic as well as metal packages the cavity formation in the base of the package is an established method. The MEMS chip is attached to the base package using low temperature solder based epoxies and baked for removal of gaseous by products of the solder or epoxy. The next step involves the placement of the top cover over the base package in a vacuum or nitrogen atmosphere. Next, hermetic sealing is done along the package rim which is performed using localized heating.

This method of packaging is expensive and is suitable for telecommunication base stations, satellites and defense systems but not for high volume applications like mobile phone handsets. Furthermore, the additional costs are mainly due to the great care with which the MEMS chips are to be handled after their release. Furthermore, standard scribing procedures cannot be used for dicing the wafer into chips since there is a high possibility of introduction of contaminants on wafer surface. These contaminants cannot be removed by mere cleaning. This cleaning will furthermore require a critical point drying for every chip which would further escalate the costs. A generic 1-level packaging is as shown in **Figure 15**.

In this thesis the focus is on die level packaging using available surface mount style RF packages. However, the whole packaging process is performed under low temperature in order to free the MEMS structures of thermally induced stress which otherwise would affect the performance of the switch. The details of the packaging process starting from the design of RF feed throughs on the Alumina substrates to the die attachment, wire bonding and hermetic sealing are discussed in details in the following sections.

1.2.2 Packaging of RF MEMS shunt capacitive switches

The packaging of RF MEMS switches involves the following steps:

- Dicing of wafer
- Design of RF feed throughs on Alumina substrate
- Attachment of the base package to Alumina substrate

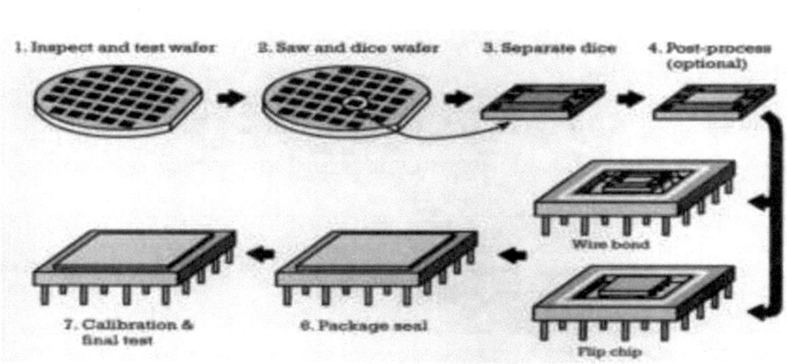

Figure 15.
Simplified one level RF MEMS packaging flow.

- Die bonding to package base
- Wire bonding
- Hermetic sealing

2. Dicing of wafer

Wafer dicing is the process by which the individual unit of dies are separated from the wafer. This process may be carried out by using mechanical sawing, scribing, breaking or laser cutting. Of the several issues and challenges of RF MEMS packaging, dicing is one of the foremost challenges. In the case of ICs, the resultant contaminants or debris due to the dicing process, on the surface of the die can be easily removed by a post-dicing cleaning process, however, in the case of MEMS devices the fragile mechanical structures on the die may get damaged by these contaminants. Dicing methods such as mechanical sawing, scribing and breaking lead to debris from the dicing process which may scatter on to the die leading to buckling or breaking of the delicate MEMS structures. Therefore, the choice of the dicing process is of utmost importance in the case of RF MEMS devices.

In order to obtain least residues from the dicing process, Chicago Laser System (CLS 960) Neodymium-doped Yttrium Aluminum Garnet (Nd:YAG) laser has been employed. This is shown in **Figure 16**. Nd:YAG lasers are one of the most common types of laser used in cutting and welding steel, semiconductors and various alloys. These lasers typically emit light with a wavelength of 1064 nm, in the infrared.

Figure 16.
Chicago laser system (CLS 960) laser dicer.

Figure 17.
Diced chips as seen under a microscope.

The wafer to be diced was mounted on the dicing platform with the alignment set. The wafer is then diced into unit chips or dies with high precision. The unit dies obtained were observed under a microscope. It was visually confirmed that the RF MEMS switches were undamaged. The samples of diced chip are as shown in **Figure 17**.

3. Design of RF feed throughs on Alumina substrate

There are several choices of substrates for packaging like Quartz, Silicon, Aluminum nitride (AlN) and Alumina (Al_2O_3) to name a few. Ceramic substrates such as Aluminum Nitride and Alumina are most commonly used packaging materials for MEMS. Alumina is the primary choice because it combines economic, physical and electrical advantages [20]. Also, Alumina is readily available in sizes that range from tiny chips to large ceramics in thicknesses from 0.25 mm to 1.5 mm and in a variety of shapes and designs. The finished substrate can be drilled or cut with diamond tools and lasers.

Some of the key properties of Alumina are as given below:

- Good thermal conductivity
- High strength and stiffness
- Resistance to strong acid and alkali attack at high temperatures
- Excellent size and shape capability
- Excellent dielectric properties from DC to GHz frequencies
- compatibility with thick film resistors and dielectrics
- Excellent adhesion with thick film conductors

Having chosen Alumina as the substrate material, the RF feedthroughs on the Alumina substrate had to be designed. The generic CPW line is as shown in **Figure 18**. The design of the CPW lines on the Alumina substrate was based on many parameters such as trace (S) and ground line (G) lengths, permittivity of Alumina (ϵ_r), material properties of conductor and the operating frequency. The designed layout for CPW lines on the Alumina substrate is as shown in **Figure 19**.

Silver Palladium paste (7474 Ag/Pd) is used to form the CPW conductors on the Alumina substrate. This paste is chosen for its excellent solderability and excellent

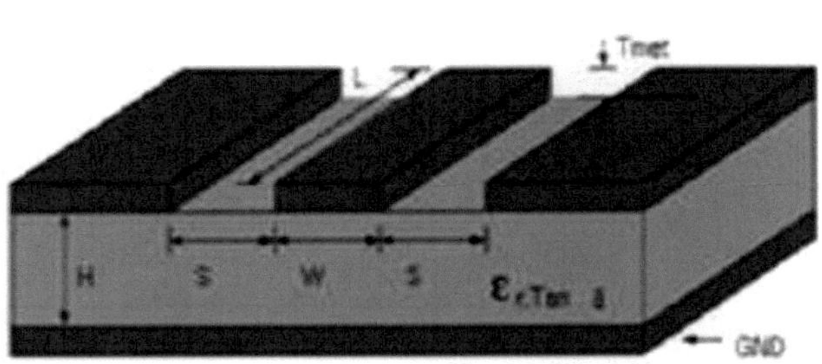

Figure 18.
Schematic of CPW line.

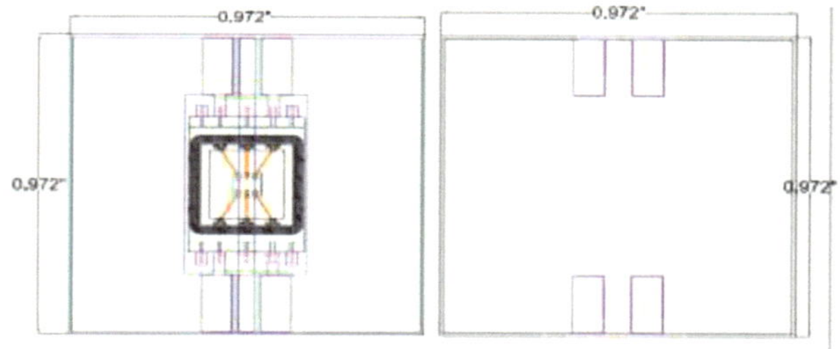

Figure 19.
CPW layout on alumina substrate. (1) Front view. (2) Back view.

aged adhesion on substrates like Alumina and it's comparatively low cost. The three steps involved in the formation of CPW lines on the Alumina substrate is as given below.

- Scribing
- Screen-Making
- Printing, Drying and Firing

i. **Scribing:** Scribing is basically designing on the substrate using laser. The blank substrate is first divided into a number of regions by scribing. The laser used in this process is a combination of Nitrogen, Carbon Dioxide (CO_2) and helium gases. At higher temperatures, the valence electrons combine to produce laser light.

The advantages of laser scribing are

1. High edging steepness
2. Small edge roughness
3. No micro cracks
4. Small thermal influence by optimized uv treatment
5. Contact free material processing
6. High precision and positional accuracy.

After the scribing process the plates are subjected to the de-burring process. De-burring is done to remove the ceramic particles that accumulate on the surface due to laser penetration. De-burring is done using another ceramic plate. The plates are cleaned in de-ionized water and then dried in an oven at 120°C.

ii. **Screen-Making:** This is the preliminary process for printing. Here, the stainless steel mesh is first stretched with hydraulic force. The frames are then attached to the mesh. The chromo difloro film, is first stuck on to the wet screen and dried in the oven after which they are exposed to UV light with the respective photo film layer. The film is developed using water. After screen making process is over, printing is performed on the Alumina substrate.

These are the following precautions to be observed in screening:

1. The screen should be free of foreign particles.

2. The screen tension should be within the specification of workmanship.

iii. **Printing:** In this process, the conductors are printed on the substrate. The conductor paste is (7474 –Palladium Silver) is screen printed. It is then dried at 150° C for 15–20 minutes in order to remove the solvents and then fired at 850° C in a fast firing furnace. At this temperature sintering takes place with a dwell time of 10 min and then ramp down takes place after which the paste starts behaving like a conductor. At the end of this process CPW lines on the Alumina substrate were formed.

4. Attachment of base package to alumina substrate

Surface mount packages are used for packaging the diced chips. The current packaging methodology proposes the use of surface mounted plastic packages supplied by Elecsys technologies, USA. These packages are suitable for DC to 18 GHz range which is also the frequency of interest of the RF MEMS switches. These packages also have their leads to be co-planar compatible. These packages have a conductive metal base attached to an Alumina ceramic ring frame with a cup shaped lid with a b-stage epoxy preform for sealing. **Figure 20** shows the layout details of the SMX series packgae used for packaging the RF MEMS shunt switches. **Figure 20(a)** shows the base of the package with leads made of Copper and Gold. **Figure 20(b)**shows the top view of the base package showing the jutting leads.

The attachment of Base Package to the substrate is achieved by using a non-conductive epoxy named 8700 K. This epoxy is a high thermal conductivity, low temperature curing, microelectronics grade adhesive. It is then cured at 150°C for 2 hours. In order to connect the base package pins to the CPW conductor lines, a conductive epoxy 84–1 LMINB1 is used. This conductive epoxy used, is a high purity silver filled die attach adhesive ideal for application by automatic dispenser.

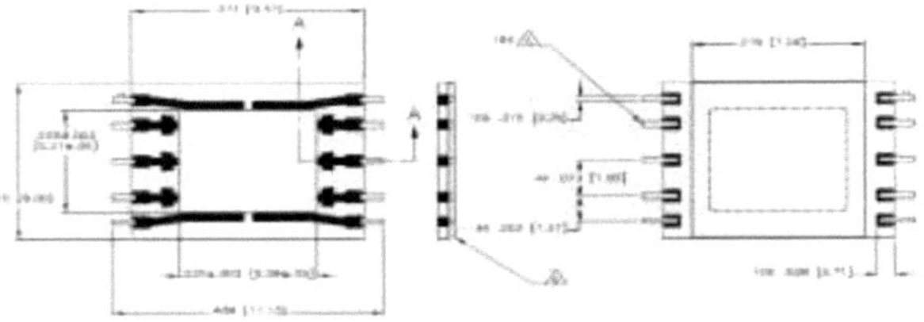

(a) (b)

Figure 20.
Layout details of the SMX series package. (a) inside view (b) top view.

5. Die bonding

Die bonding or die attach is one of the most crucial steps in the packaging process especially in the case of MEMS devices. This requires careful handling of the diced MEMS chip/die since the die contains fragile mechanical structures. The dies have to be picked from the wafer either using manual methods or by automatized grippers. They have to be then placed on the base package cavity. The choice of die bonding process depends upon package sealing strategy, operating conditions and environmental and reliability requirements. The die attach can be achieved through the following bonding methods:

1. eutectic bonding

2. solder attach

3. epoxies, silver filled glass or polymide

Eutectic bonding uses a die bonding technique with an intermediate metal layer (Au/Al) which would result in a eutectic system. The most important feature of this type of bonding is that the eutectic temperature can be much lower than the melting temperature of individual elements. Solder Attach is the most preferred type of die bonding since the solder provides for good thermal conductivity. But this type of die bond would lead to large amount of heat generation during the attachment process which may lead to a large thermal stress on the mechanical structure in the case of a MEMS device.

Epoxy bonding is achieved by attaching die to the substrate by using epoxy glue. A drop of the epoxy is first dispensed on the substrate and the die is placed on it. In order to cure the epoxy the substrate or package may need to be heated. Most commonly used adhesives are polyimide, epoxy and silver filled glass. Epoxy bonding has the following important features such as low curing temperature, used for wide range of die sizes and can be reworked easily [21, 22]. Epoxy is used for die attachment in this work.

The bare die is attached to the Base Package using non- conductive epoxy (H74 epoxy) and curing at room temperature of 25°C for 48 hours, keeping in mind the low temperature requirement for packaging in this work [23]. H74 epoxy is a thermally conductive epoxy designed for hybrid circuit assembly including die attach. The outstanding feature of this epoxy is that its curing process is fast even at low temperatures and also has a built in color change when the adhesive is cured. The adhesion of the dies is good and is confirmed by the non-destructive pull test (NDPT) and die shear test. It is passing the NDPT of greater than 16 grams and the die shear strength is greater than 6.55 kgs. The tested samples are as shown in the **Figure 21**.

Figure 21.
Die shear test.

6. Wire bonding

Wire bonding is a process by which interconnections are made between the die to the suitable location on the substrate or package. Wire bonding has the advantage of being low cost and flexible method of interconnection and is widely used to assemble majority of semiconductor packages. They also have the advantage that they can be used upto a frequency of 100 GHz if properly designed. Thus it is most suitable for RF MEMS switches.

Thermosonic bond is formed by the combination of three parameters, ultrasonic, thermal and mechanical force. A thermosonic bonding machine uses a piezoelectric transducer which converts the electrical energy to a vibratory/ultrasonic motion. This is in turn converted to an amplified oscillatory motion using a velocity transformer. This oscillatory motion is delivered to a heated bonding tip. The thermal energy and the ultrasonic motion together create a softening of the lead wire and hence its deformation leading to a required contact area using low temperature and low force.

Hence, in the proposed work thermosonic bonding has been chosen as the wire bonding technique due to its desirable properties of operation at low temperature and low force. A Kuilelle and Soffa thermosonic bonder is as shown in **Figure 22** which is used for the wire bonding process. Ball and wedge bonds of Gold wire of 2 mil are used for wire bonding between bare die to the package base pads as shown in **Figure 23(a)**. The NDPT test was also performed to ensure the strength of the wire bonds. The wire bonds were then covered with Epotek-301-2FL, a low stress adhesive especially used for glob top encapsulation over wire bonds. The curing for

Figure 22.
Kulicke and Soffa thermosonic ball bonder.

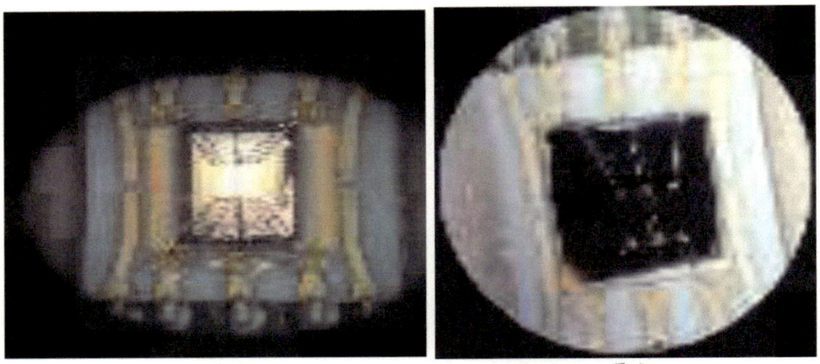

(a) (b)

Figure 23.
Wire bonding. Hermetic sealing and soldering of SMA. (a) Gold wire bonding. (b) Gel dispensed on bonding Wedge.

this adhesive was done at room temperature of 25°C for 72 hours. **Figure 23(b)** shows the adhesive covered wire bonds.

Hermetic seals are airtight seals that prevent the invasion of oxygen, moisture, humidity, and any outside contaminant to enter a sealed environment. This kind of a sealing is of utmost importance in semiconductor devices and MEMS devices. In the case of MEMS devices this is a top priority since the performance of a MEMS sensor or actuator directly depends on the ambient conditions under which they operate.

This work proposes the use of epoxy resins to seal the package lid to the base package. The top can/case is attached to the base package using a non-conductive epoxy H74 and cured at a room temperature of 25°C for 48 hours. The package with the top case attached is as shown in **Figure 24**. This method of curing was tested on samples in order to ascertain the complete curing. The specification sheets states that this epoxy requires a temperature of 150°C for 5 minutes and a temperature of 100°C for 20 minutes, for curing. However, curing at room temperature of 48 hours has led to the complete sealing. This was ascertained by performing a leak test on the packed RF MEMS switch. There are several types of leak tests to confirm the hermiticity of sealed packages. The Helium leak test was performed following a procedure as explained below. The vacuum method is the most sensitive leak detection technique. It requires that part of the package be placed under hard vacuum and the other part to be pressurized with helium. The side which is placed under vacuum is connected to the leak detector. If there is a leak, the helium that penetrates this side will be detected by the leak detector. The package under test passed the standard leak test with a test value of 5.0×10^{-8} std. atm.cc/sec.

In order to characterize the packaged MEMS switches Sub Miniature version A (SMA) connectors have been used. These connectors are designed to be used between DC and 18 GHz. The SMA's chosen for this work with part number 1367–000-G91P-35 were procured from Delta Electronics Manufacturing Corporation.

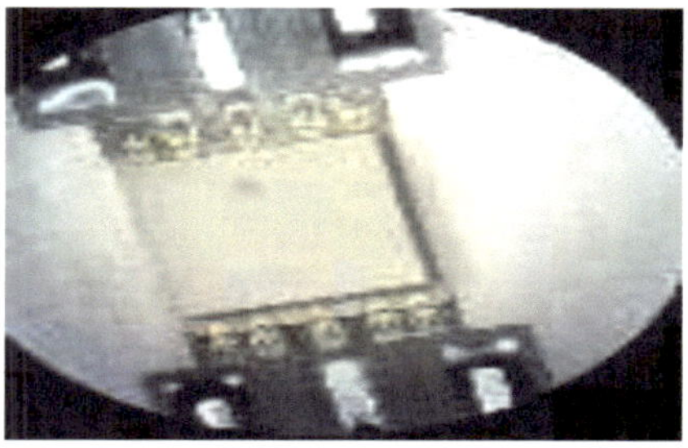

Figure 24.
Lid sealed package.

Figure 25.
Package with SMA connectors.

These connectors were soldered to the Alumina substrate using solder wire (Sn 63: Pb37) as shown in **Figure 25**.

7. Conclusions

The objective is to fabricate the simulated designs using low cost fabrication processes. Considering the ease of implementation and complications of the processes involved, it is focused to fabricate only the capacitive shunt switches using a low cost, low resistivity silicon wafer as the substrate and using only four masks for the whole process. Surface micromachining process was used to fabricate these switches. During the fabrication several challenges such as residual stress of top gold film, planarization of sacrificial layer, release of top beam were encountered and are overcome. Several rounds of optimization of unit process led to the successful fabrication of these switches. Packaging the fabricated switches, was done using low temperature methods to minimize the effect of packaging on the structure. The packaging of these switches used SMAs. The packaging involved several steps such as wafer dicing, conductor screen printing on substrate, die bonding, wire bonding and hermetic sealing. The switches are packaged and hermetically sealed by using a unique method of curing and sealing using room temperature methods, in order to avoid thermally induced stress in the fragile MEMS beams of the switches. The proposed packaging methodology has passed both the shear test and the hermeticity tests. By optimizing the fabrication process to cater to batch processing and also finding methods of CMOS compatible methods, this technology will help meet the ever growing demands of wired as well as wireless communication for low loss high performance RF switches.

Acknowledgements

My sincere thanks to Dr. Premila Manohar, Professor and Head, Department of Electronics and Instrumentation, Ramaiah Institute of Technology who has encouraged me throughout the research project towards its completion and implementation. My sincere thanks to Dr. N. Sayanu Pamidhighantam, who introduced me to the beautiful world of MEMS. His discussions on the subject were highly enlightening and thought provoking. I deeply acknowledge his guidance and advice throughout this endeavor of mine. I owe a lot to Dr. K. Natarajan, former Professor and Head, Department of Telecommunication, for support during the project based on my work which led to fabrication of my devices which I thought was a distant dream.

I think my research would have been only in the simulations stage if not for the project funded by National Program on Smart Materials and Structure (NPMASS) ADA, INDIA. This opportunity changed my perspective of my research since I was able to fabricate devices hands on in one of the best laboratories of the world.

The fabrication was carried out at the Centre for Nano Science and Engineering, IISc, Bangalore. I would like to express my deep sense of gratitude to Dr. K. N Bhat, Professor Emeritus, CeNSe, IISc, Bangalore for his time and valuable inputs during the fabrication of RF MEMS switches. With his repertoire of knowledge, he was able to guide us through difficult phases of fabrication. I also would like to thank the team at CeNSe for their co-operation throughout the project.

I owe a lot to my family who supported me in every way possible for the completion of my research work.

Author details

Lakshmi Swaminathan
Department of Electronics and Communication, M.S. Ramaiah Institute of Technology, Bangalore, Karnataka, India

*Address all correspondence to: lakshmi_ramesh@msrit.edu

References

[1] G. K. Ananthasuresh, K. J. Vinoy, S. Gopalakrishnan, K. N. Bhat, V. K. Aatre, "Micro and Smart Systems", Wiley India, 2010.

[2] Vijay K. Varadan, K.J. Vinoy, K.A. Jose "RF MEMS and Their Applications", John Wiley & Sons, 2003.

[3] Gabriel M. Rebiez " RF MEMS: Theory, Design and Technology" , John Wiley and Sons, 2003.

[4] Hyman, Daniel, Juan Lam, Brett Warneke, Adele Schmitz, T. Y. Hsu, Julia Brown, James Schaffne, "Surface-micromachined RF MEMs switches on GaAs substrates", International Journal of RF and Microwave Computer-Aided Engineering, Vol. 9, no. 4, pp.348–361, 1999.

[5] Kim, Che-Heung. "Mechanically coupled low-voltage electrostatic resistive RF multithrow switch." IEEE Transactions on Industrial Electronics, Vol.59, no. 2, pp. 1114–1122, 2012.

[6] Marc J. Madou ,"Fundamentals of Microfabrication: The Science of Miniaturization", Second Edition, CRC Press, 13-Mar-2002

[7] S. Pacheco, C. T. Nguyen, and L. P. B. Katehi, " Micromechanical electrostatic K-band switches", IEEE International Microwave Symposium Digest, MTT-S, Baltimore, USA, pp. 1569–1572, June 1998.

[8] Dimitrios Peroulis, S.P.Pacheco, K. Sarabandi,Linda P.B.Katehi.," Electromechanical Considerations in developing Low Voltage RF MEMS switches", IEEE Transactions on Microwave theory and techniques, Vol. 51 no.1, pp.259–70, Jan 2003.

[9] S. Pacheco, C. T. Nguyen, and L. P. B. Katehi," Design of low actuation voltage RF MEMS switch", International Microwave Symposium Digest, IEEE MTT-S, Boston, USA, Vol.1, pp. 165–168, 2000.

[10] D. Peroulis, S. Pacheco, and L. P. B. Katehi," MEMS devices for high isolation switching and tunable filtering", International Microwave Symposium Digest, Boston, MA, IEEE MTT-S, Vol.2, pp. 1217–1220, June 2000.

[11] Shekhar S, Vinoy KJ, Ananthasuresh GK,"Surface-Micromachined Capacitive RF Switches with Low Actuation Voltage and Steady Contact", Journal of Microelectromechanical Systems. Apr. 2017.

[12] W.Kern, Ed, Handbook of Semiconductor cleaning Technology, Noyes Publishing; Park Ridge, NJ, 1993, Ch1

[13] Jasbir, Sharma, Krishanapura Nagendra, and Das Gupta Amitava. "Fabrication of low pull-in voltage RF MEMS switches on glass substrate in recessed CPW configuration for V-band application" , Journal of Micromechanics and Microengineering, Vol.22, no.2, pp. 025001, 2012.

[14] Philippine MA, Zareie H, Sigmund O, Rebeiz GM, Kenny TW," Experimental validation of topology optimization for RF MEMS capacitive switch design", Journal of Microelectromechanical Systems. Vol.22, no.6, pp. 1296–309, Dec 2013.

[15] Fernández-Bolaños, M., Perruisseau-Carrier, J., Dainesi, P. and Ionescu, A.M," RF MEMS capacitive switch on semi-suspended CPW using low-loss high-resistivity silicon substrate", Microelectronic Engineering, Vol.85, no.5, pp.1039–1042, 2008.

[16] Jiang, Hongrui, Zhihong Li, and Norman C. Tien. "Reducing silicon-

substrate parasitics of on-chip transformers." In Micro Electro Mechanical Systems, 2002. The Fifteenth IEEE International Conference on, pp. 649–652. IEEE, 2002.

[17] Yoshikiyo Hatakeyama, Kei Onishi and Keiko Nishikawa, "Effects of sputtering conditions on formation of gold nanoparticles in sputter deposition technique", **RSC Adv.**, 2011,**1**, 1815–1821

[18] Rahman, H.U., Johnson, B.C., Mccallum, J.C, " Fabrication and characterization of PECVD silicon nitride for RF MEMS applications", Microsyst Technol 19, 131–136 2013.

[19] Ijaz H. Jafri, Heinz Busta, Steven T. Walsh, "Critical point drying and cleaning for MEMS technology", Proceedings Volume 3880, MEMS Reliability for Critical and Space Applications; 1999

[20] Khatib, M-K. El, Arnaud Pothier, and Pierre Blondy. "Packaging of RF MEMS switching functions on alumina substrate", arXiv preprint, arXi: 0711.3297, 2007.

[21] Cheng Zhao, Jing Song, Lei Han, Qing-An Huang, "Effects of thermally induced packaging stress on a distributed RF MEMS phase shifter", Microsystem Technology, Vol. 21, pp. 869–87, 2015.

[22] Y K Kim, E K Kim, S W Kim, and B K Ju, "Low temperature epoxy bonding for wafer level MEMS packaging", Sensors and Actuators A, Vol. 143, no. 2, pp. 323–8, May 2008

[23] Y K Kim, E K Kim, S W Kim, and B K Ju, "Low temperature epoxy bonding for wafer level MEMS packaging", Sensors and Actuators A, Vol. 143, no. 2, pp. 323–8, May 2008.

Chapter 16

Vacuum-Free Fabrication of Transparent Electrodes for Soft Electronics

Arshad Khan, Shawkat Ali, Saleem Khan, Moaaz Ahmed, Bo Wang and Amine Bermak

Abstract

Optoelectronic devices are advancing from existing rigid configurations to deformable configurations. These developing devices need transparent electrodes (TEs) having high mechanical deformability while preserving the high electrical conductivity and optical transparency. In agreement with these requirements, vacuum-fabricated conventional TEs based on transparent conducting oxides (TCOs) are receiving difficulties due to its low abundance, film brittleness, and low optical transmittance. Novel solution-processed TE materials including regular metal meshes, metal nanowire (NW) grids, carbon materials, and conducting polymers have been studied and confirmed their capabilities to address the limitations of the TCO-based TEs. This chapter presents a comprehensive review of the latest advances of these vacuum-free TEs, comprising the electrode material classes, the optical, electrical, mechanical and surface feature properties of the soft TEs, and the vacuum-free practices for their fabrication.

Keywords: fabrication, transparent electrodes, solution processed, soft electronics, stretchable electronics

1. Introduction

Nanofabrication means the manufacturing techniques of material or structures with critical dimensions in range of one to few hundreds of nanometers. These techniques realizes exceptionally small, features, structures, devices and systems those have applications in numerous fields of basic and applied sciences. It is comparatively a new class of manufacturing that signifies recent areas of sciences as well as creates new markets. Unlike conventional fabrication approaches, research in nanofabrication is multidisciplinary and needs combined work crosswise conventional fields. In nanofabrication, the final product is based on nanoscale materials, such as powders or fluids, and the components are realized either in "bottom up" or "top down" fashion, using various nanotechnologies. Similar to other fields, the applications of nanofabrication approaches are enormous in optoelectronic devices, [1] for instance, solar cells, [2] smart windows, [3] light-emitting diodes, [4] displays, [5] transparent sensors, [6] and touchscreens. [7] Transparent electrodes (TEs) are the key components in such optoelectronic devices. In addition to high optical transmittance and low sheet resistance [8] required for traditional TEs, next-generation soft optoelectronic devices also need decent mechanical deformability [1, 9] in TEs. Currently, the most utilized TEs are based on vacuum-processed TCOs, comprising fluorine-doped tin oxide and indium tin oxide (ITO). [10, 11] Although TCOs based TEs have demonstrated the required optoelectronic performance, several limitations, such as low abundance, [12] film brittleness, [13] low infrared transparency, [14] and failure during high temperature sintering, undermine their appropriateness for utilization in the future soft optoelectronic systems. Thus, researchers have developed novel TE materials and vacuum-free approaches for its fabrication to substitute the TCOs. [15, 16]

Novel intrinsically transparent materials including graphene, [17] carbon nanotubes (CNTs), [18] and conducting polymers [19, 20] have been explored to replace the TCOs. Besides, other promising class of soft TEs designed from metals are widely employed due to their excellent electrical, optical, and mechanical performance. This typically include metal NWs networks [21, 22] and systematic metal meshes, [23–28] and ultra-thin metal films. [29–31] In addition to the advancement of new materials for soft TEs, plenty of research is performed on the development of vacuum-free technologies for the low-cost fabrication of soft TEs. The list of these techniques is mainly consists of spin coating, [32] spray deposition, [33] inkjet printing, [34] screen printing, [35] transfer printing, [36] and slot-die coating. [37]

There have been several reviews published over the years, aiming at soft TEs from applications perspective. [1, 11, 38] However, few of them focuses on the soft TEs from the fabrication perspective. In this chapter, latest review of the vacuum-free fabricated TEs for emerging soft electronic devices is presented. The chapter begins with the discussion of key properties of TEs for soft electronics (sections 2). We then introduce the TE materials including metals, carbon materials, and transparent conducting polymers (section 3,4). Finally in section 5, the recent progress on vacuum-free methods that are typically employed for the realization of TEs, discussing their merits and demerits. We hope this chapter will enlighten the readers about the emergent soft TEs to better design and fabricate low-cost soft electronics devices.

2. Important properties of the soft transparent electrodes

2.1 Optical transmittance and electrical conductivity

Preferably, TEs must exhibit both high optical transmittance and high electrical conductance, and these are rather contrary from the physics perspective. It is due to an essential requirement for the electrical conductance of a material is the high charge density, that is restricted via the optical absorption of the free charges. [9] Figure of merit (FoM) is commonly used for evaluation of the overall performance of the transparent electrodes. FoM, which is the proportion of electrical conductivity to optical conductivity (σ_{dc}/σ_{opt}), and is measured by means of the commonly used expression, as given below: [20, 22, 25, 26].

$$\mathrm{FoM} = \frac{\sigma_{dc}}{\sigma_{opt}} = \frac{188.5}{R_s\left(\frac{1}{\sqrt{T}} - 1\right)} \qquad (1)$$

Where, T represents the optical transparency value at a wavelength of 550 nm (as it is close to most sensitive wavelength of the human eyes, [39]) and R_s represents the sheet resistance. A larger FOM value discloses a smaller sheet resistance value at a particular optical transmittance value, and vice versa. **Figure 1** presents a comparison of FoMs for metallic soft TEs reported in recent studies. Among

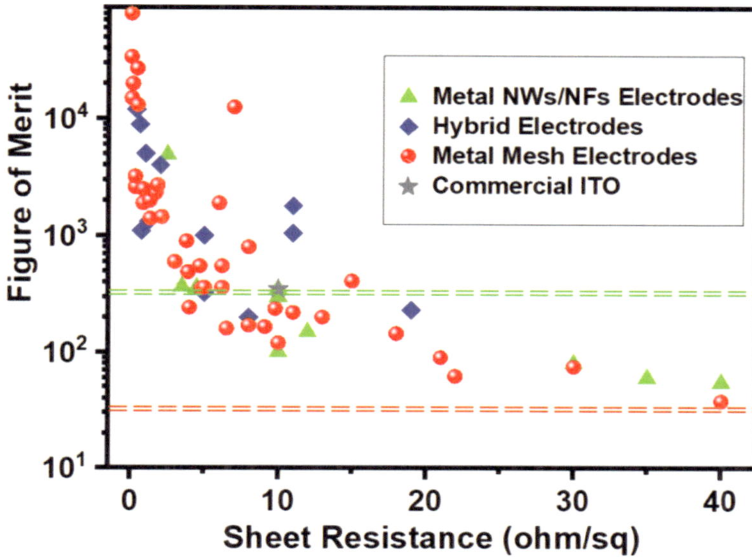

Figure 1.
Comparison of the FoMs of soft TEs (metal NW, metal mesh, and hybrid) and industrial standards. The data was acquired from the literature. [15] The dashed lines represent typical industrial standard (green line) and minimum industrial standard (red line). [40].

these classes, metal-mesh based TEs has higher FoM values, both alone and as part of the hybrid TEs. The detailed FoM values of metal based TEs are presented in **Tables 1–3**.

2.2 Mechanical stability

Mechanically resilient TEs remarkable optoelectronic properties are vital for the development of soft optoelectronic devices as without this, these systems will

R_{sh} (Ω-□$^{-1}$)	T (%)	FoM	Applications	Reference
40	85	55	Low-cost TEs	[41]
35	84	60	Low-cost TEs	[42]
6.2	85	360	Low-cost TEs	[43]
10	70	100	GaN-based LEDs	[44]
12	82	150	OLEDs	[45]
4.5	80	357	Photodetectors	[46]
10	89	300	Touch-screens	[47]
92	92	40	Touch-screens	[48]
3.5	76	366	Low-cost TEs	[49]
10	90	350	Large-area TEs	[50]
2.5	97.3	4920	High-performance TEs	[51]
130.5	92	35	Touch-screens	[52]
30	85	80	OSCs	[53]

Table 1.
Optical and electrical performance, and applications of metal NWs based soft TEs published in recent literature.

R_{sh} (Ω-□$^{-1}$)	T (%)	FoM	Applications	Reference
0.036	75	34000	DSSCs/Heaters	[15]
4	70	242	OLEDs	[54]
15	96	410	Touch-screens	[24]
21	85	90	Stretchable TEs	[55]
40	80	38	OFETs/ OLEDs/OSCs	[56]
22	78	62	OSCs	[57]
30	85	75	Touch-screens	[23]
3	82	600	Transparent Heaters	[58]
8	77	170	OSCs	[59]
0.3	70	3200	Large-area TEs	[27]
1.7	82	2700	EL displays	[60]
0.03	86	80,000	Transparent Heaters	[61]
4.8	81	355	Printed TEs	[62]
18	76	145	High-durable TEs	[26]
5	82	360	Touch-screens	[25]
9.8	85.2	237	Touch-screens	[63]
6	97	1900	Transparent Heaters	[64]
8	94	800	Printed TEs	[65]
7	96	12600	Nanofiber based TEs	[66]
0.43	97	27000	Wearable TEs	[67]
0.07	72	15000	Transparent Heaters	[16]
0.13	86	20000	Transparent Heaters	[68]
1.32	82	1400	DSSCs	[69]
0.84	84	2500	EL displays	[70]
3.8	90	900	Wearable Heaters	[71]
2.1	88.6	1450	QLEDs	[72]
13	87	200	Touch-screens	[73]
10	75	120	Solar Cells	[74]
11	86	220	Transparent Heaters	[75]
4.7	87	550	OLEDs	[76]
6.2	90	550	Transparent Heaters	[77]
3.9	84	490	Highly Bendable TEs	[78]

Table 2.
Optical and electrical performance, and applications of regular metal mesh based soft TEs published in recent literature.

not be not able to preserve electrical conductivity under significant mechanical deformation. [69] Various approaches are developed to enhance the mechanical stability of the soft TEs. For example, metal meshes are embedded and mechanically anchored into the soft polymer substrates, which significantly enhanced its adhesion with the substrate and as a result improves its mechanical stability under deformation. [15, 16, 86] In addition to the mechanical stability of the TEs, the

R_{sh} (Ω-□$^{-1}$)	T (%)	FoM	Applications	Reference
1	92	5000	High-performance TEs	[40]
2	95	4000	High-performance TEs	[40]
1.2	80	1330	OSCs	[79]
19	92	232	Perovskite Solar Cells	[80]
9.1	79	165	OSCs	[81]
5	80	325	Long-term Stable TEs	[82]
0.6	93	8900	High-performance TEs	[40]
0.36	92	12000	E-chromic Devices	[83]
0.7	65	1100	Stretchable TEs	[55]
11	98	1800	High-performance TEs	[84]
3	92	1400	High-performance TEs	[84]
60	90	60	PLEDs	[85]
11	88	1050	Stretchable TEs	[55]

Table 3.
Optical and electrical performance, and applications of hybrid soft TEs published in recent literature.

intrinsic mechanical stability of the other functional materials are equally important concerning the successful operations of the soft electronic devices.

2.3 Other surface properties

TEs Surface roughness is significant as this considerably influences the morphology and uniformity of the subsequent printed/coated layers. Though, it's hard to define a strict extreme roughness value vital for the effective production of soft electronic devices. Yet, bottom TEs with lower surface roughness value are preferred to minimize the possibility of electrical short circuiting. For instance, the roughness (root-mean-square) of a coated/printed continuous PEDOT:PSS film is normally <10 nm, which is adequately flat for most of the functional thin-films involved in fabrication of electronic devices. But, the surface roughness of metal TEs is much higher (hundreds of nm to few μm). For example, screen printed silver mesh is >2 μm thick, making the subsequent functional layer uniform deposition impossible. [87] To address this, researchers have has embedded the metal-mesh into the polymer substrates to flat the TEs top surface. [16, 69] Similarly, metal NW networks also demonstrate decent FoM as stated above, however, its high roughness resulted in poor device performance. [88] Therefore, multiple approaches have been established to flatten the metal NWs TEs by compacting the unattached networks to a dense structure or filling the openings with supplementary TE materials. [89, 90]

Chemical compatibility of the TEs/functional materials interface is another important concern for TEs. An unsteady interface can cause substandard performance and also fast deprivation of the TEs. For instance, the acidic behavior of PEDOT:PSS TEs can corrode the base ITO layer, causing the diffusion of indium at the TE/active layer boundary. Such erosion might result in critical gap conditions which further caused the degradation of device. [91] To minimize the risk of chemical/electrochemical decay of the sensitive metallic TEs, a traditional method is covering the sensitive metallic materials with a thin-film. [92] This thin-film can be either from another class of conductor, for example, graphene, [93] and less-sensitive metals, [94] or an insulating material, for instance, poly(methyl methacrylate) (PMMA) and alumina, [95] The insulating film must be ultra-thin (< few nanometers) for proficient charge transport. [96] In addition to roughness and chemical compatibility, surface energy of TEs is also an essential factor to be considered for the efficient performance of the active materials in soft electronic devices. [97]

3. Metal based soft TEs

Due to the high density of free electrons, metals demonstrate the uppermost electrical conductance among all the conductive materials. Yet, metallic materials in bulk are unable to work as TEs directly as it has high light reflection at visible wavelength. [11] Thus, shape structuring is essential for metallic materials to attain the required optoelectronic characteristics. Following are the classes of metal-based TEs frequently reported in recent years. These typically include metal nanoparticle/nanowire/nanofiber networks, regular metal meshes, and ultra-thin metal films.

3.1 Metal nanoparticle/nanowire/nanofiber networks

One of the major classes of soft TEs is prepared from the metal NPs/NWs networks [21, 22], that have exhibited enormous performance in optical transparency, electrical conductivity, and mechanical deformation. The metal NPs or NWs must be gathered to form transparent metal meshes using several vacuum-free fabrication methods to realize soft TEs. In reality, the porous arrangement of these class of TEs permit the light to go across the free spaces in the grids. Therefore, the electrical and optical conductivity of these electrodes are greatly reliant on the grid arrangement. Simply, the electrical conductance depends on the density of metallic materials, while the optical transmittance is determined by the area fraction of metal coverage. Among these, TEs prepared from the metal NWs got much attention because of their shape and that they can easily be dispersed in various solvents. Therefore, these can be processed by multiple vacuum-free techniques to create TEs having decent optoelectronic performance for soft electronic applications. **Table 1** reviews the electrical and optical performance, and applications of metal NWs based soft TEs published in recent literature. Similar to other classes, metal NWs soft TEs also suffer from quite a few difficulties such as problem in achieving smooth NWs distribution across the large-area substrates, and the NWs delamination from the substrate during deformation. [9] In addition, the dispersed NWs network cannot be employed directly as further processing steps are normally required to eliminate the polymer capping around the NWs to decrease the junction resistance. This is achieved either using selective welding, bulk heating, or chemical processes. In addition to metal NWs, nanofiber based TEs have also got great interest due to their wide range of unique capabilities. Nanofibers are fabricated by employing various approaches, however, electrospinning technique is considered to be facile and low-cost to realize nanofibers with decent reproducibility, well-controlled shape, high aspect ratio, and saleable size. Moreover, the production of nanofibers can be enhanced by means of electrospinning system with multi-nozzles. [98] Despite this potential, TEs based on nanofibers [24, 67, 84, 99] have the randomly distributed patterns and because of this, the reproducibility of placing the nanofibers in precise locations and alignments remains a foremost challenge in these TEs. [15, 100]

3.2 Regular metal meshes

Compared with metal NPs/NWs, metal-mesh based soft TEs look extra proficient as their electrical and optical conductivity can easily be adjusted in a broad assortment via changing the line width, mesh opening, and thickness. [26] Besides, numerous metals can be employed as metal-mesh based TEs to attain the desired chemical characteristics and work functions for the targeted soft electronic applications. [24] **Table 2** summarizes the electrical and optical performance, and applications of metal mesh based soft TEs published in recent literature. The presented data shows that the FoM values of metal-mesh based TEs are comparatively higher than that of metal NPs/NWs based TEs. This is mainly due to the low junction resistances, offered by the regular metal meshes. Regardless of the superior performances, rough surface topography and poor adhesion between the meshes and substrates constrained the extensive use of metal-mesh based TEs in soft electronic industry.

3.3 Transparent thin metal films

Mostly, bulk metallic films having tens to hundreds of nanometers thicknesses are utilized as back-electrodes (opaque-cathodes). But, ultra-thin metal films with only few nanometers thicknesses can also be utilized as front-electrodes (transparent-anodes). Since, these metal layers are thinner in comparison with the light visible wavelength, and thus are optically transparent to human-eye. The thickness and uniformity of the metal films determine the optoelectronic performance of these TEs for the desired soft electronic applications. Several metals having different work-functions, including silver, nickel, gold, and platinum are effectively employed as transparent electrodes in soft electronic devices. [100] However, the vacuum-free fabrication of these ultra-thin transparent metallic films over large area is difficult, and thus substantial advancements in the fabrication methods are required to efficiently mass-produce these thin metal films.

4. Other soft TEs

4.1 Carbon materials

Graphene: Graphene efficiently conducts electricity and heat, is stronger than steel (~200 times), and is nearly transparent. [101] Due to these unique characteristics, it has been suggested as a substitute soft TE material. Over the years, various vacuum-free approaches are established to produce thin films of graphene on soft substrate materials. [102–104] Recently, significant advancement has been made to enhance the optoelectronic properties of the graphene based TEs. Large-area graphene film was made-up on copper catalyst (~30 inches diagonal size), which was then accurately transferred to the target soft substrate using transfer printing technique. [105] In an ideal world, graphene has massive capability and is currently offering the assurance of being the vital transparent material for soft TEs. However as a matter of fact, uniform ultra-thin films of graphene are exceedingly challenging and are costly to produce. In addition, optoelectronic performances of graphene based TEs reduce quickly, due to the wrinkles/folds and crystallographic defects formed in these ultra-thin films during mechanical deformation. [105, 106]

Carbon Nanotubes: Similar to graphene, CNT is one of the hardest materials recognized. Due to its decent electronic and mechanical characteristics, CNTs are productively employed as TE material in soft electronic devices. [107–110]

Numerous vacuum-free approaches, for example, spin coating [111] and transfer printing, [112] are developed to produce CNTs based soft TEs. While, CNTs based TEs have attractive characteristics, such as higher optical transmittance and superior mechanical deformation capability, these have typically poor electrical conductivity. This limitation makes CNTs less suitable for large-area commercial soft electronics.

4.2 Transparent conducting polymers

PEDOT:PSS: Few transparent polymers, having intrinsically poor electrical conductivity, are transformed into conducting polymers via addition of conductive dopants into their iterating chains. Poly (3,4-ethylenedioxythiophene) polystyrene sulfonate (PEDOT:PSS) is one of the classic model of such conducting polymers. In PEDOT: PSS unit chain, PEDOT acts as the conducting polymer, while the PSS plays the role of a dopant, enhancing its electrical conductance via significantly increasing the charge carriers. Since, PEDOT:PSS has no visible absorptive resonances, therefore it is routinely used as TEs in small scale soft electronic devices. Yet, a number of concerns, for example, instable molecular structure and high water solubility have limited the use of PEDOT:PSS in large-scale soft electronics. [113, 114]

Other Conducting Polymers: Besides PEDOT:PSS, other conducting polymers comprising poly(p-phenylene-vinylene) (PPV), polyaniline (PANI), polyfuran (PF), polypyrrole (PPy), are utilized as TE materials for several soft electronic devices, due to their decent electrical and optical conductivity. [115, 116]

As discussed above, each class of soft TEs offers unique set of favorable properties, and also has some disadvantages. Researchers have combined different classes of TEs into a single electrode structure to fabricate hybrid soft TEs. The objectives of developing this new class TEs are: (1) take advantage of the benefits offered by individual electrode. (2) overcome those challenges associated with the electrode once employed individually. **Table 3** summarizes, the optical and electrical performance, and applications of hybrid soft TEs published in recent literature.

5. Vacuum-free fabrication approaches for soft TEs

Vacuum-free thin film fabrication techniques are favored by soft electronic industry because of low cost, low material waste and high output as compared with conventional vacuum fabrication processes. Yet, accomplishing equivalent quality solution-processed TEs is a challenging job due to several reasons, including the substrate/TE adhesion, the solvent volatility, surface wettability, and solution rheology need to be accustomed. Following are the most commonly reported vacuum-free printing and coating approaches for the fabrication of soft TEs.

5.1 Spin coating

It is a simple technique used to coat continuous thin films onto rigid flat surfaces. Typically a small amount of coating material is put on the substrate's center, that is ideally spinning at low speed. The substrate is then rotated at high speed (max ~10 k rpm) to uniformly spread the coat-material utilizing the centrifugal force, as schematically illustrated in **Figure 2a**. One main benefit of the spin coating process is its capacity of dense coating of uniform and thin films onto rigid flat surfaces. This ability is quite attuned along the requirement of excellent TEs, as the thickness of TEs needs to be optimized. It is an attractive method to fabricate transparent thin

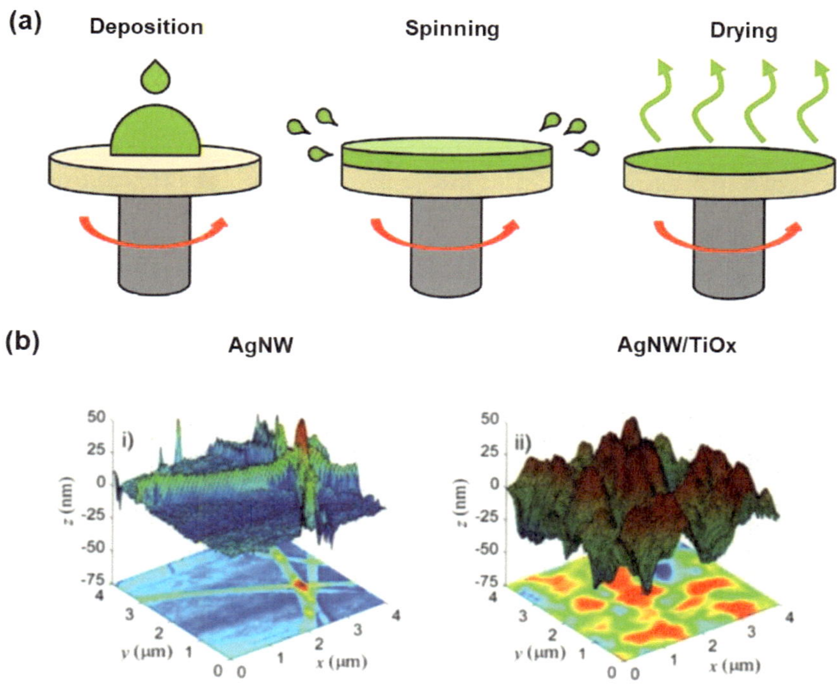

Figure 2.
(a) Schematic illustration of spin coating process. (b) AFM topographies of silver NW TEs without (left) and with the TiOx buffer layer (right). Reproduced with permission from Ref. [32].

graphene films (few nanometer-thick), as the optical transparency of these films will decline considerably with increase in thickness. For instance, each graphene layer absorbs 2.3% of white light. [117] Therefore, graphene-based TEs needs to be ultra-thin to obtain appropriate optical transmittance. Thin (3.1 nm) graphene TEs are fabricated using spin-coated for realizing OSCs. [118] Occasionally, the smoothness of spin-coated TEs is not perfect because of the material properties itself. For instance, silver NWs have decent dispersion in isopropanol, water, and few other frequently employed solvents and therefore can be easily spin coated on several substrates for the fabrication of TEs. Conversely, the spin coated silver NWs typically creates a nano-mesh (with certain thickness) on the substrates, making roughness for the subsequent processing, and therefore limits the applications of bottom TEs. In addition to the roughness concern, the weak silver NW/substrates adhesion causes mechanical failure of the devices, particularly in soft electronics. [88] This issue is resolved by spin coating a TiO_x buffer layer (~200 nm) over the silver NWs to get a comparatively uniform film, as displayed in **Figure 2b**. [32] Despite such potential, spin coating process have few limitations for the realizing of soft TEs. First, flatness of the spin-coated TEs is typically sensitive to spin speed, humidity, and substrate cleanness, which make the processing difficult to reproduce in ambient environment. Second, spin coating on large-area substrates is precisely difficult as it is challenging to clamp a hefty substrate and keep it stable at a high rotating speed. As a result, the spin coated films thickness is spatially different over a large substrate due to the variation of the localized centrifugation speed. Third, majority of the material is spun-off the substrate in spin coating, making this material-wasting approach. Bearing in mind, major portion of the total price of the raw materials of the soft electronic devices comprises of the material cost of TEs alone. Therefore, this wastage of material by spin coating is not financially viable for industrial mass-production, even though partially this may be reused.

5.2 Spray deposition

It is a coating process that uses a spray of particles or droplets to deposit a material onto a substrate using a nozzle, as schematically illustrated in **Figure 3a**. The spray nozzle creates a spray that comprises small drops of TE material and leads the materials transportation to the substrate by the help of carrier gas or electric charge. [119] Compared with other vacuum-free deposition techniques, the main benefit of spray coating is its capability of uniform coating of materials on non-flat substrates. **Figure 3b** displays organic photodetector (fiber-based) using PEDOT:PSS TE, that was realized using spray coating. it difficult to coat smooth PEDOT:PSS film though spin coating on the curved optical fiber surface. [120] It is also useful for subsequent processing, for instance, to spray coat on uneven surfaces, for instance, metal NWs, metal mesh coated substrates, as spin-coating of solutions can create non-continuous surface coverage. [33] Besides condense and smooth TCO-free films, spray coating has also the capability to deposit TCO films. **Figure 3c** shows

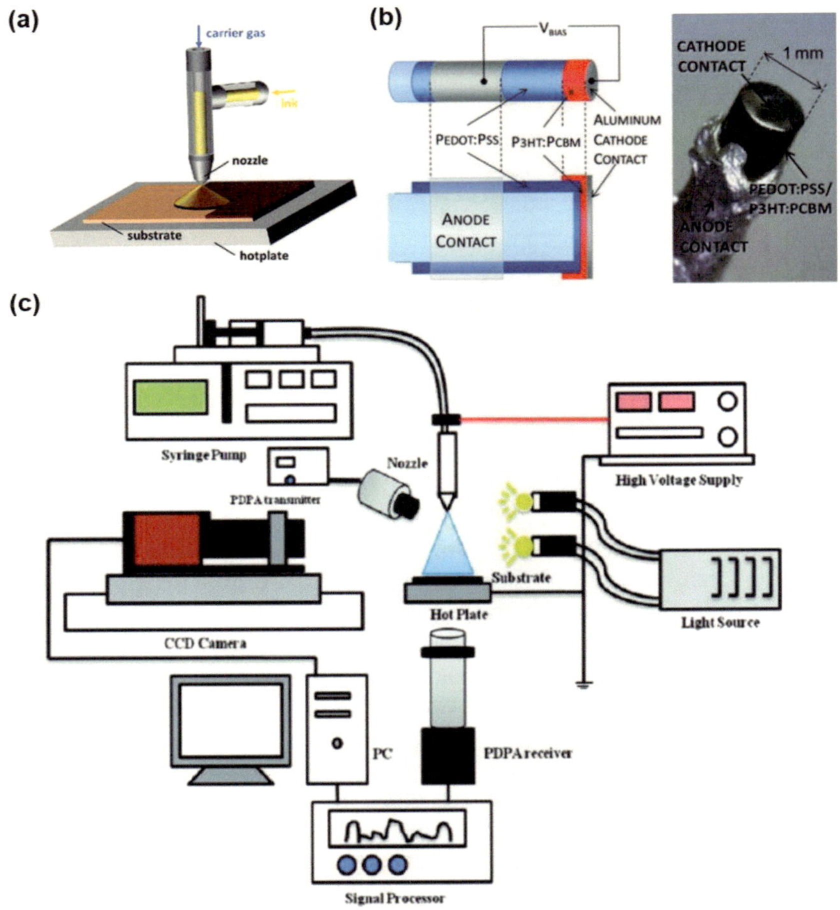

Figure 3.
(a) Schematic illustration of spray coating process. (b) Schematic (left) and photo (right) of fiber-based organic photodetector produced by spray coating. Reproduced with permission from Ref. [120] (c) schematic illustration of electrospray system. Reproduced with permission from Ref. [121].

the electrospray setup, utilized for deposition of zinc oxide (ZnO) and aluminum doped zinc oxide (AZO) films. [121] Despite such capabilities, spray processed TEs has the scalability problem, much more prominent as compared with other vacuum-free coating techniques. This limitation of low throughput has hammered its widespread adoption for production of large area soft TEs.

5.3 Inkjet printing

It is another highly used technique for making soft TEs. Inkjet printing is devised from dispenser printing where ink droplets exit the nozzles by a vibrant practice. By controlling the contraction expansion of the piezoelectric actuator, discrete ink drops are ejected from the nozzle making the anticipated design on top of the substrate, as schematically illustrated in **Figure 4a**. It is direct printing technique for high-resolution patterning, without the need of lithography other advantage key advantages that the printed design can be easily changed by modifying the digital pattern that controls the actuator. [122] Inkjet printing is an effective approach to producing large area soft TEs. **Figure 4b** displays a large-area organic solar cell (OSC) having silver current collecting mesh fabricated by inkjet-printing. The printed silver mesh consisted only small portion (~8%) of the total substrate area due to the mesh relatively small line width (~160 μm). The thickness of the printed silver mesh lines was >2 μm, which caused large height variation for the subsequent processing i.e. spin-coating of PEDOT:PSS and other active materials of the solar cell. This problem was resolved by embedding the silver mesh into an extra barrier film. The large-area OSCs having flexible Ag/PEDOT:PSS mesh TEs shown excellent performance as compared to that of TCO-solar cell, due to the high conductivity (sheet resistance ~1 Ω/□) of the silver mesh. [87] Inkjet printing processes based on mechanisms other than piezoelectric actuation are also utilized

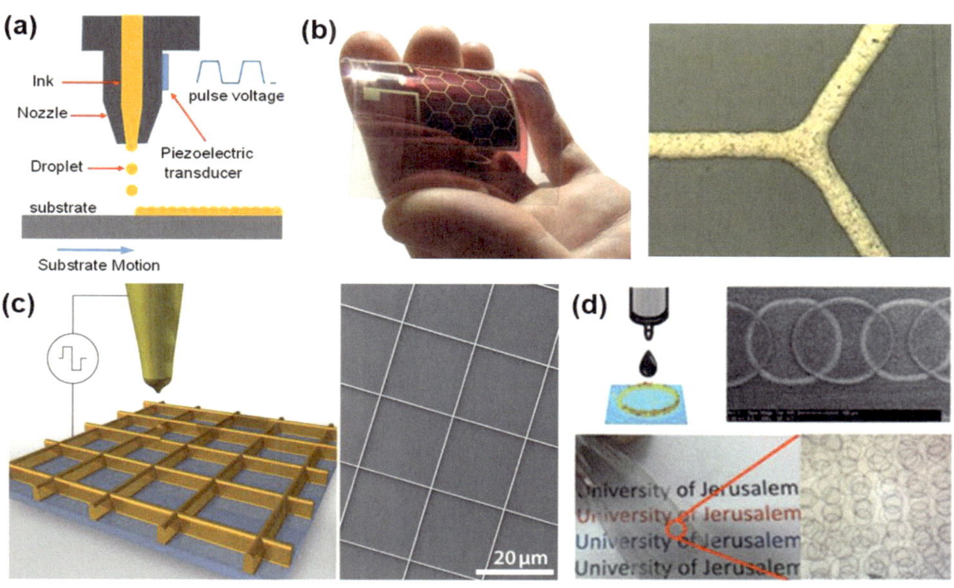

Figure 4.
(a) Schematic illustration of inkjet printing process. Reproduced with permission from Ref. [34]. (b) Photographs of inkjet-printed silver current mesh for large-area OSCs. Reproduced with permission from Ref. [87]. (c) Schematic illustration of high-aspect ratio metal grid along with electrohydrodynamic inkjet printing and the SEM image of the printed gold metal mesh electrode. Reproduced with permission from Ref. [65] (d) schematic diagram, SEM image, and photographs of the inkjet printed CNTs based TEs by "coffee ring effect". Reproduced with permission from Ref. [123].

for fabricating TEs. For instance, electrohydrodynamic inkjet process, as shown in **Figure 4c**, enabled the printing of high resolution gold meshes (feature size line 80 to 500 nm line widths) for realizing high performance TEs (8 Ω/□ at 94% optical transmittance), that can be custom-made for the application in different electronic devices. [65]

One major obstacle in attaining uniform inkjet printed structures is the coffee-ring effect, that initiates because of the capillary flow in the solvent evaporation step. [124] Though, this effect is also occasionally useful for making TEs with particular ring shapes. [125] **Figure 4d** demonstrates a CNTs based TE having joined ring patterns, that was made through inkjet printing the CNT ink on top of a pre-heated PET film. The height and diameter of the rings were the functions of applied temperature. Post heat-curing further lowered the sheet resistance of the CNT coatings. [123]

5.4 Screen printing

It is one of the attractive methods used to print soft TEs. In this technique, viscous inks are forced across stencils or patterned mesh (typically used as the template) using a scraper as shown in **Figure 5a**. The density of the used mesh and ink viscosity define the printing resolution and thickness of the pattern. [35] This handy and relatively simple technique is utilized mainly for graphene and PEDOT:PSS, however metals can also be printed. The resolution of conventional screen printing processes is not high, however, it can be improved to tens of micrometer using an improved screen-offset approach. [126] **Figure 5b** displays the screen printed graphite oxide (GO) arrays on PET film, which was afterwards reduced to rGO using hydriodic acid (HI) in modest environments. This technique developed an easy way to manufacture large-area graphene TEs (patterned), having thickness of few hundred nanometers. [127] Similar to other screen printable materials, mesh-patterned PEDOT:PSS TEs can be realized with various width/period ratios by adjusting the wire diameter, mesh size, and photoresist thickness. [128] Beside graphene and PEDOT:PSS, screen printing is also utilized for the patterning of metallic inks. **Figure 5c** shows the schematic illustration of the structure of OSC having printed silver mesh as TEs. This work relates the screen printed hybrid TEs having PEDOT:PSS on top of silver mesh with various other printing approaches for

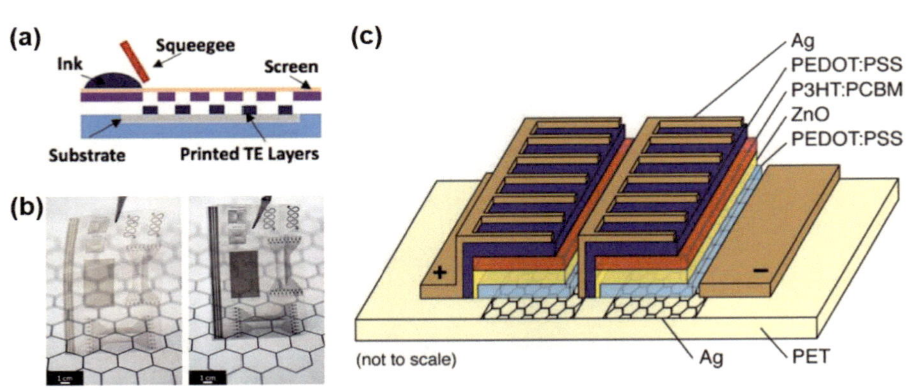

Figure 5.
(a) Schematic illustration of screen printing process. Reproduced with permission from Ref. [35]. (b) Photographs of GO (left) and rGO (right) films, fabricated by screen printing. Reproduced with permission from Ref. [127]. (c) Schematic illustration of the OSCs containing the layers of P3HT:PCBM, ZnO, PEDOT:PSS, and silver electrodes. The back silver electrode is printed by various processes including screen printing. Reproduced with permission from Ref. [129].

the vacuum-free and TCO free OSCs. It concludes that the uniformity of screen-printed silver meshes was superior as compared to inkjet printed and flexographic printed TEs, which were damaged by de-wetting in the subsequent PEDOT:PSS film processing. Consequently, the OSCs having screen-printed silver TEs showed better performance equated with inkjet printed and flexographic printed solar cells. [129]

5.5 Transfer printing

Transfer printing is an emergent method for fabrication of soft TEs, that empowers the processing of various materials into the chosen useful shapes. This produces manufacturing prospects in the field of soft electronics with comparable performance to that of traditional wafer-based processes, however with capacity to be deformed. In this technique, first the materials structures are fabricated on the conventional donor substrate and then wisely transferred onto unconventional soft substrates, as described in **Figure 6a**. [36] For instance, graphene ultra-thin films are first coated on Ni or Cu foils using the standard chemical vapor deposition (CVD) technique. [130] In order to be used as TEs, the this graphene has to be transferred directly to top of the devices or transparent substrates. There are two different transfer approaches (wet transfer and dry transfer) to transfer CVD graphene onto various soft substrates, as shown in **Figure 6b**. In wet transfer,

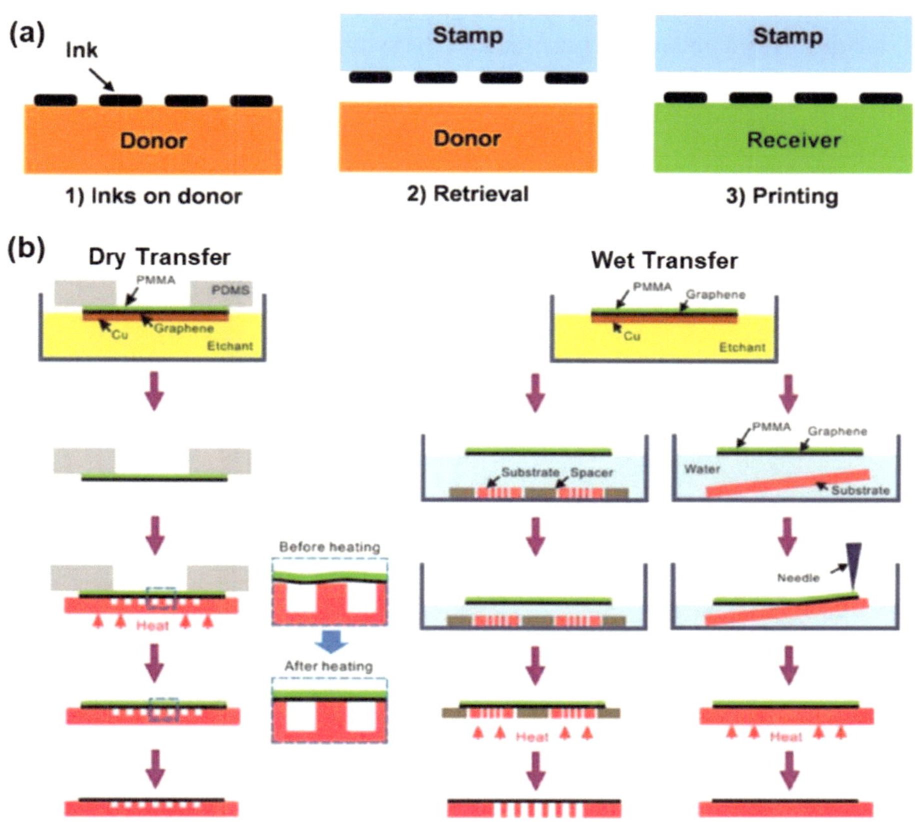

Figure 6.
(a) Schematic illustration of the transfer printing technique. Reproduced with permission from Ref. [36]. (b) Schematics of wet (right) and dry (lift) transfer printing for graphene soft TEs fabrication. Reproduced with permission from Ref. [131].

the graphene was initially covered by a PMMA thin film. Next, the underneath Cu film was removed by an etching step in $FeCl_3$. The graphene film covered by PMMA was then lifted-off either using a PDMS stamp for transfer, or directly picked up using the target substrate itself. [131] To enhance the throughput and production speed, transfer printing has been integrated with R2R process for the fabrication of large-area graphene (30-in) soft TEs. [105] Despite such potential, wet transfer has a limitation for the fabrication of top graphene-based TEs for the soft thin-film devices as the functional materials used in these devices are sensitive to moisture. To overcome this a dry transfer approach is developed, where a the film is directly coated on the PDMS stamps before transfer. [132] Besides graphene, other major transfer printable material for soft TEs is the metal nanowire/mesh films. These films typically have weak adhesion with the transfer substrates. This poor adhesion between the transfer substrates and metal films makes it easier to lift these films up with the PDMS, or another sticky polymeric stamp/target substrate. [60, 69] The high optical transmittance and superior conductivity of fabricated soft TEs using transfer printing ensure the high performance of soft electronic devices. [15, 60, 69, 70]

5.6 Slot-die coating

It is an effective process for printing one-dimensional structures. Slot-die coating is typically integrated with the R2R system for rapid production of soft electronic devices. As shown in **Figure 7a** and **b**, the solution is pushed out of the slot-die using a pneumatic scheme, and the solution is printed laterally in the

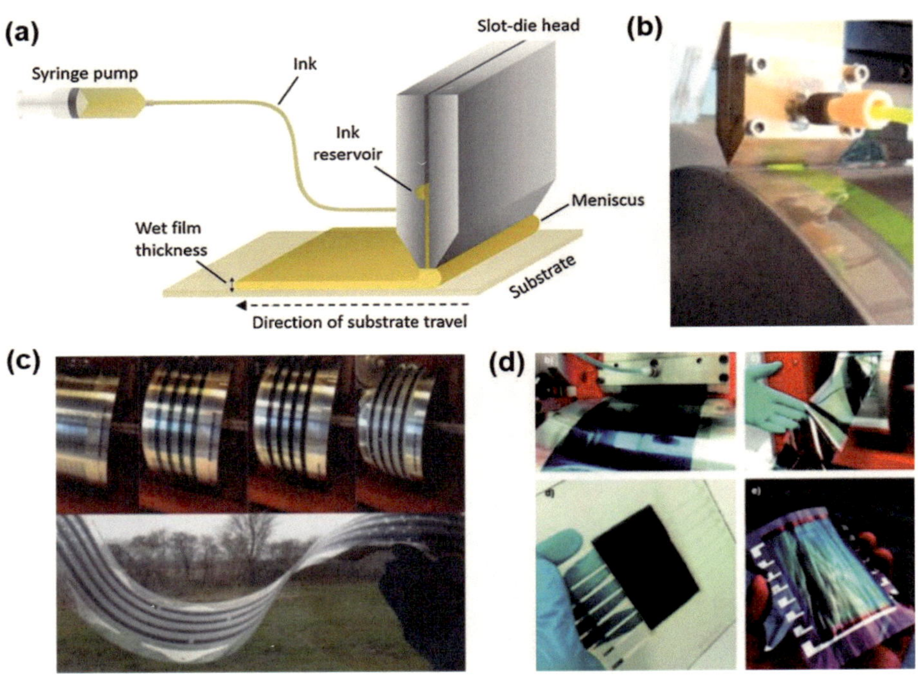

Figure 7.
(a) Schematic illustration of the slot-die coating technique. Reproduced with permission from Ref. [37] (b) photograph of the slot-die coating system. Reproduced with permission from Ref. [133] (c) photographs of the large-area soft OCSs having PEDOT:PSS TEs fabricated by slot-die coating. Reproduced with permission from Ref. [134] (d) photographs of the OCSs with high geometric fill factor. The employed PEDOT:PSS TEs are fabricated by slot-die coating. Reproduced with permission from Ref. [135].

direction of the moving head. The thickness of the printed structure is typically determined through the solution's concentration and it's flow rate, while the head speed controls the speed of printing. PEDOT:PSS is the most commonly processed TE material for slot-die coating. **Figure 7c** and **d** display flexible large-area OSCs, where, the PEDOT:PSS TEs and the organic active material were both printed by slot-die coating. [134] Key benefit of using slot-die coating is its capability to print on large-area substrates, as in slot-die coated films the center-to-edge thickness difference is negligible. Therefore, large-area OSCs having a high geometric fill factor (98.5%) were realized through integrating laser patterning with slot-die coating. [135] Besides PEDOT:PSS, slot-die coating has been effectively utilized for other conductive inks including silver NWs, [136] CNTs, [137] and graphene. [138] Similar to other processes, slot-die coating has also few limitations including, the harsh requirements regarding inks rheology for high quality coatings [139] and the existence of high density printing defects such as ribbing and rivulet. [140]

6. Conclusions

Recent progress of the development of vacuum-free TEs for soft electronics has been promising. This chapter presents a detailed overview on the latest advances of the vacuum-free soft TEs, comprising the introduction of electrode materials classes, the optical, electrical, mechanical, and surface features of the soft TEs. The chapter summarizes the vacuum-free techniques for the fabrication of soft TEs. Regardless of all the shortcomings discussed, we are optimistic that the vacuum-free TEs be going to play vital roles in soft electronic industries in the future.

Acknowledgements

This work was supported by the Qatar National Research Fund (a member of Qatar Foundation) under grant NPRP11S-0110-180246. The findings herein reflect the work and are solely the responsibility of the authors.

Conflict of interest

Authors declare no conflict of interests.

Author details

Arshad Khan*, Shawkat Ali, Saleem Khan, Moaaz Ahmed, Bo Wang and Amine Bermak
Division of Information and Computing Technology, College of Science and Engineering, Hamad Bin Khalifa University, Qatar Foundation, Doha, Qatar

*Address all correspondence to: arkhan4@hbku.edu.qa

References

[1] Hecht DS, Hu L, Irvin G. Emerging Transparent Electrodes Based on Thin Films of Carbon Nanotubes, Graphene, and Metallic Nanostructures. Advanced Materials. 2011;23:1482-513.

[2] Yu Z, Li L, Zhang Q, Hu W, Pei Q. Silver Nanowire-Polymer Composite Electrodes for Efficient Polymer Solar Cells. Advanced Materials. 2011;23:4453-7.

[3] deb SK, Lee S, Edwin Tracy C, Roland Pitts J, Gregg BA, Branz HM. Stand-alone photovoltaic-powered electrochromic smart window. Electrochimica Acta. 2001;46: 2125-30.

[4] Wu J, Agrawal M, Becerril HA, Bao Z, Liu Z, Chen Y, et al. Organic Light-Emitting Diodes on Solution-Processed Graphene Transparent Electrodes. ACS Nano. 2010;4:43-8.

[5] Blake P, Brimicombe PD, Nair RR, Booth TJ, Jiang D, Schedin F, et al. Graphene-Based Liquid Crystal Device. Nano Letters. 2008;8:1704-8.

[6] Cai J, Zhang C, Khan A, Liang C, Li W-D. Highly transparent and flexible polyaniline mesh sensor for chemiresistive sensing of ammonia gas. RSC Advances. 2018;8:5312-20.

[7] Wang J, Liang M, Fang Y, Qiu T, Zhang J, Zhi L. Rod-Coating: Towards Large-Area Fabrication of Uniform Reduced Graphene Oxide Films for Flexible Touch Screens. Advanced Materials. 2012;24:2874-8.

[8] Gordon RG. Criteria for Choosing Transparent Conductors. MRS Bulletin. 2000;25:52-7.

[9] Ellmer K. Past achievements and future challenges in the development of optically transparent electrodes. Nat Photon. 2012;6:809-17.

[10] Minami T. Transparent conducting oxide semiconductors for transparent electrodes. Semiconductor Science and Technology. 2005;20:S35-S44.

[11] Zhang Y, Ng S-W, Lu X, Zheng Z. Solution-Processed Transparent Electrodes for Emerging Thin-Film Solar Cells. Chemical Reviews. 2020;120:2049-122.

[12] Kumar A, Zhou C. The Race To Replace Tin-Doped Indium Oxide: Which Material Will Win? ACS Nano. 2010;4:11-4.

[13] Cairns DR, Witte RP, Sparacin DK, Sachsman SM, Paine DC, Crawford GP, et al. Strain-dependent electrical resistance of tin-doped indium oxide on polymer substrates. Applied Physics Letters. 2000;76:1425-7.

[14] Bel Hadj Tahar R, Ban T, Ohya Y, Takahashi Y. Tin doped indium oxide thin films: Electrical properties. Journal of Applied Physics. 1998;83:2631-45.

[15] Khan A, Liang C, Huang Y-T, Zhang C, Cai J, Feng S-P, et al. Template-Electrodeposited and Imprint-Transferred Microscale Metal-Mesh Transparent Electrodes for Flexible and Stretchable Electronics. Advanced Engineering Materials. 2019;21:1900723.

[16] Khan A, Lee S, Jang T, Xiong Z, Zhang C, Tang J, et al. High-Performance Flexible Transparent Electrode with an Embedded Metal Mesh Fabricated by Cost-Effective Solution Process. Small. 2016;12:3021-30.

[17] Bonaccorso F, Sun Z, Hasan T, Ferrari AC. Graphene photonics and optoelectronics. Nat Photon. 2010;4:611-22.

[18] Zhang M, Fang S, Zakhidov AA, Lee SB, Aliev AE, Williams CD, et al.

Strong, Transparent, Multifunctional, Carbon Nanotube Sheets. Science. 2005;309:1215-9.

[19] Kirchmeyer S, Reuter K. Scientific importance, properties and growing applications of poly(3,4-ethylenedioxythiophene). Journal of Materials Chemistry. 2005;15:2077-88.

[20] Vosgueritchian M, Lipomi DJ, Bao Z. Highly Conductive and Transparent PEDOT:PSS Films with a Fluorosurfactant for Stretchable and Flexible Transparent Electrodes. Advanced Functional Materials. 2012;22:421-8.

[21] De S, Higgins TM, Lyons PE, Doherty EM, Nirmalraj PN, Blau WJ, et al. Silver Nanowire Networks as Flexible, Transparent, Conducting Films: Extremely High DC to Optical Conductivity Ratios. ACS Nano. 2009;3:1767-74.

[22] van de Groep J, Spinelli P, Polman A. Transparent Conducting Silver Nanowire Networks. Nano Letters. 2012;12:3138-44.

[23] Hong S, Yeo J, Kim G, Kim D, Lee H, Kwon J, et al. Nonvacuum, Maskless Fabrication of a Flexible Metal Grid Transparent Conductor by Low-Temperature Selective Laser Sintering of Nanoparticle Ink. ACS Nano. 2013;7:5024-31.

[24] Wu H, Kong D, Ruan Z, Hsu P-C, Wang S, Yu Z, et al. A transparent electrode based on a metal nanotrough network. Nat Nano. 2013;8:421-5.

[25] Han B, Pei K, Huang Y, Zhang X, Rong Q, Lin Q, et al. Uniform Self-Forming Metallic Network as a High-Performance Transparent Conductive Electrode. Advanced Materials. 2014;26:873-7.

[26] Kim H-J, Lee S-H, Lee J, Lee E-S, Choi J-H, Jung J-H, et al. High-Durable AgNi Nanomesh Film for a Transparent Conducting Electrode. Small. 2014;10:3767-74.

[27] Kiruthika S, Gupta R, Rao KDM, Chakraborty S, Padmavathy N, Kulkarni GU. Large area solution processed transparent conducting electrode based on highly interconnected Cu wire network. Journal of Materials Chemistry C. 2014;2:2089-94.

[28] Choi H-J, Choo S, Jung P-H, Shin J-H, Kim Y-D, Lee H. Uniformly embedded silver nanomesh as highly bendable transparent conducting electrode. Nanotechnology. 2015;26:055305.

[29] Zhijun W, Shufen C, Huishan Y, Yi Z, Jingying H, Shiyong L. Top-emitting organic light-emitting devices based on silicon substrate using Ag electrode. Semiconductor Science and Technology. 2004;19:1138.

[30] Peng H, Zhu X, Sun J, Xie Z, Xie S, Wong M, et al. Efficient organic light-emitting diode using semitransparent silver as anode. Applied Physics Letters. 2005;87:173505.

[31] Gu D, Zhang C, Wu Y-K, Guo LJ. Ultrasmooth and Thermally Stable Silver-Based Thin Films with Subnanometer Roughness by Aluminum Doping. ACS Nano. 2014;8:10343-51.

[32] Leem D-S, Edwards A, Faist M, Nelson J, Bradley DDC, de Mello JC. Efficient Organic Solar Cells with Solution-Processed Silver Nanowire Electrodes. Advanced Materials. 2011;23:4371-5.

[33] Choi DY, Kang HW, Sung HJ, Kim SS. Annealing-free, flexible silver nanowire–polymer composite electrodes via a continuous two-step spray-coating method. Nanoscale. 2013;5:977-83.

[34] Lin X, Kavalakkatt J, Lux-Steiner MC, Ennaoui A. Inkjet-

Printed Cu2ZnSn(S, Se)4 Solar Cells. Advanced Science. 2015;2:1500028.

[35] Shin S, Kumar R, Roh JW, Ko D-S, Kim H-S, Kim SI, et al. High-Performance Screen-Printed Thermoelectric Films on Fabrics. Scientific Reports. 2017;7:7317.

[36] Linghu C, Zhang S, Wang C, Song J. Transfer printing techniques for flexible and stretchable inorganic electronics. npj Flexible Electronics. 2018;2:26.

[37] Patidar R, Burkitt D, Hooper K, Richards D, Watson T. Slot-die coating of perovskite solar cells: An overview. Materials Today Communications. 2020;22:100808.

[38] Groenendaal L, Jonas F, Freitag D, Pielartzik H, Reynolds JR. Poly(3,4-ethylenedioxythiophene) and Its Derivatives: Past, Present, and Future. Advanced Materials. 2000;12:481-94.

[39] Wald G. The Receptors of Human Color Vision. Action spectra of three visual pigments in human cones account for normal color vision and color-blindness. 1964;145:1007-16.

[40] Gao T, Li Z, Huang P-s, Shenoy GJ, Parobek D, Tan S, et al. Hierarchical Graphene/Metal Grid Structures for Stable, Flexible Transparent Conductors. ACS Nano. 2015;9:5440-6.

[41] Zhang D, Wang R, Wen M, Weng D, Cui X, Sun J, et al. Synthesis of Ultralong Copper Nanowires for High-Performance Transparent Electrodes. Journal of the American Chemical Society. 2012;134:14283-6.

[42] Rathmell AR, Bergin SM, Hua Y-L, Li Z-Y, Wiley BJ. The Growth Mechanism of Copper Nanowires and Their Properties in Flexible, Transparent Conducting Films. Advanced Materials. 2010;22:3558-63.

[43] Tokuno T, Nogi M, Jiu J, Sugahara T, Suganuma K. Transparent Electrodes Fabricated via the Self-Assembly of Silver Nanowires Using a Bubble Template. Langmuir. 2012;28:9298-302.

[44] Guo H, Lin N, Chen Y, Wang Z, Xie Q, Zheng T, et al. Copper Nanowires as Fully Transparent Conductive Electrodes. Scientific Reports. 2013;3:2323.

[45] Yu Z, Zhang Q, Li L, Chen Q, Niu X, Liu J, et al. Highly Flexible Silver Nanowire Electrodes for Shape-Memory Polymer Light-Emitting Diodes. Advanced Materials. 2011;23:664-8.

[46] Wang J, Yan C, Kang W, Lee PS. High-efficiency transfer of percolating nanowire films for stretchable and transparent photodetectors. Nanoscale. 2014;6:10734-9.

[47] Lee J, Lee P, Lee H, Lee D, Lee SS, Ko SH. Very long Ag nanowire synthesis and its application in a highly transparent, conductive and flexible metal electrode touch panel. Nanoscale. 2012;4:6408-14.

[48] Suh YD, Hong S, Lee J, Lee H, Jung S, Kwon J, et al. Random nanocrack, assisted metal nanowire-bundled network fabrication for a highly flexible and transparent conductor. RSC Advances. 2016;6:57434-40.

[49] Sciacca B, van de Groep J, Polman A, Garnett EC. Solution-Grown Silver Nanowire Ordered Arrays as Transparent Electrodes. Advanced Materials. 2016;28:905-9.

[50] Lee J-G, Kim D-Y, Lee J-H, Sinha-Ray S, Yarin AL, Swihart MT, et al. Production of Flexible Transparent Conducting Films of Self-Fused Nanowires via One-Step Supersonic Spraying. Advanced Functional Materials. 2017;27:1602548-n/a.

[51] Liu J-W, Wang J-L, Wang Z-H, Huang W-R, Yu S-H. Manipulating

Nanowire Assembly for Flexible Transparent Electrodes. Angewandte Chemie International Edition. 2014;53:13477-82.

[52] Gong S, Zhao Y, Yap LW, Shi Q, Wang Y, Bay JAPB, et al. Fabrication of Highly Transparent and Flexible NanoMesh Electrode via Self-assembly of Ultrathin Gold Nanowires. Advanced Electronic Materials. 2016;2:1600121-n/a.

[53] Wu H, Hu L, Rowell MW, Kong D, Cha JJ, McDonough JR, et al. Electrospun Metal Nanofiber Webs as High-Performance Transparent Electrode. Nano Letters. 2010;10:4242-8.

[54] Kang MG, Guo LJ. Nanoimprinted Semitransparent Metal Electrodes and Their Application in Organic Light-Emitting Diodes. Advanced Materials. 2007;19:1391-6.

[55] Guo CF, Sun T, Liu Q, Suo Z, Ren Z. Highly stretchable and transparent nanomesh electrodes made by grain boundary lithography. Nat Commun. 2014;5.

[56] Park JH, Lee DY, Kim Y-H, Kim JK, Lee JH, Park JH, et al. Flexible and Transparent Metallic Grid Electrodes Prepared by Evaporative Assembly. ACS Applied Materials & Interfaces. 2014;6:12380-7.

[57] Kang M-G, Joon Park H, Hyun Ahn S, Jay Guo L. Transparent Cu nanowire mesh electrode on flexible substrates fabricated by transfer printing and its application in organic solar cells. Solar Energy Materials and Solar Cells. 2010;94:1179-84.

[58] Kiruthika S, Rao KDM, Ankush K, Ritu G, Kulkarni GU. Metal wire network based transparent conducting electrodes fabricated using interconnected crackled layer as template. Materials Research Express. 2014;1:026301.

[59] Kang M-G, Kim M-S, Kim J, Guo LJ. Organic Solar Cells Using Nanoimprinted Transparent Metal Electrodes. Advanced Materials. 2008;20:4408-13.

[60] Zhang C, Khan A, Cai J, Liang C, Liu Y, Deng J, et al. Stretchable Transparent Electrodes with Solution-Processed Regular Metal Mesh for an Electroluminescent Light-Emitting Film. ACS Applied Materials & Interfaces. 2018;10:21009-17.

[61] Chen X, Nie S, Guo W, Fei F, Su W, Gu W, et al. Printable High-Aspect Ratio and High-Resolution Cu Grid Flexible Transparent Conductive Film with Figure of Merit over 80 000. Advanced Electronic Materials. 2019;5:1800991.

[62] Yonghee J, Jihoon K, Doyoung B. Invisible metal-grid transparent electrode prepared by electrohydrodynamic (EHD) jet printing. Journal of Physics D: Applied Physics. 2013;46:155103.

[63] Jang S, Jung W-B, Kim C, Won P, Lee S-G, Cho KM, et al. A three-dimensional metal grid mesh as a practical alternative to ITO. Nanoscale. 2016;8:14257-63.

[64] Huang Y, Bai X, Zhou M, Liao S, Yu Z, Wang Y, et al. Large-Scale Spinning of Silver Nanofibers as Flexible and Reliable Conductors. Nano Letters. 2016;16:5846-51.

[65] Schneider J, Rohner P, Thureja D, Schmid M, Galliker P, Poulikakos D. ElectrohydrodynamicNanoDripPrinting of High Aspect Ratio Metal Grid Transparent Electrodes. Advanced Functional Materials. 2016;26:833-40.

[66] Bai X, Liao S, Huang Y, Song J, Liu Z, Fang M, et al. Continuous Draw Spinning of Extra-Long Silver Submicron Fibers with Micrometer Patterning Capability. Nano Letters. 2017;17:1883-91.

[67] An S, Jo HS, Kim D-Y, Lee HJ, Ju B-K, Al-Deyab SS, et al. Self-Junctioned Copper Nanofiber Transparent Flexible Conducting Film via Electrospinning and Electroplating. Advanced Materials. 2016;28:7149-54.

[68] Xian Z, Han B, Li S, Yang C, Wu S, Lu X, et al. A Practical ITO Replacement Strategy: Sputtering-Free Processing of a Metallic Nanonetwork. Advanced Materials Technologies. 2017;2:1700061-n/a.

[69] Khan A, Huang Y-T, Miyasaka T, Ikegami M, Feng S-P, Li W-D. Solution-Processed Transparent Nickel-Mesh Counter Electrode with in-Situ Electrodeposited Platinum Nanoparticles for Full-Plastic Bifacial Dye-Sensitized Solar Cells. ACS Applied Materials & Interfaces. 2017;9:8083-91.

[70] Zhang C, Cai J, Liang C, Khan A, Li W-D. Scalable Fabrication of Metallic Nanofiber Network via Templated Electrodeposition for Flexible Electronics. Advanced Functional Materials. 2019;29:1903123.

[71] An BW, Gwak E-J, Kim K, Kim Y-C, Jang J, Kim J-Y, et al. Stretchable, Transparent Electrodes as Wearable Heaters Using Nanotrough Networks of Metallic Glasses with Superior Mechanical Properties and Thermal Stability. Nano Letters. 2016;16:471-8.

[72] Chen X, Guo W, Xie L, Wei C, Zhuang J, Su W, et al. Embedded Ag/Ni Metal-Mesh with Low Surface Roughness As Transparent Conductive Electrode for Optoelectronic Applications. ACS Applied Materials & Interfaces. 2017;9:37048-54.

[73] Jang J, Im H-G, Jin J, Lee J, Lee J-Y, Bae B-S. A Flexible and Robust Transparent Conducting Electrode Platform Using an Electroplated Silver Grid/Surface-Embedded Silver Nanowire Hybrid Structure. ACS Applied Materials & Interfaces. 2016;8:27035-43.

[74] Iwahashi T, Yang R, Okabe N, Sakurai J, Lin J, Matsunaga D. Nanoimprint-assisted fabrication of high haze metal mesh electrode for solar cells. Applied Physics Letters. 2014;105:223901.

[75] Gupta R, Rao KDM, Srivastava K, Kumar A, Kiruthika S, Kulkarni GU. Spray Coating of Crack Templates for the Fabrication of Transparent Conductors and Heaters on Flat and Curved Surfaces. ACS Applied Materials & Interfaces. 2014;6:13688-96.

[76] Zhou L, Xiang H-Y, Shen S, Li Y-Q, Chen J-D, Xie H-J, et al. High-Performance Flexible Organic Light-Emitting Diodes Using Embedded Silver Network Transparent Electrodes. ACS Nano. 2014;8:12796-805.

[77] Kang J, Jang Y, Kim Y, Cho S-H, Suhr J, Hong BH, et al. An Ag-grid/graphene hybrid structure for large-scale, transparent, flexible heaters. Nanoscale. 2015;7:6567-73.

[78] Hak-Jong C, Soyoung C, Pil-Hoon J, Ju-Hyeon S, Yang-Doo K, Heon L. Uniformly embedded silver nanomesh as highly bendable transparent conducting electrode. Nanotechnology. 2015;26:055305.

[79] Mao L, Chen Q, Li Y, Li Y, Cai J, Su W, et al. Flexible Silver Grid/PEDOT:PSS Hybrid Electrodes For Large Area Inverted Polymer Solar Cells. Nano Energy. 2014;10:259-67.

[80] Im H-G, Jeong S, Jin J, Lee J, Youn D-Y, Koo W-T, et al. Hybrid crystalline-ITO/metal nanowire mesh transparent electrodes and their application for highly flexible perovskite solar cells. Npg Asia Materials. 2016;8:e282.

[81] Zou J, Yip H-L, Hau SK, Jen AK-Y. Metal grid/conducting polymer hybrid

transparent electrode for inverted polymer solar cells. Applied Physics Letters. 2010;96:203301.

[82] Mayousse C, Celle C, Fraczkiewicz A, Simonato J-P. Stability of silver nanowire based electrodes under environmental and electrical stresses. Nanoscale. 2015;7:2107-15.

[83] Hsu P-C, Wang S, Wu H, Narasimhan VK, Kong D, Ryoung Lee H, et al. Performance enhancement of metal nanowire transparent conducting electrodes by mesoscale metal wires. Nat Commun. 2013;4.

[84] Bao W, Wan J, Han X, Cai X, Zhu H, Kim D, et al. Approaching the limits of transparency and conductivity in graphitic materials through lithium intercalation. Nat Commun. 2014;5.

[85] Ok JG, Kwak MK, Huard CM, Youn HS, Guo LJ. Photo–Roll Lithography (PRL) for Continuous and Scalable Patterning with Application in Flexible Electronics. Advanced Materials. 2013;25:6554-61.

[86] Au - Khan A, Au - Lee S, Au - Jang T, Au - Xiong Z, Au - Zhang C, Au - Tang J, et al. Scalable Solution-processed Fabrication Strategy for High-performance, Flexible, Transparent Electrodes with Embedded Metal Mesh. JoVE. 2017:e56019.

[87] Galagan Y, J.M. Rubingh J-E, Andriessen R, Fan C-C, W.M. Blom P, C. Veenstra S, et al. ITO-free flexible organic solar cells with printed current collecting grids. Solar Energy Materials and Solar Cells. 2011;95:1339-43.

[88] Lee J-Y, Connor ST, Cui Y, Peumans P. Solution-Processed Metal Nanowire Mesh Transparent Electrodes. Nano Letters. 2008;8:689-92.

[89] Yim JH, Joe S-y, Pang C, Lee KM, Jeong H, Park J-Y, et al. Fully Solution-Processed Semitransparent Organic Solar Cells with a Silver Nanowire Cathode and a Conducting Polymer Anode. ACS Nano. 2014;8:2857-63.

[90] Hauger TC, Al-Rafia SMI, Buriak JM. Rolling Silver Nanowire Electrodes: Simultaneously Addressing Adhesion, Roughness, and Conductivity. ACS Applied Materials & Interfaces. 2013;5:12663-71.

[91] Meng Y, Hu Z, Ai N, Jiang Z, Wang J, Peng J, et al. Improving the Stability of Bulk Heterojunction Solar Cells by Incorporating pH-Neutral PEDOT:PSS as the Hole Transport Layer. ACS Applied Materials & Interfaces. 2014;6:5122-9.

[92] Huo D, Kim MJ, Lyu Z, Shi Y, Wiley BJ, Xia Y. One-Dimensional Metal Nanostructures: From Colloidal Syntheses to Applications. Chemical Reviews. 2019;119:8972-9073.

[93] Dou L, Cui F, Yu Y, Khanarian G, Eaton SW, Yang Q, et al. Solution-Processed Copper/Reduced-Graphene-Oxide Core/Shell Nanowire Transparent Conductors. ACS Nano. 2016;10:2600-6.

[94] Chen Z, Ye S, Stewart IE, Wiley BJ. Copper Nanowire Networks with Transparent Oxide Shells That Prevent Oxidation without Reducing Transmittance. ACS Nano. 2014;8:9673-9.

[95] Hsu P-C, Wu H, Carney TJ, McDowell MT, Yang Y, Garnett EC, et al. Passivation Coating on Electrospun Copper Nanofibers for Stable Transparent Electrodes. ACS Nano. 2012;6:5150-6.

[96] Sajjad MT, Park J, Gaboriau D, Harwell JR, Odobel F, Reiss P, et al. CuSCN Nanowires as Electrodes for p-Type Quantum Dot Sensitized Solar Cells: Charge Transfer Dynamics and Alumina Passivation. The Journal of Physical Chemistry C. 2018;122:5161-70.

[97] Bulliard X, Ihn S-G, Yun S, Kim Y, Choi D, Choi J-Y, et al. Enhanced Performance in Polymer Solar Cells by Surface Energy Control. Advanced Functional Materials. 2010;20:4381-7.

[98] Khan A, Rahman K, Hyun M-T, Kim D-S, Choi K-H. Multi-nozzle electrohydrodynamic inkjet printing of silver colloidal solution for the fabrication of electrically functional microstructures. Appl Phys A. 2011;104:1113-20.

[99] Jo HS, An S, Lee J-G, Park HG, Al-Deyab SS, Yarin AL, et al. Highly flexible, stretchable, patternable, transparent copper fiber heater on a complex 3D surface. Npg Asia Materials. 2017;9:e347.

[100] Khan A. Novel Embedded Metal-mesh Transparent Electrodes: Vacuum-free Fabrication Strategies and Applications in Flexible Electronic Devices. Singapore: Springer Nature; 2020.

[101] Mattevi C, Eda G, Agnoli S, Miller S, Mkhoyan KA, Celik O, et al. Evolution of Electrical, Chemical, and Structural Properties of Transparent and Conducting Chemically Derived Graphene Thin Films. Advanced Functional Materials. 2009;19:2577-83.

[102] Park S, Lee K-S, Bozoklu G, Cai W, Nguyen ST, Ruoff RS. Graphene Oxide Papers Modified by Divalent Ions—Enhancing Mechanical Properties via Chemical Cross-Linking. ACS Nano. 2008;2:572-8.

[103] Reina A, Jia X, Ho J, Nezich D, Son H, Bulovic V, et al. Large Area, Few-Layer Graphene Films on Arbitrary Substrates by Chemical Vapor Deposition. Nano Letters. 2009;9:30-5.

[104] De S, King PJ, Lotya M, O'Neill A, Doherty EM, Hernandez Y, et al. Flexible, Transparent, Conducting Films of Randomly Stacked Graphene from Surfactant-Stabilized, Oxide-Free Graphene Dispersions. Small. 2010;6:458-64.

[105] Bae S, Kim H, Lee Y, Xu X, Park J-S, Zheng Y, et al. Roll-to-roll production of 30-inch graphene films for transparent electrodes. Nat Nano. 2010;5:574-8.

[106] Bae S-Y, Jeon I-Y, Yang J, Park N, Shin HS, Park S, et al. Large-Area Graphene Films by Simple Solution Casting of Edge-Selectively Functionalized Graphite. ACS Nano. 2011;5:4974-80.

[107] Baughman RH, Zakhidov AA, de Heer WA. Carbon Nanotubes--the Route Toward Applications. Science. 2002;297:787-92.

[108] Iijima S. Helical microtubules of graphitic carbon. Nature. 1991;354:56-8.

[109] Tasis D, Tagmatarchis N, Bianco A, Prato M. Chemistry of Carbon Nanotubes. Chemical Reviews. 2006;106:1105-36.

[110] Wu Z, Chen Z, Du X, Logan JM, Sippel J, Nikolou M, et al. Transparent, Conductive Carbon Nanotube Films. Science. 2004;305:1273-6.

[111] Jo JW, Jung JW, Lee JU, Jo WH. Fabrication of Highly Conductive and Transparent Thin Films from Single-Walled Carbon Nanotubes Using a New Non-ionic Surfactant via Spin Coating. ACS Nano. 2010;4:5382-8.

[112] Pint CL, Xu Y-Q, Moghazy S, Cherukuri T, Alvarez NT, Haroz EH, et al. Dry Contact Transfer Printing of Aligned Carbon Nanotube Patterns and Characterization of Their Optical Properties for Diameter Distribution and Alignment. ACS Nano. 2010;4:1131-45.

[113] Reuter AESKWLoUMK. PEDOT: Principles and Applications of an Intrinsically Conductive Polymer. FL: CRC Press,; 2010.

[114] Lipomi DJ, Lee JA, Vosgueritchian M, Tee BCK, Bolander JA, Bao Z. Electronic Properties of Transparent Conductive Films of PEDOT:PSS on Stretchable Substrates. Chemistry of Materials. 2012;24:373-82.

[115] Das TK, Prusty S. Review on Conducting Polymers and Their Applications. Polymer-Plastics Technology and Engineering. 2012;51:1487-500.

[116] Gangopadhyay R, De A. Conducting Polymer Nanocomposites: A Brief Overview. Chemistry of Materials. 2000;12:608-22.

[117] Nair RR, Blake P, Grigorenko AN, Novoselov KS, Booth TJ, Stauber T, et al. Fine Structure Constant Defines Visual Transparency of Graphene. Science. 2008;320:1308-.

[118] Wu J, Becerril HA, Bao Z, Liu Z, Chen Y, Peumans P. Organic solar cells with solution-processed graphene transparent electrodes. Applied Physics Letters. 2008;92:263302.

[119] Khan A, Rahman K, Kim DS, Choi KH. Direct printing of copper conductive micro-tracks by multi-nozzle electrohydrodynamic inkjet printing process. Journal of Materials Processing Technology. 2012;212:700-6.

[120] Binda M, Natali D, Iacchetti A, Sampietro M. Integration of an Organic Photodetector onto a Plastic Optical Fiber by Means of Spray Coating Technique. Advanced Materials. 2013;25:4335-9.

[121] Mahmood K, Swain BS, Jung HS. Controlling the surface nanostructure of ZnO and Al-doped ZnO thin films using electrostatic spraying for their application in 12% efficient perovskite solar cells. Nanoscale. 2014;6:9127-38.

[122] Khan A, Roo JS, Kraus T, Steimle J. Soft Inkjet Circuits: Rapid Multi-Material Fabrication of Soft Circuits using a Commodity Inkjet Printer. Proceedings of the 32nd Annual ACM Symposium on User Interface Software and Technology. New Orleans, LA, USA: Association for Computing Machinery; 2019. p. 341-54.

[123] Shimoni A, Azoubel S, Magdassi S. Inkjet printing of flexible high-performance carbon nanotube transparent conductive films by "coffee ring effect". Nanoscale. 2014;6:11084-9.

[124] Yunker PJ, Still T, Lohr MA, Yodh AG. Suppression of the coffee-ring effect by shape-dependent capillary interactions. Nature. 2011;476:308-11.

[125] Layani M, Gruchko M, Milo O, Balberg I, Azulay D, Magdassi S. Transparent Conductive Coatings by Printing Coffee Ring Arrays Obtained at Room Temperature. ACS Nano. 2009;3:3537-42.

[126] Fukuda K, Someya T. Recent Progress in the Development of Printed Thin-Film Transistors and Circuits with High-Resolution Printing Technology. Advanced Materials. 2017;29:1602736.

[127] Overgaard MH, Kühnel M, Hvidsten R, Petersen SV, Vosch T, Nørgaard K, et al. Highly Conductive Semitransparent Graphene Circuits Screen-Printed from Water-Based Graphene Oxide Ink. Advanced Materials Technologies. 2017;2:1700011.

[128] Zhou L, Yu M, Chen X, Nie S, Lai W-Y, Su W, et al. Screen-Printed Poly(3,4-Ethylenedioxythiophene):Poly(Styrenesulfonate) Grids as ITO-Free Anodes for Flexible Organic Light-Emitting Diodes. Advanced Functional Materials. 2018;28:1705955.

[129] Hösel M, Søndergaard RR, Angmo D, Krebs FC. Comparison of Fast Roll-to-Roll Flexographic, Inkjet, Flatbed, and Rotary Screen Printing of

Metal Back Electrodes for Polymer Solar Cells. Advanced Engineering Materials. 2013;15:995-1001.

[130] Novoselov KS, Fal'ko VI, Colombo L, Gellert PR, Schwab MG, Kim K. A roadmap for graphene. Nature. 2012;490:192-200.

[131] Suk JW, Kitt A, Magnuson CW, Hao Y, Ahmed S, An J, et al. Transfer of CVD-Grown Monolayer Graphene onto Arbitrary Substrates. ACS Nano. 2011;5:6916-24.

[132] You P, Liu Z, Tai Q, Liu S, Yan F. Efficient Semitransparent Perovskite Solar Cells with Graphene Electrodes. Advanced Materials. 2015;27:3632-8.

[133] Sandström A, Dam HF, Krebs FC, Edman L. Ambient fabrication of flexible and large-area organic light-emitting devices using slot-die coating. Nature Communications. 2012;3:1002.

[134] Cheng P, Bai H, Zawacka NK, Andersen TR, Liu W, Bundgaard E, et al. Roll-Coated Fabrication of Fullerene-Free Organic Solar Cells with Improved Stability. Advanced Science. 2015;2:1500096.

[135] Lucera L, Machui F, Kubis P, Schmidt HD, Adams J, Strohm S, et al. Highly efficient, large area, roll coated flexible and rigid OPV modules with geometric fill factors up to 98.5% processed with commercially available materials. Energy & Environmental Science. 2016;9:89-94.

[136] Kim D-J, Shin H-I, Ko E-H, Kim K-H, Kim T-W, Kim H-K. Roll-to-roll slot-die coating of 400 mm wide, flexible, transparent Ag nanowire films for flexible touch screen panels. Scientific Reports. 2016;6:34322.

[137] Shin K, Park J, Lee C. A 250-mm-width, flexible, and continuous roll-to-roll slot-die coated carbon nanotube/silver nanowire film fabrication and a study on the effect of anti-reflective overcoat. Thin Solid Films. 2016;598:95-102.

[138] Hu X, Meng X, Xiong J, Huang Z, Yang X, Tan L, et al. Roll-to-Roll Fabrication of Flexible Orientated Graphene Transparent Electrodes by Shear Force and One-Step Reducing Post-Treatment. Advanced Materials Technologies. 2017;2:1700138.

[139] Jakubka F, Heyder M, Machui F, Kaschta J, Eggerath D, Lövenich W, et al. Determining the coating speed limitations for organic photovoltaic inks. Solar Energy Materials and Solar Cells. 2013;109:120-5.

[140] Nam J, Carvalho MS. Flow visualization and operating limits of tensioned-web-over slot die coating process. Chemical Engineering and Processing: Process Intensification. 2011;50:471-7.

Chapter 17

Calculation of the Electronic Properties and Reactivity of Nanoribbons

Pedro Navarro-Santos, Rafael Herrera-Bucio, Judit Aviña-Verduzco and Jose Luis Rivera

Abstract

It has been demonstrated that matter at low dimensionality exhibits novel properties, which could be used in promising applications. An effort to understand their behavior is being through the application of computational methods providing strategies to study structures, which present greater experimental challenges. It is proven that thin and narrow carbon-based nanostructures, such as, nanoribbons show promising tunable electronic properties, particularly when they are substitutionally functionalized. This chapter is proposed as a guidance to help the readers to apply conceptual density functional theory to calculate helpful intrinsic properties, e. g., energetic, electronic and reactivity of one-dimension nanomaterial's, such as, carbon nanoribbons. As a case of study, it is discussed the effect of boron atoms on the properties of pristine carbon nanoribbons concerning the main aspect and considerations must take into account in their computational calculations.

Keywords: DFT, band structure, DOS, MEP

1. Introduction

Carbon nanoribbons (CNRs) are strips of graphene whose edges symmetry, width and cut orientation give them specific electronic properties. These carbon nanostructures have attracted the attention in both experimental and theoretical fields because of their peculiar properties, which have been studied widely in the last decade as a function of topology, width, as well as doping. [1–5] Depending the chain-type along the periodic direction, carbon nanoribbons are commonly classified either armchair carbon nanoribbons (ACNR) when these grow through dimer chains, or zigzag carbon nanoribbons (ZCNR) if those have zigzag type chains along the periodic direction. **Figure 1** shows a pristine ACNR and ZCNR respectively, their distances between their C – C edged lengths are 13.44 and 24.19 Å respectively, although there could be named referring their length and width (MxN), in such case, both CNRs shown in **Figure 1** are 12x2 size.

Through different experimental techniques, it is possible to obtain carbon nanoribbons. [6–8] However, these techniques have not succeeded in controlling the edges shape of carbon nanoribbons. For example, Cai et al. [9] have proposed a chemical technique which is able to synthesize narrow nanoribbons having symmetric edges, so that, it is possible to obtain experimentally carbon nanoribbons with perfect edges

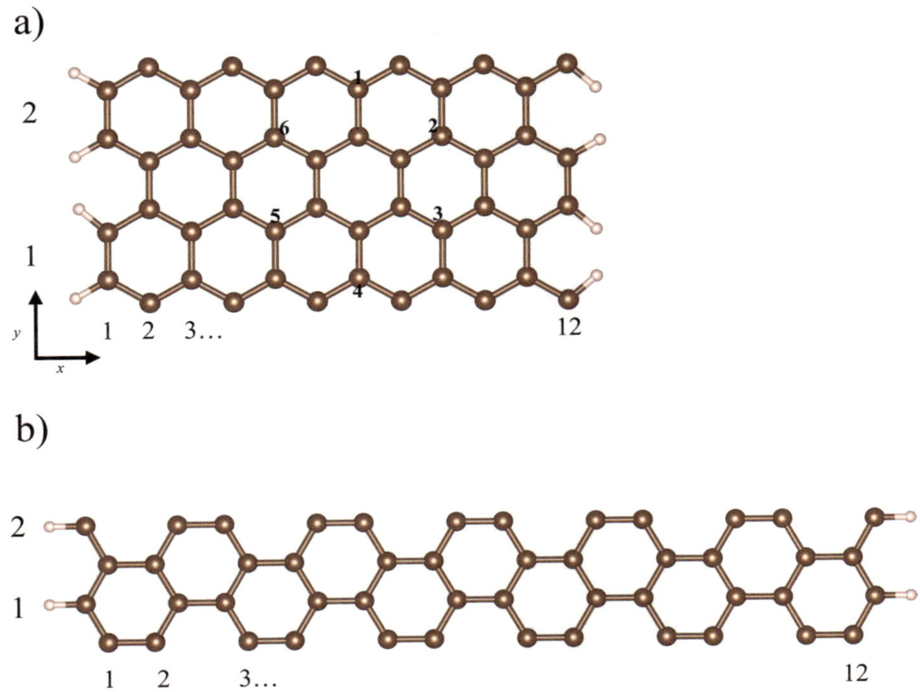

Figure 1.
Optimized structure of bare (a) ACNR and (b) ZCNR of size 12x2.

and specific topology. To date, succeeding methods to obtain CNRs come from two different strategies, namely, top-down, which refers to break down large performed carbon-base structures, i. e., CNTs and multiwall CNTs (MWCNTs) and bottom-up, i. e., using several chemical reactions to tailor building-blocks into a complex structure. **Table 1** shows a comparative chart representing synthetic strategies to obtain CNRs, employed characterization techniques, advantages and disadvantages.

Because of their finite dimension, at nanoscale, CNRs have peculiar properties associated to their electronic states close the edges, playing an important role on the reactivity. [17–22] Several theoretical models, e. g., tight binding, all electron techniques, density functional theory (DFT), etc., have been applied to explore the electronic properties, magnetic states or band structure of carbon nanoribbons. [1, 5, 23] Some of them, have focused on the zigzag topology because they intrinsically have dangling bonds at the edges. This behavior provides active sites for chemical reactions. Moreover, ZCNRs have peculiar properties, e.g., theoretical calculations have shown that ZCNRs have localized electrons largely on the edge C atoms close to the Fermi level. [4, 22] This large contribution of electronic states forms two-fold degenerate flat band at Fermi level, such that, the ground state has spin coupling of each edge ferromagnetic whereas between edges antiferromagnetic. Despite zigzag edges of synthesized carbon nanoribbons have been observed, [8] there is not direct experimental evidence about the magnetic states of ZCNRs. It was theoretically suggested that magnetism of ZCNR could be destroyed substituting defects or vacancies directly on carbon edges. [24]

On the other hand, all hydrogen-passivated ACNRs are semiconducting [22]. However, ACNRs are expected to reach the graphene limit of zero band gap for sufficiently large widths. [25]

Concerning these fascinating properties, CNRs may fit for promising technological applications, mainly if the presence of donor or acceptor impurities bring

Strategies	Method	Characteristics	Advantages	Disadvantages
Top-down	Unzipping CNTs by chemical oxidation [7]	Using potassium permanganate and sulfuric acid	The production of 100 nm wide nanoribbons	
	Etching of graphene [10]		It is possible to narrow the ribbons down to <10 nm using gas phase etching chemistry	Complicated procedures, CNRs with edges abnormalities
	Treating multiwalled CNTs [11, 12]	Longitudinal splitting of MWCNTs using hydrothermal approach	This procedure gives highly conducting CNRs with over 80% yield at low cost fabrication.	
		Intercalating lithium and ammonia into MWCNTs followed by thermal expansion	Graphene flakes attractive for many applications	Structures with large number of non-symmetric edge atoms
	Chemical Procedure Organic synthesis	Mechanical exfoliation of highly oriented pyrolyzed graphite	Produce micrometer length involving non complicated procedures	
		Chemical oxidation/ exfoliation of graphite followed by reduction of the resulting nanomaterial [13]		Resulting graphene oxide
		Cross coupling building blocks followed by dehydrogenation		Complicated procedures poorly defined edges
		Conversion of precursors inside CNTs	Because CNTs impose spatial limitations on the structure of the product, may obtain narrow CNRs	
Bottom-up	Organic synthesis [14–16]	Surface assisted polymerization followed by dehydrogenation in an ultra-high vacuum environment	Defined edge type and narrow widths, potential techniques for scale-up	Depends on the precursor's nature, which defines the ribbon's dimension

Table 1.
Comparative chart of synthetic methods to obtain nanoribbons and their advantages or disadvantages.

specific reactive properties. [26, 27] So that, this chapter is proposed as a guidance to help the readers to apply conceptual density functional theory to calculate helpful intrinsic properties, e. g., energetic, electronic and reactivity of one-dimension nanomaterial's, such as, carbon nanoribbons in order to predict or tune their properties; particularly when they are substitutional doped.

2. Structural and energetic properties

To give insights about the structural stability of nanostructures, firstly, it is suggested to evaluate if the proposal unit cell may array forming a stable crystalline state. Usually, a structural analysis is carried out computing the cohesive energy per atoms o per unit cell. The cohesive energy (*EC*) is the energy required to disassemble a molecular system into its constituent parts. From a physical point of view, a bound (stable) system has a positive value of *EC*, which represents the energy gained during the formation of the bound state. To calculate the *EC* of ACNRs, it is necessary to obtain the optimized energy of the unit cell being aware of the well converged energy with respect to the *k*-points and the cutoff energy for the planewave basis set, evaluating the impact of the exchange-correlation functional used and its ability to accurately describe both the atom and bulk phase.

Although *EC* is a reference to know the stability of bulk materials, it differs from a nanoparticle. [28–30] At nanoscale, size effects on the cohesive energy of nanoparticles has been demonstrated, which decreases with decreasing the particle size. [31] However, slight differences of *EC* are found when nuclei radii of constituent are similar, which do difficult to analyze or find a trend, e. g., the effect of the relative position of dopants along the NRs. For example, **Table 2** shows the calculated values of *EC* of armchair carbon nanoribbons (ACNR) doped with boron atoms in randomly (ACNR-R) and forming one B nanoisland (ACNR-I) arrangements compared with those pristine ACNRS of size 16x2, 20x2, 16x2 and 20x4 respectively. [32] The arrangement of the nanoisland (ACNR-I) explained in this section is shown in part (a) of **Figure 1** numbering from 1 to 6 the C atoms are substituted for impurities. Note that, B doping slightly reduces the cohesive energy of ACNRs compared with the pristine ones with similar *EC* values mainly found in the largest B-ACNRs. However, at lower doping concentrations, i. e. in the case of the largest ACNR (20x4) very close values of *EC* are obtained which makes difficult to observe a trend.

Because of these CNRs has 3 different chemical species, *EC* does not provide a suitable way to evaluate the relative stability. **Table 2** also shows in brackets the calculated values of the Gibbs free formation energy to take into account the chemical composition of ACNRs. The relative thermodynamic stability that is considered to evaluate the relative stability of multicomponent systems. This approach has been used in binary and tertiary phase thermodynamics and nanostructures other than NRs. [25, 33, 34] it can be calculated by using the following expression:

$$\delta G = E(x) + \sum_{i=1}^{n} x_i \mu_i \quad (1)$$

MxN	Pristine	ACNR -R	ACNR-I
16x2	7.224 (0.003)	6.992 (−0.272)	7.003 (−0.291)
20x2	7.338 (0.002)	7.143 (−0.224)	7.158 (−0.239)
16x4	7.225 (0.003)	7.112 (−0.142)	7.116 (−0.147)
20x4	7.338 (0.002)	7.249 (−0.119)	7.250 (−0.130)

Table 2.
Cohesive per atom (Gibbs free) energy in eV of pristine and B-doped ACNRs of randomly (ANCR-R) and forming a B-nanoisland (ACNR-I) [32].

where $E(x)$ is the binding energy per atom of the B-ACNR for the example shown in **Table 1**, x_i corresponds to the molar fraction of the conformant components (H, N, B, C) which satisfies $\sum x_i = 1$, where $x_i = \frac{n_i}{n_T}$, being n_i the number of atoms of specie i in the unit cell and n_T the total number of atoms conforming the unit cell. The chemical potential (μ_i) can be approximate as the binding energy per atom of the singlet ground state of the H_2, the triplet ground state of the B_2 molecule and the cohesive energy per atom of the graphene sheet respectively. Note that positive values of δG represent a metastable structure with respect to the conforming constituents, whereas negative values of δG refer to stable structures in accordance with their constituents. As we can observe in **Table 2**, δG suggests that the arrangement of B-nanoisland leads to stabilize energetically the pristine carbon nanoribbons more than the randomly cases.

3. Electronic properties of nanoribbons

The electronic properties of nanoribbons can be inferred from the band structure and total and local density of states (DOS and LDOS) respectively. For the case of NRs, these calculations are relatively simple because they are computed sampling the Brillouin zone only in one direction, i. e., the grown direction from 0 to gamma point. We recommend to use a denser grid than the case of the total energy calculations, including a Gaussian smearing (of width 0.01 eV) to improve the convergence of the integrals of the energy levels for the band structure calculations, for DOS calculations, to use the tetrahedron method with Blöchl corrections. [35, 36]

Pristine CNRs with hydrogen passivated armchair edges passivated are direct bandgap semiconducting, which decreases as their width increases. The edges of ACNRs play an important role on their electronic properties and reactivity because of quantum confinement gaps, which can be characterized by $\Delta_{Na} \sim w_a^{-1}$. [19, 23, 37]

In order to evaluate the electronic nature of nanoribbons, firstly, spin-polarized and non-spin polarized solutions of the Kohn-Sham equations must be taken into account to evaluate possible magnetic configurations, as found in zigzag carbon nanoribbons, [38] that implies the magnetic state is the most stable. For armchair ribbons, the non-magnetic state is always the most stable [22] even for ACNRs doped with boron atoms, [32] so that, for simplicity, we consider the armchair topology as a case of study.

The electronic behavior of ACNRs can be tuned for the influence of substitutional dopants. To illustrate this fact, we think about a unit cell containing one pristine CNR with even number of electrons of valence. If we replace only one carbon atom (with 4 valence electrons) for B (3 valence electrons) or N (5 valence electrons) such change gives one unit cell with odd number of valence electrons, in such cases is necessary to search for spin polarized solutions of the Kohn Sham equation, i. e, to evaluate if there are significant differences with respect to the non-spin polarized solution.

Figure 2 presents the band structure, total density of states and local density of states of dopants (shown in line red) of the 12x2 ACNR pristine, B-doped and N-doped substituting two dopants on positions 3 and 4 using the numbering shown in **Figure 1**. Note that, the pristine ACNR is a semiconducting in agreement with the literature [22] and the positive doping caused for the B moves the Fermi level (EF) to lower energies meanwhile the negative doping related with the N moves the EF

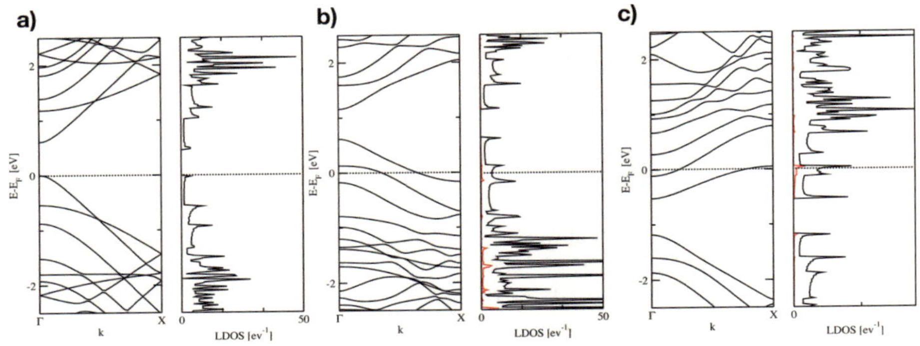

Figure 2.
Band structure and DOS of (a) pristine, (b) B-doped and (c) N-doped ACNRs of size 12x2.

to higher energies with respect to the pristine one. In both cases, the closest energy bands to EF are partially unoccupied and occupied respectively giving rise to metallic behavior in both cases.

4. Reactivity of nanoribbons

In order to explore the reactivity of 1D nanomaterial's, such as nanoribbons, it is mandatory to use appropriate reactivity descriptor. However, there is not a well stablish criteria to accomplish this task without prior knowing of an adsorption mechanism or experimental evidence, particularly for doped nanoribbons.

This is why, it is suggested the employment of two reactivity descriptors that are able to cover covalent and non-covalent interactions. The first one is the electrostatic potential, defined as:

$$V(\mathbf{r}) = \sum_A \frac{Z_A}{|\mathbf{R}_A - \mathbf{r}|} - \int \frac{\rho(\mathbf{r}')d\mathbf{r}'}{|\mathbf{r}' - \mathbf{r}|} \tag{2}$$

Where Z_A and R_A are the atomic number of nucleus A and its position respectively, $|R_A - r|$ is its distance from the point r and $\rho(r')$ is the electron density in each volume element. This descriptor provides the response of electron density when a positive unit charge is approaching, which is commonly plotted in a color scheme. Because of the electrostatic potential $V(\mathbf{r})$, is a local property, it has one value for each $\mathbf{r}$ point in the space surrounding a molecule or unit cell, so that, depending the nature of the ions (for instance positive or negative nature), the electrostatic potential will depend on the radial distance $\mathbf{r}$ from the nucleus. Commonly is followed that the contour of the electrostatic potential is plotted on the isovalue of the molecular electron density, for example, see the Bader's suggestion. [39] Be aware that, the chosen outer electron density contour depends on the Van der Waals radii of involving ions, which should reflect the molecular properties we want to observe, e. g., lone pairs, strained bonds, conjugated π systems, etc.

To illustrate, **Figure 3** shows the electrostatic potential of pristine, B-doped and N-doped carbon nanoribbons of size 12x2 plotted on the electron density surface of value 0.001 au, computed by using the generalized gradient approximation (GGA) in the form proposed by Perdew et al. for exchange-correlation

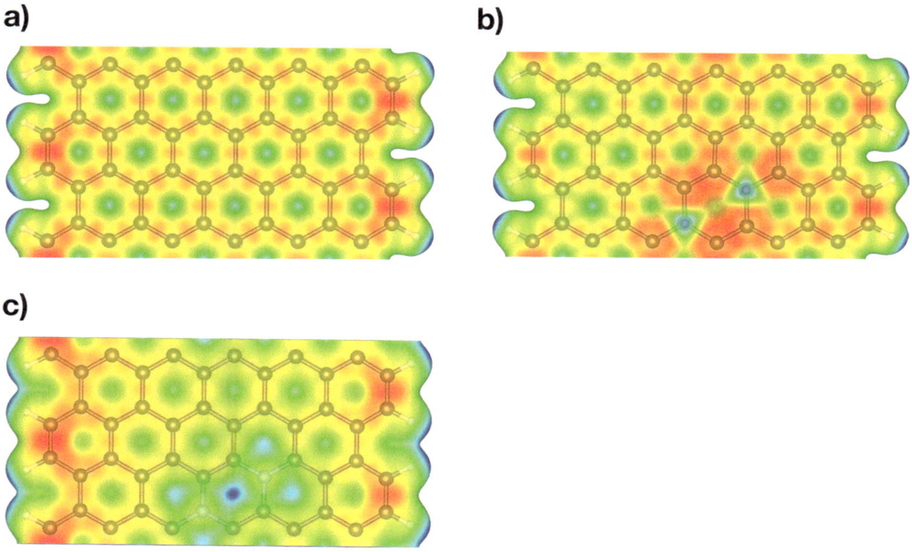

Figure 3.
Molecular electrostatic potential of the nanoribbons (a) pristine (b) B-doped and (c) N-doped of size 12x2.

functional. **Figure 3** was built in the software VESTA [40] plotting the charge file and then, adding the cube file containing the local potential. The color scheme used in **Figure 3** represents in blue, regions where a positive charge may repel each other, unlike in red, it represents regions where a positive charge, ion or chemical group can interact.

We can observe from **Figure 3a** that, hydrogens are weak positive meanwhile the delocalized charge is distributed along the carbon atoms, particularly found in the edged carbons, which is in agreement with the DOS of pristine ACNRs. The lacking of π electrons of the boron atoms is particularly observed in **Figure 3b**, which influences over their neighbor carbon atoms finding localized charge in such region. On the other hand, the N doping influences over the edges with more negative electrostatic potential than the pristine carbon nanoribbon.

The second reactivity descriptor is the Fukui or frontier functions (Ffs), helpful chemical reactivity descriptors for process controlled by electron transfer. Ffs were introduced for the first time by Parr and Yang, [41], which is convinced from the area of research so-called conceptual Density Functional Theory given by Geerlings in a comprehensive way. [42] Fukui functions play an important role linking the Molecular Orbital Theory with the HSAB principle, [43] they are defined as the change of the electronic density with respect to number of electrons (N), considering the nuclei position fixed, i.e. constant external potential $v(r)$:

$$f(r) = \left(\frac{\partial \rho(\mathbf{r})}{\partial N} \right)_{v(r)} \tag{3}$$

Due to the discontinuity of the above equation with respect to N, in a finite difference approximation three functions can be defined as:

$$f_{v,N}^{+}(r) = \rho_{v,N+1}(r) - \rho_{v,N}(r) \tag{4}$$

$$f_{v,N}^{-}(r)=\rho_{v,N}(r)-\rho_{v,N-1}(r) \tag{5}$$

$$f_{v,N}^{0}(r)=0.5\left(\rho_{v,N+1}(r)-\rho_{v,N-1}(r)\right) \tag{6}$$

Where $\rho_{v,N+1}(r)$, $\rho_{v,N}(r)$, and $\rho_{v,N-1}(r)$, are the electronic densities of the system with N + 1, N, and N–1 electrons, respectively, all with the ground state geometry of the N electron system. Expressions 4–6 concern the Fukui function for: nucleophilic attack, the chemical change where a molecule gains an electron; electrophilic attack, when a molecule loses charge; and for free radical attacks. [44]

Although, the finite difference approximation to the Fukui functions works for a specific set of configurations whilst for others is worthless to implement (i.e., full configuration interaction), [45] in most cases they are considered a reliable descriptor to indicate how the electron (incoming or outcoming) is redistributed in regions of the molecules. [46] Chemical reactivity is based on the assumption that, when molecules A and B interact in order to form a product AB, occurs a molecular densities-perturbation. [47] As the electronic density contains all sort of information, the chemical reactivity has to be reflected within its sensitivity to infinitesimal electron changes at constant external potential $v(r)$. Calculation of the frontier orbitals (HOMO or LUMO) are unambiguously defined. Within the frozen orbital approximation, [48] Ffs can be written in terms of the Kohn-Sham orbitals as follows:

$$f^{+}(r)=\left|\phi_{v,N}^{LUMO}(r)\right|^{2}+\sum_{i=1}^{N}\left(\frac{\partial\left|\phi_{i}(r)\right|}{\partial N}\right)\approx\rho_{v,N}^{LUMO}(r) \tag{7}$$

$$f^{-}(r)=\left|\phi_{v,N}^{HOMO}(r)\right|^{2}+\sum_{i=1}^{N}\left(\frac{\partial\left|\phi_{i}(r)\right|}{\partial N}\right)\approx\rho_{v,N}^{HOMO}(r) \tag{8}$$

In molecules, the relaxation term is usually very small for the discrete nature of Kohn Sham orbitals. So that, if Eqs. 7 and 8 neglect the second-order variations in the electron density, Ffs may approximate to the electron densities of its frontier orbitals.

On the other hand, referring to periodic systems, it is difficult to identify one frontier state because of the continuous character of the Blöch states, which makes difficult to compute the Fukui functions in DFT of the solid state. Although there is scarcely literature on this topic, a very useful reference for the numerical calculation of the condensed Ffs in periodic boundary conditions within the DFT applied to oxide bulk and surfaces is found here. [49]

One qualitative way to obtain Ffs for delocalized periodic systems, such as, the carbon nanoribbons is to extract its electron density and evaluate it by using the Eq. (7) and (8) respectively. From the electronic structure of these nanomaterials we can observe that only one occupied electronic band contributes below and above the Fermi level.

Figure 3 depicts the Ffs evaluated for electrophilic attacks respectively for B-doped and N-doped armchair carbon nanoribbons of size 12x2 with doping made on positions 3 and 4 using the numbering shown in **Figure 1**.

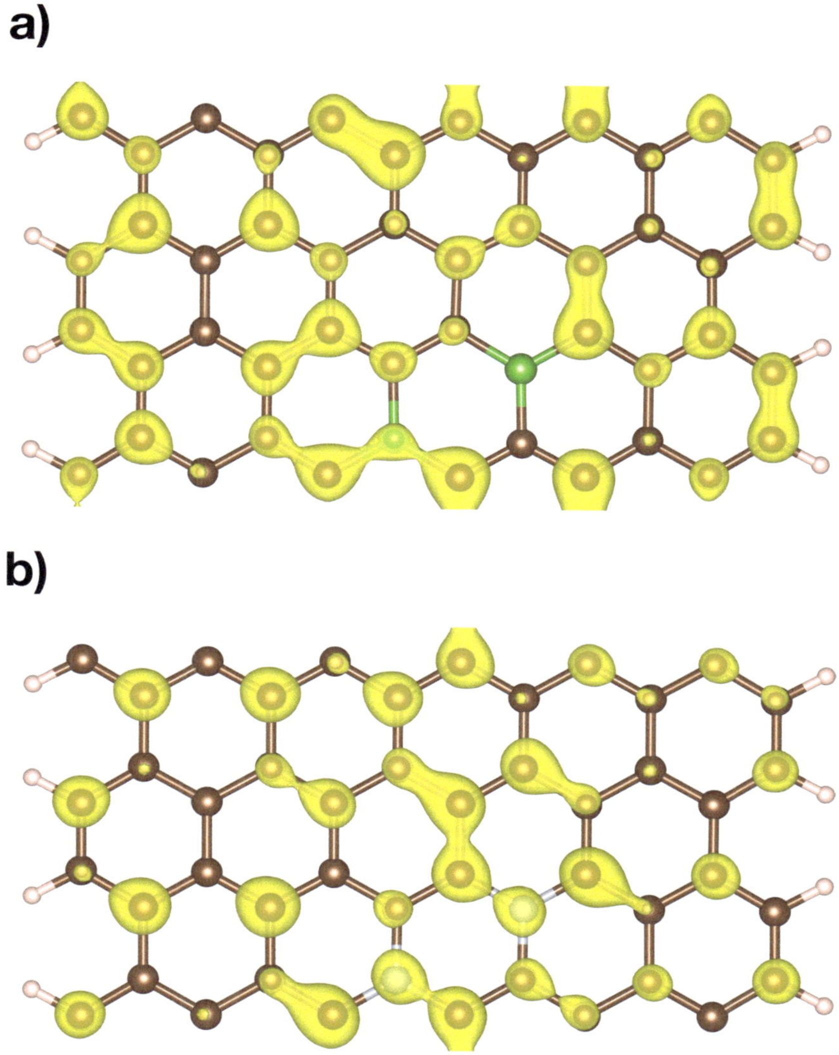

Figure 4.
Fukui functions for nucleophilic attack of (a) B-doped and (b) N-doped ACNRs of size 12x2.

We observe from **Figure 4** that the B atoms contributes to form regions where an electrophilic attack can occur on the doped nanoribbons, i. e. larges values of f^- mean regions where the ACNR will lose charge to stabilize it in a chemical change.

The electrostatic potential and the Fukui functions provide information on the local selectivity for donor-acceptor interactions. In here, the electrostatic potential describes the long-range non-covalent interactions. [50] The evaluation of the incoming charge distribution on nanoribbons states that "The Fukui function is strong while regions of a molecule are chemically softer than the regions where the Fukui function is weak. By invoking the hard and soft acid and bases (HSAB) principle [51] in a local sense, it is possible to establish the behavior of the different sites as function of hard or soft reagents (adsorbates)". [32, 52–54] **Figure 4** shows the Fukui functions for electrophilic attack, calculated by using Eq. (8), we observe the contribution of doping particularly on the neighboring carbon atoms. Indeed, from parts (b) and (c) of **Figure 2** is observed the electronic states of dopants contributing near the Fermi level.

5. Conclusions

In this chapter is presented how the energetic, electronic and reactivity of can be calculated for 1D nanomaterial's, such as, carbon nanoribbons. Although the carbon nanoribbons are used as case of study, this methodology can be applied for other kind of chemical compositions, in our experience we have explore the reactivity and stability of doped boron nitride at nanoscale. It is worthy to mention that, the evaluation of Fukui functions in periodic boundary conditions is limited in the usual computational approaches, so that, we suggest to support and compare such analysis with others e. g., charge analysis, global reactivity descriptors depending the nature of the involving chemical species.

Acknowledgements

PNS thanks to CONACYT for grant number 252239 and Cátedras CONACYT for Research Fellow.

Conflict of interest

The authors declare no conflict of interest.

Author details

Pedro Navarro-Santos[1*], Rafael Herrera-Bucio[2], Judit Aviña-Verduzco[2] and Jose Luis Rivera[3]

1 CONACYT-Universidad Michoacana de San Nicolás de Hidalgo, Edif. B-1, Francisco J. Múgica, s/n, Morelia 58030, Michoacán, Mexico

2 Instituto de Investigaciones Quimico-Biologicas, Universidad Michoacana de San Nicolás de Hidalgo, Edif. B-1, Francisco J. Múgica, s/n, Morelia 58030, Michoacán, Mexico

3 Facultad de Ciencias Fisico-Matematicas, Universidad Michoacana de San Nicolás de Hidalgo, Francisco J. Múgica, s/n, Morelia 58030, Michoacán, Mexico

*Address all correspondence to: pnavarrosa@conacyt.mx

References

[1] Nakada K, Fujita M, Dresselhaus G, Dresselhaus MS. Edge state in graphene ribbons: Nanometer size effect and edge shape dependence. Phys Rev B. 1996;**54**(24):17954-17961

[2] Martins TB, Miwa RH, da Silva AJR, Fazzio A. Electronic and Transport Properties of Boron-Doped Graphene Nanoribbons. Phys Rev Lett. 2007;98(19):196803.

[3] Jiang D-e, Sumpter BG, Dai S. Unique chemical reactivity of a graphene nanoribbon's zigzag edge. J Chem Phys. 2007;126(13):-.

[4] Yu SS, Zheng WT, Wen QB, Jiang Q. First principle calculations of the electronic properties of nitrogen-doped carbon nanoribbons with zigzag edges. Carbon. 2008;46(3):537-43.

[5] Dutta S, Manna AK, Pati SK. Intrinsic Half-Metallicity in Modified Graphene Nanoribbons. Phys Rev Lett. 2009;102(9):096601.

[6] Berger C, Song Z, Li X, Wu X, Brown N, Naud C, et al. Electronic Confinement and Coherence in Patterned Epitaxial Graphene. Science. 2006;312(5777):1191-6.

[7] Kosynkin DV, Higginbotham AL, Sinitskii A, Lomeda JR, Dimiev A, Price BK, et al. Longitudinal unzipping of carbon nanotubes to form graphene nanoribbons. Nature. 2009;458(7240):872-6.

[8] Tapasztó L, Dobrik G, Lambin P, Biró LP. Tailoring the atomic structure of graphene nanoribbons by scanning tunnelling microscope lithography. Nature Nanotechnology. 2008;3(7):397-401.

[9] Cai J, Ruffieux P, Jaafar R, Bieri M, Braun T, Blankenburg S, et al. Atomically precise bottom-up fabrication of graphene nanoribbons. Nature. 2010;466(7305):470-3.

[10] Wang X, Dai H. Etching and narrowing of graphene from the edges. Nature Chemistry. 2010;2(8):661-5.

[11] Cano-Márquez AG, Rodríguez-Macías FJ, Campos-Delgado J, Espinosa-González CG, Tristán-López F, Ramírez-González D, et al. Ex-MWNTs: Graphene Sheets and Ribbons Produced by Lithium Intercalation and Exfoliation of Carbon Nanotubes. Nano Lett. 2009;9(4):1527-33.

[12] Shinde DB, Majumder M, Pillai VK. Counter-ion Dependent, Longitudinal Unzipping of Multi-Walled Carbon Nanotubes to Highly Conductive and Transparent Graphene Nanoribbons. Scientific Reports. 2014;4(1):4363.

[13] Stankovich S, Dikin DA, Dommett GHB, Kohlhaas KM, Zimney EJ, Stach EA, et al. Graphene-based composite materials. Nature. 2006;442(7100):282-6.

[14] Vo TH, Shekhirev M, Kunkel DA, Orange F, Guinel MJF, Enders A, et al. Bottom-up solution synthesis of narrow nitrogen-doped graphene nanoribbons. Chem Commun. 2014;50(32):4172-4.

[15] Sakaguchi H, Kawagoe Y, Hirano Y, Iruka T, Yano M, Nakae T. Width-Controlled Sub-Nanometer Graphene Nanoribbon Films Synthesized by Radical-Polymerized Chemical Vapor Deposition. 2014;26(24):4134-8.

[16] Yang X, Dou X, Rouhanipour A, Zhi L, Räder HJ, Müllen K. Two-Dimensional Graphene Nanoribbons. J Am Chem Soc. 2008;130(13):4216-7.

[17] Cervantes-Sodi F, Csányi G, Piscanec S, Ferrari AC. Edge-functionalized and substitutionally

doped graphene nanoribbons: Electronic and spin properties. Phys Rev B. 2008;77(16):165427.

[18] Ezawa M. Peculiar width dependence of the electronic properties of carbon nanoribbons. Phys Rev B. 2006;73(4):045432.

[19] Fujita M, Wakabayashi K, Nakada K, Kusakabe K. Peculiar Localized State at Zigzag Graphite Edge. J Phys Soc Jpn. 1996;65(7):1920-3.

[20] Lee Y-L, Lee Y-W. Ground state of graphite ribbons with zigzag edges. Phys Rev B. 2002;66(24):245402.

[21] Miyamoto Y, Nakada K, Fujita M. First-principles study of edge states of H-terminated graphitic ribbons. Phys Rev B. 1999;60(23):16211-.

[22] Son Y-W, Cohen ML, Louie SG. Energy Gaps in Graphene Nanoribbons. Phys Rev Lett. 2006;97(21):216803.

[23] Wakabayashi K, Fujita M, Ajiki H, Sigrist M. Electronic and magnetic properties of nanographite ribbons. Phys Rev B. 1999;59(12):8271-82.

[24] Huang B, Liu F, Wu J, Gu B-L, Duan W. Suppression of spin polarization in graphene nanoribbons by edge defects and impurities. Phys Rev B. 2008;77(15):153411.

[25] Barone V, Hod O, Scuseria GE. Electronic Structure and Stability of Semiconducting Graphene Nanoribbons. Nano Lett. 2006;6(12):2748-54.

[26] Novoselov KS, Geim AK, Morozov SV, Jiang D, Zhang Y, Dubonos SV, et al. Electric Field Effect in Atomically Thin Carbon Films. Science. 2004;306(5696):666-9.

[27] Zhang Y, Tan Y-W, Stormer HL, Kim P. Experimental observation of the quantum Hall effect and Berry's phase in graphene. Nature. 2005;438(7065):201-4.

[28] Jiang Q, Aya N, Shi FG. Nanotube size-dependent melting of single crystals in carbon nanotubes. Appl Phys A. 1997;64(6):627-9.

[29] David TB, Lereah Y, Deutscher G, Kofman R, Cheyssac P. Solid-liquid transition in ultra-fine lead particles. Philos Mag A. 1995;71(5):1135-43.

[30] Lamber R, Wetjen S, Jaeger NI. Size dependence of the lattice parameter of small palladium particles. Phys Rev B. 1995;51(16):10968-71.

[31] Qi WH, Wang MP. Size effect on the cohesive energy of nanoparticle. J Mater Sci Lett. 2002;21(22):1743-5.

[32] Navarro-Santos P, Ricardo-Chávez JL, Reyes-Reyes M, Rivera JL, López-Sandoval R. Tuning the electronic properties of armchair carbon nanoribbons by a selective boron doping. J Phys: Condens Matter. 2010;22(50):505302.

[33] Dumitrică T, Hua M, Yakobson BI. Endohedral silicon nanotubes as thinnest silicide wires. Phys Rev B. 2004;70(24):241303.

[34] Kan E-j, Li Z, Yang J, Hou JG. Half-Metallicity in Edge-Modified Zigzag Graphene Nanoribbons. J Am Chem Soc. 2008;130(13):4224-5.

[35] Kresse G, Joubert D. From ultrasoft pseudopotentials to the projector augmented-wave method. Phys Rev B. 1999;59(3):1758-75.

[36] Blöchl PE. Projector augmented-wave method. Phys Rev B. 1994;50(24):17953-79.

[37] Abanin DA, Lee PA, Levitov LS. Spin-Filtered Edge States and Quantum Hall Effect in Graphene. Phys Rev Lett. 2006;96(17):176803.

[38] Magda GZ, Jin X, Hagymási I, Vancsó P, Osváth Z, Nemes-Incze P, et al. Room-temperature magnetic order on zigzag edges of narrow graphene nanoribbons. Nature. 2014;514(7524):608-11.

[39] Bader RFW, Carroll MT, Cheeseman JR, Chang C. Properties of atoms in molecules: atomic volumes. J Am Chem Soc. 1987;109(26):7968-79.

[40] Momma K, Izumi F. VESTA 3 for three-dimensional visualization of crystal, volumetric and morphology data. J Appl Crystallogr. 2011;44(6):1272-6.

[41] Yang W, Parr RG. Hardness, softness, and the fukui function in the electronic theory of metals and catalysis. 1985;82(20):6723-6.

[42] Geerlings P, De Proft F, Langenaeker W. Conceptual Density Functional Theory. Chem Rev. 2003;103(5):1793-874.

[43] Li Y, Evans JNS. The Fukui Function: A Key Concept Linking Frontier Molecular Orbital Theory and the Hard-Soft-Acid-Base Principle. J Am Chem Soc. 1995;117(29):7756-9.

[44] Yang W, Mortier WJ. The use of global and local molecular parameters for the analysis of the gas-phase basicity of amines. J Am Chem Soc. 1986;108(19):5708-11.

[45] Ayers PW, De Proft F, Borgoo A, Geerlings P. Computing Fukui functions without differentiating with respect to electron number. I. Fundamentals. J Chem Phys. 2007;126(22):224107.

[46] Chermette H, Boulet P, Stefan P. Reviews of Modern Quantum Chemistry: A Celebration of the Contributions of Robert G. Parr **Parr**. Singapore: World Scientific; 2002.

[47] L. GJ. Structure and Bonding. Berlin: Springer-Verlag; 1993. 268 p.

[48] Ayers PW, Yang W, Bartolotti LJ. Chemical Reactivity Theory: A Density Functional View: CRC Press; 2009. 610 p.

[49] Cerón ML, Gomez T, Calatayud M, Cárdenas C. Computing the Fukui Function in Solid-State Chemistry: Application to Alkaline Earth Oxides Bulk and Surfaces. The Journal of Physical Chemistry A. 2020;124(14):2826-33.

[50] Politzer P, Murray JS, Peralta-Inga Z. Molecular surface electrostatic potentials in relation to noncovalent interactions in biological systems. Int J Quantum Chem. 2001;85(6):676-84.

[51] Pearson RG. Hard and Soft Acids and Bases. J Am Chem Soc. 1963;85(22):3533-9.

[52] Morales-Palacios FG, Navarro-Santos P, Beiza-Granados L, Rivera JL, García-Gutiérrez HA, Herrera-Bucio R. Conjugate addition between syringol and a captodative olefin catalyzed by BF3. 2019;32(12):e4011.

[53] Rivera JL, Navarro-Santos P, Guerra-Gonzalez R, Lima E. Interaction of Refractory Dibenzothiophenes and Polymerizable Structures. International Journal of Polymer Science. 2014;2014:11.

[54] Rivera JL, Navarro-Santos P, Hernandez-Gonzalez L, Guerra-Gonzalez R. Reactivity of Alkyldibenzothiophenes Using Theoretical Descriptors. J Chem. 2014;2014:8